Technology Entrepreneurship

Technology Entrepreneurship
Taking Innovation to the Marketplace

Second Edition

Thomas N. Duening, Ph.D
El Pomar Chair of Business and Entrepreneurship
Director, Center for Entrepreneurship
College of Business
University of Colorado at Colorado Springs

Robert D. Hisrich, Ph.D
Garvin Professor of Entrepreneurship
Director, Walker Center for Global Entrepreneurship,
Thunderbird School of Global Management

Michael A. Lechter, Esq., CLP
CEO, Michael Lechter, PC
CEO, TechPress Inc.
Adjunct Professor, Fulton School of Engineering,
Arizona State University

AMSTERDAM • BOSTON • HEIDELBERG • LONDON
NEW YORK • OXFORD • PARIS • SAN DIEGO
SAN FRANCISCO • SINGAPORE • SYDNEY • TOKYO
Academic Press is an imprint of Elsevier

Academic Press is an imprint of Elsevier
32 Jamestown Road, London NW1 7BY, UK
525 B Street, Suite 1800, San Diego, CA 92101-4495, USA
225 Wyman Street, Waltham, MA 02451, USA
The Boulevard, Langford Lane, Kidlington, Oxford OX5 1GB, UK

Second edition 2015

Copyright © 2015, 2010 Thomas N. Duening, Robert D. Hisrich, Michael A. Lechter.
Published by Elsevier Inc. All rights reserved.

No part of this publication may be reproduced, stored in a retrieval system or transmitted in any form or by any means electronic, mechanical, photocopying, recording or otherwise without the prior written permission of the publisher Permissions may be sought directly from Elsevier's Science & Technology Rights Department in Oxford, UK: phone (+44) (0) 1865 843830; fax (+44) (0) 1865 853333; email: permissions@elsevier.com. Alternatively you can submit your request online by visiting the Elsevier web site at http://elsevier.com/locate/permissions, and selecting Obtaining permission to use Elsevier material.

Notice
No responsibility is assumed by the publisher for any injury and/or damage to persons or property as a matter of products liability, negligence or otherwise, or from any use or operation of any methods, products, instructions or ideas contained in the material herein. Because of rapid advances in the medical sciences, in particular, independent verification of diagnoses and drug dosages should be made.

Library of Congress Cataloging-in-Publication Data
Duening, Thomas N.
  Technology entrepreneurship: taking innovation to the marketplace/by Thomas N. Duening, Robert D. Hisrich, Michael A. Lechter.
    pages cm
  Revised edition of the authors' Technology entrepreneurship : creating, capturing, and protecting value, published in 2010.
  Includes bibliographical references and index.
  ISBN 978-0-12-420175-0 (alk. paper)
 1. Technological innovations–Economic aspects. 2. High technology industries–Management. 3. Entrepreneurship. 4. Intellectual property. 5. New business enterprises–Management. I. Hisrich, Robert D. II. Lechter, Michael A. III. Title.
  HC79.T4D84 2015
  658.5'14–dc23
                                                                                2014025056
A catalog record for this book is available from the Library of Congress

British Library Cataloguing in Publication Data
A catalogue record for this book is available from the British Library

For information on all **Academic Press** publications
visit our web site at store.elsevier.com

Transferred to Digital Printing in 2014

Dedication

To my mentor, John M. Ivancevich, and to my wife, Charlene.

Thomas N. Duening

To my wife, Tina Hisrich; daughters, Kary, Katy, and Kelly; son-in-law, Rich; and grandchildren, Rachel, Andrew, and Sarah.

Robert D. Hisrich

To my wife, Sharon L. Lechter, and children and grandchildren, Phillip, Angela, Kristian, and Dylan; Shelly, Jeff, and Emma; and William Eric, ever the Marine, may he rest in peace.

Michael A. Lechter

Contents

ACKNOWLEDGMENTS xvii
ABOUT THE AUTHORS xix
PREFACE xxiii

Part 1   You Are Here: X

CHAPTER 1   Technology Entrepreneurship Today: Trends, Opportunities, Challenges 3
   1.1   Introduction 4
   1.2   Trends and Opportunities in Technology Entrepreneurship 6
   1.3   A Word about Global Markets 11
   1.4   Foundations of this Book 12
   1.5   Chapter Summary 13
   Keyterms 15
   Additional Reading 15
   Web Resources 16
   Endnotes 16

CHAPTER 2   Five Pillars of Technology Entrepreneurship 17
   2.1   Introduction 17
   2.2   Pillar #1: Value Creation 18
   2.3   Pillar #2: The Lean Startup 20
   2.4   Pillar #3: Customer Discovery and Validation 23
   2.5   Pillar #4: The Business Model Canvas 26
   2.6   Pillar #5: The Entrepreneurial Method 29
         2.6.1   Principle #1: Expert Technology Entrepreneurs Believe Value Creation is the Primary Purpose of their Business 30

vii

		2.6.2	Principle #2: Expert Technology Entrepreneurs Rebound Personally and Professionally from Failure 30

        2.6.2 Principle #2: Expert Technology Entrepreneurs Rebound Personally and Professionally from Failure 30
        2.6.3 Principle #3: Expert Technology Entrepreneurs Respect Private Property and Uphold Contractual Obligations 31
        2.6.4 Principle #4: Expert Technology Entrepreneurs Respect the Judgment of the Marketplace 31
  2.7 Deliberate Practice 32
        2.7.1 When to Start Practicing 34
  2.8 Chapter Summary 35
  Keyterms 36
  Additional Reading 37
  Web Resources 37
  Endnotes 37

CHAPTER 3 Technology Venture Idea Generation 39
  3.1 Introduction 39
  3.2 Fundamental Venture Types 41
        3.2.1 Business-to-Business 43
        3.2.2 Business to Consumer 44
        3.2.3 Business to Government 46
  3.3 The Idea Generation Process 47
        3.3.1 Step 1: Identify a Point of Pain 47
        3.3.2 Step 2: Innovate a Product or Service 49
        3.3.3 Step 3: Test Business Models 51
        3.3.4 Step 4: Explore How to Acquire Customers 53
  3.4 The Opportunity Register 54
  3.5 Nontraditional Idea Sources 56
        3.5.1 Read Alternative Literature 57
        3.5.2 Visit New Places and Experience New Things 58
        3.5.3 Meet Thought Leaders 58
        3.5.4 Team Up 59
  3.6 Chapter Summary 59
  Keyterms 60
  Additional Reading 61
  Web Resources 61
  Endnotes 61

Part 2 Countdown to Launch

CHAPTER 4 Markets and Product or Service Development 65
- 4.1 Introduction 66
- 4.2 Product Planning and Development 66
- 4.3 The Idea Development Process 68
  - 4.3.1 Describe the Idea and Competition 68
  - 4.3.2 Determine the Need 69
  - 4.3.3 Modify and Validate 69
  - 4.3.4 Develop a Marketing Plan 70
  - 4.3.5 Develop the Final Product or Service 70
  - 4.3.6 Launch the Idea 70
- 4.4 The Concept of Newness 70
  - 4.4.1 Newness to the Consumer 71
  - 4.4.2 Newness to the Organization 72
  - 4.4.3 Newness to the Distribution System 72
- 4.5 Opportunity Assessment Plan 73
- 4.6 Disruptive Technology 75
- 4.7 The Market 75
  - 4.7.1 Market Segmentation 77
  - 4.7.2 Target Market and Positioning 79
- 4.8 Chapter Summary 80
- Keyterms 80
- Additional Readings 80
- Web Resources 81
- Endnotes 81

CHAPTER 5 Protecting Your Intellectual Property 83
- 5.1 Introduction 84
- 5.2 IP And Technology Ventures 84
  - 5.2.1 IP Protection 87
- 5.3 Recognizing IP 89
- 5.4 Record Keeping 90
  - 5.4.1 Record Keeping Procedures 91
  - 5.4.2 Guidelines for Record Keeping 92
- 5.5 Trade Secrets 95
  - 5.5.1 Procedures for Trade Secret Protection 95
- 5.6 Patents 97
  - 5.6.1 The Patent Application 99
  - 5.6.2 Written Description 100
  - 5.6.3 Claims 102

		5.6.4	Exclusive Right	102

		5.6.4	Exclusive Right 102
		5.6.5	Patentability 103
		5.6.6	The Patent Examination Process 106
		5.6.7	Patent Pending 106
		5.6.8	Patent Ownership 107
		5.6.9	International Patents 107
	5.7	Copyrights 108
		5.7.1	Considerations with Respect to Software 109
		5.7.2	Copyrights and the Internet 110
		5.7.3	Notice 111
		5.7.4	Copyright Registration 111
		5.7.5	Copyright Ownership 111
	5.8	Mask Works 112
	5.9	Trademarks 112
		5.9.1	Acquiring Trademark Rights 113
		5.9.2	Registering a Trademark 114
		5.9.3	Trademark Notice Symbols 116
		5.9.4	International Protection of Trademarks 116
		5.9.5	The Strength of a Mark 116
		5.9.6	Choosing a Mark 119
	5.10	Chapter Summary 119
	Keyterms 120
	Additional Reading 122
	Web Resources 123
	Endnotes 123

CHAPTER 6 Legal Structure and Equity Distribution 127
    6.1 Introduction 128
    6.2 Ownership and Liability Issues 129
        6.2.1 Limited Versus Unlimited Liability 129
        6.2.2 The Extent of Limited Liability 131
    6.3 Choice of Legal Structure 134
        6.3.1 Sole Proprietorship 135
        6.3.2 General Partnership 137
        6.3.3 Limited Partnership 139
        6.3.4 Corporation 141
        6.3.5 Limited Liability Company 146
        6.3.6 Limited Liability Entities—A Comparison ... 148

CHAPTER 13 Valuing and Exiting Your Venture 309
    13.1 Introduction 309
    13.2 Due Diligence 311
        13.2.1 Finances 312
        13.2.2 Product/Service Line 312
        13.2.3 Synergy 312
        13.2.4 Markets and Customers 313
        13.2.5 Research and Development and Intellectual Property 313
        13.2.6 Operations 314
        13.2.7 Management and Key Personnel 314
    13.3 Valuation 315
        13.3.1 Valuation Techniques 315
        13.3.2 Multiples Technique 315
        13.3.3 Discounted Cash Flow Technique 316
    13.4 Exit Via Succession 318
        13.4.1 Advantages of Exit Via Succession 320
        13.4.2 Disadvantages of Exit Via Succession 320
    13.5 Exit Via Acquisition 321
        13.5.1 The Acquisition Deal 323
        13.5.2 Advantages of Exit Via Acquisition 323
        13.5.3 Disadvantages of Exit Via Acquisition 324
    13.6 Exit Via Merger 325
        13.6.1 Advantages of Exit Via Merger 326
        13.6.2 Disadvantages of Exit Via Merger 327
    13.7 Exit Via Initial Public Offering 328
        13.7.1 Timing 328
        13.7.2 Selecting an Investment Bank 329
    13.8 Registration Statement and Timetable 331
        13.8.1 The Prospectus 331
        13.8.2 The Red Herring 332
        13.8.3 Reporting Requirements 332
        13.8.4 Advantages of Exit Via IPO 333
        13.8.5 Disadvantages of Exit Via IPO 334
    13.9 Chapter Summary 335
    Keyterms 336
    Additional Reading 337
    Web Resources 337
    Endnotes 338

Appenidx I: Example of a Generic Confidentiality Agreement 339
Appendix II: Example Executive Summary 343

Appendix III: Sample Development Agreement 347
Appendix IV: Example of an Employment Agreement 353

INDEX 359

Acknowledgments

We are grateful for the support, encouragement, and feedback we received during the production of this text. Several people deserve special mention for their unwavering and tireless support of this project. Carol Pacelli assisted in the preparation of many of the chapters. Jonathan Beckley and Jason Babbel provided substantial research support and assistance in developing some of the chapters. Of course, any remaining errors, either of omission or commission, remain the sole responsibility of the authors.

In addition, a number of teachers and scholars participated in reviewing this text during its development. In addition to the ones who wish to remain anonymous, these reviewers include:

- Frank Hoy, Worcester Polytechnic Institute
- Thomas E. Kaplan, Wittenberg University

About the Authors

Dr. Thomas N. Duening is the El Pomar Chair of Business & Entrepreneurship and Director of the Center for Entrepreneurship in the College of Business at the University of Colorado, Colorado Springs. He is a 1991 graduate of the University of Minnesota with a PhD in Higher Education Administration and an MA in Philosophy of Science. He began his academic career as the Assistant Dean for the University of Houston's College of Business Administration. There, he was also a visiting faculty member in the Marketing Department and a co-founder of its Center for Entrepreneurship & Innovation.

Dr. Duening launched his first venture while a graduate student. His international consulting firm served the electric utility industry with information products centered on the issue of health effects associated with electric and magnetic fields from high-voltage power lines. Duening and his partner launched the venture in 1984. He left in 1991 upon completion of his doctorate to assume the Assistant Dean position in Houston.

After his 9-year stint as Assistant Dean, Dr. Duening founded several more companies. With a partner, he co-founded US Learning Systems in 1998. The firm provided e-learning content to providers around the country. US Learning Systems was acquired in December 1999 by Aegis Learning. Aegis provided e-learning services to corporations around the world. Dr. Duening left Aegis in 2002 to launch the Applied Management Sciences Institute (AMSI). The organization was created to develop educational products for business students. As part of this firm, Professor Duening co-wrote three business textbooks, which now enjoy wide circulation around the world.

Dr. Duening next founded INSYTE Business Services Group and launched a project to study best practices in business process outsourcing. The result of this effort was two trade books: *Business Process Outsourcing: The Competitive Advantage* and *The Essentials of Business Process Outsourcing*. Both books were published by John Wiley & Sons in 2004 and 2005, respectively. As he was conducting the research for these books, Dr. Duening co-founded INSYTE InfoLabs India, Pvt. Ltd., a business process outsourcing firm based in Bangalore, India. The firm provides outsourcing services to a wide range of companies, enabling them to reduce their cost structure. Infolabs was acquired by ANSRSource in 2004.

In 2004, Dr. Duening accepted a position as Director of its Entrepreneurial Programs Office with Arizona State University's Ira A. Fulton School of Engineering. In that role, Dr. Duening taught courses in Technology Entrepreneurship to engineers at the graduate and undergraduate levels. He also co-authored a textbook titled *Technology Entrepreneurship: Creating, Capturing, and Protecting Value* published by Elsevier.

He is currently a senior advisor to ANSRSource (www.ansrsource.com) and a founding partner in a Bangalore, India-based global accelerator called Kyron (www.kyron.me).

Dr. Robert D. Hisrich is the Garvin Professor of Global Entrepreneurship and Director of the Walker Center for Global Entrepreneurship at Thunderbird School of Global Management. Dr. Hisrich received his BA from DePauw University, his MBA and PhD degrees from the University of Cincinnati, and honorary doctorate degrees from Chuvash State University (Russia) and the University of Miskolc (Hungary). He was a Fulbright Professor at the International Management Center in 1989 and a Fulbright Professor at the Foundation for Small Enterprise Economic Development in Budapest, Hungary, from 1990 to 1991. He has held or now holds visiting professorships at the University of Ljubljana (Slovenia), the Technical University of Vienna (Austria), Donau University (Austria), Queensland University of Technology (Australia), and Jilin University (China).

Dr. Hisrich has authored or co-authored 34 books, including: *Marketing for Entrepreneurs and SMEs: A Global Perspective* (2014), *Managing Innovation and Entrepreneurship* (2014), *Entrepreneurship: Starting, Developing, and Managing a New Enterprise*, 9th edition (2013—translated into 13 languages), *Governpreneurship: Establishing a Thriving Entrepreneurial Spirit in Government* (2013), *Corporate Entrepreneurship* (2012), *International Entrepreneurship: Starting, Developing, and Managing a Global Venture*, 2nd edition (2012), *Technology Entrepreneurship: Value Creation, Protection, and Capture* (2010), and *The 13 Biggest Mistakes That Derail Small Businesses and How to Avoid Them*, and over 300 articles on entrepreneurship, international business management, and venture capital.

Dr. Hisrich has been involved in the creation of numerous ventures including: H&B Associates (consulting), Illuminations (rainbow stickers and ornaments), Jameson Inns (hotels), La Bella Terre (organic sugars), Noteworthy Medical Systems (medical software), Polymer Technology (Boston Lens) and Xeta Technologies (telecommunications).

Michael A. Lechter is an attorney, certified licensing professional and entrepreneur, CEO of TechPress Inc., a publishing and literary agency company, CEO of Michael Lechter PC, and Adjunct Professor in the Entrepreneurial Program in the Ira A. Fulton School of Engineering at Arizona State University. Lechter is also the best-selling author of *OPM, Other People's Money, How to Attract Other People's Money for Your Investments—The Ultimate Leverage* (2005, 2nd edition 2010) and *Protecting Your #1 Asset, Creating Fortunes from Your Ideas* (2001, 2nd edition 2014).

An internationally known expert in the field of intellectual property, his clients have included breweries, fast food companies, casinos, professional sports teams, major software companies, semiconductor and medical device manufacturers, start-ups, venture capitalists, and Fortune 100 companies. When asked what he does for a living, he typically replies, "I build forts and fight pirates."

Michael has been the architect of strategies for building businesses, using both conventional and unconventional forms and sources of "Other People's Money and Resources." His experience in representing both venture capitalists and start-up and emerging businesses, and experience as an angel investor himself, provides a unique perspective to the subject of building a business.

Michael is also the author of *The Intellectual Property Handbook*, TechPress (1994), coordinating editor of *Successful Patents and Patenting for Engineers and Scientists*, IEEE Press (1995), and contributing author to the *Encyclopedia of Electrical and Electronics Engineering*, Wiley (1999), and *Licensing Best Practices: The LESI Guide to Strategic Issues and Contemporary Realities*, Wiley (2002). Over the years, he has also written monthly columns for INC.com, IEEE-USA Today's Engineer, and Washington Technology.

Michael has been an active member of the Licensing Executives Society (LES) USA/Canada, serving as a trustee (1996-2000), and as Computer and Electronics Industry Sector Chair (1992-1996). He has been a LES USA/Canada delegate to LES International since 2001 and has served as Chair or Vice Chair of a number of LESI committees. Since 2013, he has been counsel to the LESI board.

He has lectured extensively throughout the world on intellectual property law and entrepreneurship. Upon request of the House Judiciary Committee he has submitted testimony to the Congress of the United States and has participated in various United Nations and foreign government proceedings on intellectual property law and technology transfer.

Michael is also the owner of Cherry Creek Lodge LLC, a resort/guest ranch in the Tonto National Forest of Northern Arizona (www.CherryCreekLodge.com), a study in rustic elegance where modern comfort meets The Old West. The Cherry Creek Lodge specializes in corporate and family retreats.

This figure is a useful illustration of the difference between the way problems are approached by most businesspeople (managers and business leaders) and the way they are approached by expert entrepreneurs. Most business people think in terms of setting a clear goal and then aggregating the resources necessary to pursue that singular goal, as illustrated on the left side of Exhibit 1.1. Indeed, the mark of an "expert manager" is the ability to set goals, persuade the organization to allocate the necessary resources, and then rally the troops to pursue the goal with single-mindedness and determination.

By way of contrast, the expert entrepreneur is aware that bringing new products to the market and building successful ventures is fraught with uncertainty. As we mentioned above, the expert entrepreneur begins the venture building process by taking stock of the resources available and currently controlled. This is illustrated on the right side of the diagram in Exhibit 1.1. This figure highlights the fact that expert entrepreneurs create value for target customers with the resources available to them and that any one of a number of different and varied "imagined ends" could count as a successful outcome. Clearly, this "entrepreneurial" way of creating value differs radically from the way corporate managers are trained to create value.

The experienced technology entrepreneur knows that customers may react in unexpected and unpredictable ways to the products they bring to the market. As such, the entrepreneur must approach the market with an open mind, a willingness to listen, and an ability to pivot to new business models and/or offerings as warranted by market response. One currently popular phrase that illustrates this perspective is: "No business plan survives first contact with the customer."[1]

A startup venture is not merely a "small" version of a large company. The startup is different from an established business in a number of important ways. For example, the startup venture has no customers. Thus, the startup cannot simply execute a proven business model because it hasn't yet determined which model will work best with its target customers. This contradicts a common understanding that entrepreneurship is all about execution. In fact, in the early days of a startup there is nothing really to execute because the entrepreneur has not yet found a way to consistently attract and deliver value to target customers. The goals of a startup are often different from those of an established business. For example, the immediate goal of a startup may be to become credible, show proof of customer demand, or market viability, as opposed to demonstrating sales and revenue.

The focus of the startup technology entrepreneur is to run experiments with products, features, and customers to *discover* a scalable, repeatable business model. The mantra for technology entrepreneurs during this startup phase is: "Fail often, fail fast." In other words, run experiments that expose your products to the market, gather feedback, and refine your offering based on that feedback until you have a viable product that customers want to buy. We will explore the process of experimenting to find a business model in greater detail in Chapter 2.

The product that the startup brings to market in these early days is not necessarily its ultimate product, but rather what is referred to as the minimal viable product (MVP). The MVP is a product having only those features (1) necessary to get the product into the hands of early adopters and (2) sufficient to demonstrate future benefit. Successive versions of MVPs are introduced into the market to test fundamental business hypotheses as part of an iterative product development process.

One outcome of this fail often, fail fast revolution in technology startups is the advent of a new type of venture capital firm referred to as an accelerator. Ycombinator was one of the first of this new genre, which has now become global. An accelerator generally invests small amounts of capital—usually less than $50 K—in a large number of promising ideas and entrepreneurial teams. Typically, following the investment, the accelerator will require their portfolio companies to attend a venture development "boot camp" that builds the team and the product. For example, Kyron Global Accelerator, based in Bangalore, India, invests less than $20 K in startup technology ventures. Companies receiving investment must participate in Kyron's four-month boot camp at its Bangalore headquarters. There, teams interact with mentors, develop their products, and undertake a series of tasks designed to help the founding teams gain knowledge of their customers and their markets.

The concepts of "effectuation" and the "minimal viable product" are just two of the more important new ways of understanding how technology entrepreneurs should build successful ventures. We will be exploring these and other concepts in more detail in Chapter 2. Let's next turn our attention to some leading trends and challenges facing technology entrepreneurs.

1.2 TRENDS AND OPPORTUNITIES IN TECHNOLOGY ENTREPRENEURSHIP

As the story that opens this chapter indicates, technology entrepreneurs must think globally and act locally to succeed. Hike is a very young startup in India that built its core technology—an SMS messaging service—to serve the unique problems faced by Indians in the fragmented local telecom market. Although Hike was the beneficiary of a sizable $7 million early-stage investment, it will succeed or fail based on its ability to serve the massive opportunity that the Indian market presents. This is a fundamental lesson for all technology entrepreneurs.

Although big technology names dominate the media and have global presence, there is always room for innovation that creates value for local customers. Hike competes against some of the biggest names in the tech industry—Google, Facebook, and others—and yet it has achieved remarkably rapid growth because it addresses local issues that the big players have overlooked.

And so, despite the fact that markets are dominated by big companies in many technology industries, it is still a good time to be a technology entrepreneur. Disruptive innovations across the spectrum of technology industries are changing the way people interact, shop, vacation, work, and play. For example, some of the more powerful transformative Internet technology trends in recent years include:

- *Big Data*: Companies are benefiting from the troves of data they have been warehousing. Big data enables companies to track customer buying habits and make adjustments quickly if necessary. This has been described as "knowing the now." In other words, companies use big data not simply to understand what happened, but to know what is happening right now—in real time.[2]
- *The Quantified Self*: The quantified self is a function of numerous converging factors: aging Baby Boomers who want to maintain their youth and vitality into their senior years; the miniaturization of sensors and their ability to be embedded in the body and in exercise gear; and the ubiquitous presence of smart phones and apps that can communicate with the embedded sensors. Nike, for example, is sponsoring a technology accelerator in Beaverton, Oregon that develops companies whose technologies complement Nike's quantified self-innovations. The Nike+ website is a great example of how the company is recruiting developers to add value to its portfolio of athletic products.
- *Collaborative Commerce*: It was only a matter of time before entrepreneurs realized that people are becoming more comfortable sharing things via the Internet. What began as social networking and the sharing of personal information has now evolved to widespread sharing of nearly everything. For example, 99dresses.com allows people to sell old dresses for "buttons"—a form of online currency that allows them to buy more dresses from others. Or consider EatFeastly.com. This site enables people to prepare meals in their homes and then find others who are interested in paying for that meal and coming to their home to dine together.
- *Context Awareness*: The ever decreasing cost of embedding sensors and microchips in objects, people, pets, and so on, has opened up a new world of what is called "context awareness." People can now maintain e-connectivity with their possessions and loved ones via a network of embedded sensors and chips. Entrepreneurs are exploiting this explosion of e-connectivity by designing applications that keep track of children

- or pets, monitor the home, and much more. Embedded sensors are also enabling a strong link between e-health and e-connectedness. Applications can now monitor key health indications in the elderly, for example, and provide immediate, real-time updates to caregivers and concerned loved ones.
- *Cloud Computing*: So-called cloud computing is the longest-running of the trends cited here. The notion that the "network is the computer" was put forward by Sun Microsystems's Scott McNealy as a founding motto of the company in 1985. Since then, cloud computing has become more important with increasing bandwidth, storage, and network speeds. Today, companies and consumers alike are comfortable with storing important information, pictures, and other digital content in the cloud. Startups such as Box.com and Dropbox are examples of companies that enable people to store digital assets in the cloud. In fact, Dropbox was an early graduate of a Ycombinator cohort and has since grown into a massive online storage and interaction platform.
- *Internet of Things*: This emerging category builds on some of the technologies listed above, but is not necessarily human-centric. That is, the embedded sensors that enable people to keep track of home appliances, children, pets, and other things also enable nonsentient "things" to talk to and keep track of one another. For example, the coffeemaker could tell the toaster when the coffee is ready, cuing the toaster to warm up the breakfast roll. When appliances talk to one another, the home owner doesn't need to attempt to calibrate the timing on each appliance so that he or she can enjoy hot coffee and a warm roll at the same time. The Internet of things would manage that process without the home owner needing to be involved. Of course, this is a banal application of what certainly will be a revolutionary change in how people and their "things" are oriented to one another.

The items listed above are only a few of the technology innovations that are proving disruptive to the status quo. There are important technological advances occurring beyond the realm of the Internet that will have profound implications for people around the world in the decades to come. Some of the more important of these innovations that are emerging as this chapter is being written include:

- Bitcoin and the advent of new digital currencies
- E-medical records and the advent of major changes in health care provision
- Cyber security and the need to ensure the integrity of data and transactions
- Nano-technology breakthroughs in medicine and other areas

- Health informatics and the ability to analyze massive amounts of health data
- Aging research that portends lengthened life spans
- Robotics and the increasing role of robots in the workplace and the home
- Brain science advances that will alter mental health care

Technology entrepreneurs will continue to drive the global economy and disrupt industries that are not evolving quickly enough to serve changing markets. The role of innovation in technology entrepreneurship cannot be underestimated. **Innovation** is defined in this book as the transformation of new ideas, inventions, and processes into value for a market. This definition makes it clear that innovation is different from merely having an idea or creating an invention. Good ideas can come from anyone. You've probably had ideas that you believed were breakthrough ideas and would be incredibly useful if they became reality. For example, it may be a good idea for electric vehicles to become more ubiquitous and affordable. However, as an *idea* there is no impact on society. It is the *innovator* who takes an idea and devises a way to create value with the idea via a product or service. The *entrepreneur* is the one who knows how to take the innovation to the marketplace. In graphic form, these distinctions are presented in Exhibit 1.2.

As you can see, innovators are people who take ideas and attempt to create value through product development, designing, iterating, and refining. Innovators often are also entrepreneurs, meaning those people who also understand how to organize a startup venture, develop a business model around the innovation, serve customers, and grow the venture. The innovator is not always a skilled entrepreneur, yet innovation and entrepreneurship are inseparable. The entrepreneur without the innovator has no value to bring to

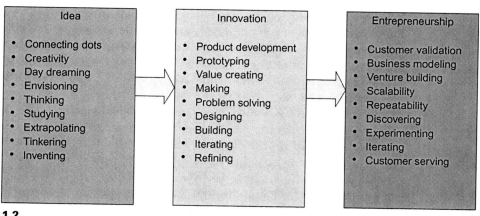

EXHIBIT 1.2
Distinguishing ideas, innovation, and entrepreneurship.

the market, and the innovator without the entrepreneur has value but doesn't possess the skills to serve markets and build a venture. However, it is not necessary that a single individual possess all of the skills necessary for success when starting a venture. Building a technology venture is a team sport, which will be discussed in more detail in Chapter 6.

The Mini-Case below is an example of someone who possesses all three of the key abilities noted above. Loren Brichter is an idea person, innovator, and successful entrepreneur.

MINI-CASE
Loren Brichter on Developing Popular Apps

Tweetie founder Loren Brichter says that he discards about 80% of his ideas to focus on those that he thinks will become "sticky" for users. Tweetie is the popular Twitter app that became a standard for mobile device users. He has since gone on to create a new app called "LetterPress," a multiplayer word game that prompts players to make words from a five-by-five grid of letters. There are several tips that aspiring app developer entrepreneurs should use to come up with products that users will desire:

1. Narrow your idea to something genuinely useful.
2. Build features that provide feedback to users so that they know what is happening.
3. Test your app wisely before releasing it to the marketplace.
4. Cross promoting your app with other popular apps is a good way to build market buzz.

Sources: Jessica E. Lessin, "Experts Share Their Dos and Don'ts for Building an App," *The Wall Street Journal*, March 4, 2013, p. B7; Seth Fiegerman, "Tweetie Creator Loren Brichter Discusses His Long-Awaited Follow-Up," *Mashable*, October 24, 2012.

When you are studying the material in this book, think about your own capacities as an idea person, an innovator, and an entrepreneur. It's possible that you possess all three skill sets like Loren Brichter does. However, you may be more of an innovator than an entrepreneur, or vice versa. Knowing yourself is a key starting point in the effectual approach to technology entrepreneurship. After all, two of the key resources that you always possess are your own unique talents and your skill sets. Knowing your personal strengths and weaknesses enables you to focus on what you do best and to find the talented people who can compensate for any of your weaknesses. For example, if you are a strong innovator but a weak entrepreneur, then you will need to find someone who possesses the skills of the entrepreneur. Likewise, skilled entrepreneurs who are not good innovators must find an innovation or an innovator around which to build a venture. By far the most likely scenario is that you'll need to work with other talented people to realize your technology venture dreams. It is rare to have all the talents and resources necessary to launch, grow, and exit a venture alone.

Technology entrepreneurship is truly a team sport that leverages the talents and contributions of multiple individuals striving together to reach the singular goal of venture success.

1.3 A WORD ABOUT GLOBAL MARKETS

All technology entrepreneurship today is global in nature. Innovation, competition, and disruptive technologies can emerge anywhere on the globe and rapidly disseminate to markets around the world. These technologies are in part driven by the Internet, which has not only enabled people to connect easily via the World Wide Web, but has also enabled rapid dissemination of knowledge and information. As a result, it is exceedingly difficult for innovators to build something that is not also being built somewhere else in the world.

The pressure on technology entrepreneurs to move quickly from innovation to market development is greater than ever. Technology entrepreneurs must develop their markets while iterating through multiple releases of their products, and during this process, their products are exposed to the global marketplace and the prying eyes of other ambitious entrepreneurs. This cannot be helped. Technology entrepreneurs cannot build their products in a vacuum (the proverbial "garage"), but must be willing to iterate through these product releases in the interest of rapidly developing their customers and their markets.

Of course, intellectual property will play a role in protecting the interests of technology entrepreneurs and their investors. One challenge technology entrepreneurs inevitably face is whether to spend more of their precious resources (and in particular, cash) on intellectual property development or on market development. Immediate needs inevitably clash with the need to lay a foundation for the future. The need to lay a foundation for intellectual property protection is discussed in Chapter 5. The priorities chosen and the path taken to solve this type of problem will vary depending on what type of technology venture is involved. For example, in many cases, entrepreneurs pursuing some type of Internet venture will likely want to spend more of their limited cash resources on market development. In contrast, a technology entrepreneur developing a medical device may want to spend more cash on intellectual property development.

It is important to realize that all global markets inevitably are local, and technology entrepreneurs who aspire to reach global markets with their products will need to take local customer needs and desires into consideration. Technology products that are created to serve markets in the United States, for example, may not be a good fit for markets in, say, India without significant modification. India is a unique and attractive market because it has an emerging middle class, is the second most populous nation on earth, and is technology savvy. However,

Internet entrepreneurs must realize that the World Wide Web is available only to 10% of the Indian population.[3] Yet, India currently has more than 900 million telephone subscribers, 96% of whom are on mobile phones. Mobile phones are used by 75% of the Indian population.[4] Internet entrepreneurs who want to develop a presence in the Indian market must tailor their offerings to take advantage of the extensive mobile phone penetration there. Offering a web-only solution to the India market right now will not be as potentially lucrative as offering a phone-based solution.

1.4 FOUNDATIONS OF THIS BOOK

This book is written for anyone who aspires to take a technology from idea to the market. We decided to write a practical guide for those intrepid souls who desire to be their own boss, build their own business ventures, and create value for global customers through technology. As such, this book is not for the faint-of-heart. Building a technology venture from scratch for the first time is daunting, stressful, and physically and psychologically taxing--—and perhaps the most exciting thing you'll ever do in your life.

We have attempted to create a practical guide that any first-time entrepreneur will find useful and easy to read and, at the same time, a book that will be helpful also to serial technology entrepreneurs (those who may be on their second, third, or fourth venture). We have written this book with three primary objectives that we used to guide us:

1. *Rules and Principles*: We have attempted to incorporate the most important time-tested rules and principles for technology venture success. Even though no two ventures will be alike, there are rules and principles that apply across every venture and we've tried to highlight the important ones throughout this book. For example, new ventures are often challenged by cash flow issues, and the savvy entrepreneur quickly learns to track and manage cash. Because each of the three authors of this text is also a successful entrepreneur, we will relate the rules and principles that we've followed in our own ventures.
2. *Global Focus*: Our global economy is changing rapidly, and technology trends and opportunities are also changing. Any book is written and released at a single point in time. To avoid becoming obsolete before publication, we have integrated discussions of emerging, tracked trends throughout this book. We believe that successful technology entrepreneurs follow a general principle summarized by hockey legend Wayne Gretzky: "skate to where the puck is going," not to where it is. It is important for aspiring technology entrepreneurs to be in tune with the trends that are in place now so that they are able to seize the opportunities these trends will create in the future.

3. *Research and Theory*: The word "theory" is anathema to many practicing entrepreneurs. That's unfortunate because, in truth, we all operate in a world of theory all the time. That is, our theories about how the world works and our role in it define all of our actions and responses to the actions of others. We confine our discussions of theory to what we believe are the most important and revealing research topics emerging from top entrepreneurship scholars and practitioners. We have also been careful to translate theory into practical implications for practicing entrepreneurs.

The first section of this book is titled "You Are Here X." Literally, it means that you have to start your entrepreneurial journey where you are. Refer back to Exhibit 1.2, where we highlighted the distinction between causal logic and effectual logic. The expert technology entrepreneur doesn't wait for all the necessary resources to be in place before creating the venture and pursuing economic opportunity. Instead, the expert technology entrepreneur starts right here, at "X," and proceeds to create value.

This book is based on the model shown in Exhibit 1.3 below. This model mirrors the steps that technology entrepreneurs take to bring their products to the global marketplace.

1.5 CHAPTER SUMMARY

This chapter was designed to provide insights into the structure and underlying philosophies of this book, and to help the reader gain a basic understanding of the steps involved in building a successful venture. We introduced the concept of "effectuation" and the notion that successful entrepreneurs begin their ventures by assessing the resources they currently control; that is, rather than waiting for all the appropriate pieces to come together—which may never happen—expert entrepreneurs push forward with the resources they control and attempt to create value for chosen markets.

We also reviewed some of the top trends emerging in technology entrepreneurship on a global scale. The trends we identified are evolving rapidly as this book goes to press and likely will continue to change while new trends also emerge. The point is to alert aspiring entrepreneurs to the fluid nature of opportunity and the need to take effectual action as soon as possible and get to market quickly.

Global markets have become accessible to nearly anyone, anywhere on the planet. As such, the nature of competition has changed dramatically from just a quarter of a century ago. Global brands can emerge overnight on the Internet, and technology entrepreneurs must be alert and build unique advantages

CHAPTER 1 Technology Entrepreneurship Today

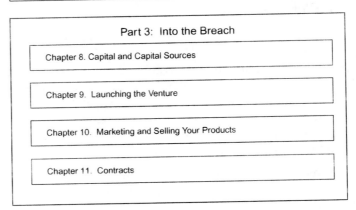

EXHIBIT 1.3
The structure of this book.

into their offerings. For example, it was highlighted that global brands can be beaten competitively by entrepreneurs who know the local market better than the global brand. Local nuances that are poorly served by a one-size-fits-all global brand can be a potent and sustainable form of competitive advantage to local firms that address those nuances.

Finally, we provided the reader with the basic information that should provide insight into the intellectual foundations of this book. This is a practical guide to technology entrepreneurship that draws inspiration from the personal experiences of the authors and other practicing and successful entrepreneurs. We also draw from the global pool of successful entrepreneurs to build awareness of the many different ways entrepreneurs around the world are writing their own success stories, and we draw inspiration from the tremendous research that continues to discover new insights into what makes successful entrepreneurs tick and how they build successful companies.

In Chapter 2, we will explore what we call the five pillars of technology entrepreneurship:

1. Value Creation
2. The Lean Startup
3. Customer Development & Validation
4. The Entrepreneurial Method
5. The Business Model Canvas

These five pillars are based on both the emerging research and on what successful technology entrepreneurs actually do to build successful ventures.

KEYTERMS

Effectuation The process of beginning a new venture by taking stock of who you are, what you know, and the resources you control.

Minimal viable product A version of an ultimate final product that is introduced to the market as an experiment to test market reaction and gather feedback for future iterations of the product.

Accelerator A new genre of venture capital firm that invests small amounts of capital into a large number of ventures.

Innovation The transformation of new ideas, inventions, and processes into value for a market.

ADDITIONAL READING

Feld, B., Cohen, D., 2010. Do More Faster: TechStars Lessons to Accelerate Your Startup. John Wiley and Sons, Hoboken, NJ.

Sarasvathy, SD., 2009. Effectuation: Elements of Entrepreneurial Expertise. Edward Elgar Publishing, Northhampton, MA.

Duening, T., Stock, G., 2012 The Entrepreneurial Method. Kendall-Hunt Publishing, Dubuque, IA.

WEB RESOURCES

Effectuation.org: This is the website of the Effectuation Society, a worldwide group of scholars who are exploring how expert entrepreneurs actually achieve their success and how those discoveries can be translated for the next generation of entrepreneurs.

TechCrunch.com: TechCrunch is an excellent online resource for tracking emerging and prevailing trends in technology entrepreneurship. The site reports on new companies, the entrepreneurs behind them, and the investors who back them.

Ycombinator.com: As discussed in the chapter, Ycombinator is one of a new breed of venture capital firms called "accelerators." This website will introduce this new style of venture investing and provide examples of the types of ventures in which they invest.

ENDNOTES

1. S. Blank and B. Dorf. 2012. The Startup Owner's Manual: The Step-by-Step Guide for Building a Great Company (Pescadero, CA: K&S Ranch Press).
2. Belicove, M.A. "Discovering Buried Treasure". *Entrepreneur*, May 2013, p. 40.
3. "Indian Technology Firms: Looking for India's Zuckerberg". *The Economist*, March 16, 2013.
4. Chotiner, I. "When a Subcontinent Goes Cellular". *The Wall Street Journal*, April 12, 2013.

CHAPTER 2

Five Pillars of Technology Entrepreneurship

iWANAMAKER EXPLORES ITS MARKET OPPORTUNITY

iWanamaker is a live-scoring golf app that is designed to allow golfers to keep track of their scores on their smartphones. When founder Doyle Heisler first came up with the idea for the app, he envisioned that golfers and golf clubs would be as excited about the app as he was. When he took the app to clubs for them to use in their tournaments, he learned that club pros were reluctant to purchase it because of the low margins in the golf industry and their general aversion to innovation and change. Doyle next decided to market his product directly to the charities that run golf tournaments as fundraisers. He thought they would be interested in the product as it makes golf more fun for players and offers sponsor opportunities within the app itself. Again, however, he learned that charities are also averse to spending additional money on their golf tournaments and are indifferent to the new technology.

As a result of these two initial failures to penetrate the lucrative golf market, Doyle decided to pivot to a new business model. Realizing that the more golfers that use the product the more he can charge for advertising, he decided to experiment with a "freemium" business model where he would virtually give the app to golf clubs and charities, and sell advertising to local and national brands that wanted exposure to golfers. As this book is going to press, iWanamaker 2.0 is about to be released under this revised business model. Time will tell if this is the model that leads to venture growth. The important thing is that Doyle was willing to change his business model based on what the market was telling him about his product and its features.

Author Duening was an investor in and board member to iWanamaker at the time this book was going to press.

2.1 INTRODUCTION

Chapter 1 examined the focus and purpose of this book and some of the leading trends and challenges for technology entrepreneurs in the modern global economy. We hope that your appetite is now whetted for a lifetime of technology entrepreneurship and that you are ready to begin learning some of the practical tools of entrepreneurial expertise. Before we begin, we must warn those of you who are novices that developing entrepreneurial expertise takes time and practice. In fact, research into what is required to become expert in anything suggests that it can take 10 years or more of what is called *deliberate practice* to achieve expert-level competence.[1] We address the process of deliberate practice in detail later in this chapter and provide you with some suggestions about how you can develop your own entrepreneurial expertise.

The central theme of this chapter is that expert technology entrepreneurs have become adept at certain specific skills and ways of thinking about products and new ventures. These skill sets are referred to as the "five pillars of entrepreneurial expertise." These five pillars are:

1. Value Creation
2. The Lean Startup
3. Customer Discovery and Validation
4. The Business Model Canvas
5. The Entrepreneurial Method

Below we discuss each of these pillars in detail, beginning with the fundamental skill that all technology entrepreneurs must possess: The ability to create value for customers.

2.2 PILLAR #1: VALUE CREATION

Expert technology entrepreneurs know intuitively that value creation is the purpose of business. In fact, it doesn't matter if you are a technology entrepreneur or a fast-food entrepreneur (or any other kind of entrepreneur). Your products and services must create value for customers. There are probably as many ways to create value as there are people on the planet.

Consider the case of three individuals in Menlo Park, California, who set out to create a new type of Internet company in 2005. Menlo Park is located in the heart of Silicon Valley, which has been the birthplace of some of the most rapidly growing technology companies in history. These three individuals were veterans of technology companies, having previously been principals at PayPal. From their garage in Menlo Park, the entrepreneurs created YouTube, one of the fastest growing companies of all time. By July 2006, YouTube reported that more than 100 million videos were being watched and as many as 50,000 videos were being added to the site each day. In October 2006, a mere 10 months after it was launched, YouTube was acquired by Google for $1.65 billion.[2]

The concept of "value" has myriad definitions. Value is defined as whatever customers believe it to be. Technology entrepreneurs can develop successful ventures based on widely different *value propositions*. A value proposition is what a venture tells its customers about the value it intends to provide to them. For example, the value proposition for YouTube is: "Broadcast yourself." That simple statement, while not necessarily appealing to everyone, is the foundation of the online video-sharing revolution.

Creating value requires vision, passion, and an ability to adjust to customer needs and constantly evolving economic, social, and technological trends and conditions. Successful technology entrepreneurs realize that steadily advancing technologies and technological form factors must be taken into consideration in their product development and design processes. For example, Rovio was a developer of games for mobile phones which were sold at retail. It had developed 50 such products, but none of them became a big hit with users. Nearing the end of its cash flow, Rovio realized that the advent of smartphones, touch-screen technologies, and Apple's new App Store would enable a new breed of games and distribution opportunities. Rovio decided to pivot from its retail-based business model to developing apps for the smart devices that were becoming increasingly ubiquitous. Their breakout product, sold through Apple's App Store, was the popular game "Angry Birds." Rovio reported that the game has been downloaded over two billion times.[3]

Value propositions are important to a venture. They help to communicate the value the venture provides to customers. Value propositions also help guide the venture's internal decision making. For example, the value proposition for well-known consumer products company Procter & Gamble is "Touching lives, improving life."[4] This value proposition tells P&G scientists and product developers how to structure their investment of research and development resources. P&G introduces hundreds of new products to markets around the world each year. The firm's value proposition guides internal decision making about which new products to pursue through multiyear development cycles.[5]

Fundamentally, creating value for customers seems too obvious to mention. Yet, a review of why new ventures fail indicates that the most common reason is because they fail to create appropriate value for customers. Instead, the failed ventures were guided by the founders' vision of the product and its features, with no guidance from customers. Products were designed, built, and released without regard to what customers really want. Of course you know by now that customers, not entrepreneurs, determine what is valuable. Customers don't always know what they want, but they always know what they don't want. Technology entrepreneurs are well-advised always to remember that customers are the ultimate judges of value and determiners of the venture's success.

2.3 PILLAR #2: THE LEAN STARTUP

The Lean Startup was conceived and developed by serial entrepreneur Eric Ries.[6] After a failed technology venture, Ries and his partners launched another technology venture called IMVU. IMVU is an instant-messaging platform that includes a novel feature that no other platform at the time was offering—3D avatars. The founders of IMVU were all technologically savvy—Ries himself is an expert programmer—so there was no question about whether they could build and deliver a working product. The question that perplexed Ries and that he wanted to solve in this new venture was "How do we get customers to buy our product?"

In his previous venture, Ries was convinced that he and his team had built a world-class technology that provided benefits to customers, but not enough customers bought the product after it was launched. Ries understood that there was no lack of effort in the work that he and his partners put into the failed venture. There were also plenty of features and benefits built into the product to attract their target customers. Ultimately, he realized, the problem was that the company ran out of money before it was able to deliver a product that attracted enough customers to generate sufficient cash flow to survive on its own.

In IMVU, Ries wanted to avoid these same problems from the previous venture. When he was between ventures, Ries decided to learn all he could about startup ventures and why they often failed as a result of having too few customers. In particular, he studied the management philosophies and tactics used by companies that excel in manufacturing. He studied Toyota and the world-renowned "Toyota Production System" (TPS). TPS is based on continuous improvement, waste and cost reduction, just-in-time inventory systems, and many other things. In addition, Ries studied the companies that had adopted a technique called "lean manufacturing," which includes many of the same elements of TPS. Lean manufacturing changes the way supply chains and production systems operate. It emphasizes empowering individuals to solve problems as they arise to promote continuous improvement and organizational learning. Lean manufacturing also emphasizes small batch sizes, just-in-time inventory systems, and accelerated cycle times. In short, lean manufacturing is about experimenting, learning, and constantly improving.

Ries realized that the same techniques that are used to continuously improve processes, products, and systems in large manufacturing organizations could be applied to the product development process in startups. Thus, the "Lean Startup" concept was born, and Ries decided to apply that concept aggressively to IMVU's product development during its startup phase. A short overview of IMVU's startup is provided in the Mini-Case below.

MINI-CASE
IMVU Uses Lean Startup Methods to Succeed

When they launched IMVU Ries and his partners thought they had devised a winning strategy. They would make their avatar-based instant messaging technology work on all of the major existing IM platforms, of which there were at least six. Unfortunately, when they launched, they did not attract any customers. So they decided to bring individuals from their target market into their headquarters to find out what was wrong. Talking directly to their target customers revealed several critical pieces of information. Number one, they learned that customers did not want to use the IMVU platform because none of their friends were on it. Further, they didn't want to invite their friends because they were not sure if the system worked—they didn't want to invite their friends to a bad experience. To solve that problem, the IMVU team built a "Chat Now" function which connected the user to a random stranger's avatar so they could meet, chat, and hang out. Customers liked that. When the IMVU team suggested they invite strangers they liked to their existing buddy lists they ran into problem number two. Users did not want to invite total strangers to their existing IM platform buddy lists.

As a result of these customer interactions, Ries and his team learned several important things that changed their marketing strategy:

- Customers loved using their 3D avatars to meet new people
- Customers did not mind adopting another stand-alone IM platform
- Customers would build new buddy lists including some from their other IM platforms

These revelations would have taken too much time to discover if the IMVU team had not brought customers to headquarters for direct interactions. Customers don't often know what they want or need until they are presented with options. Ventures using lean startup methods use experimentation, iteration, and learning to reduce the time and cost involved in discovering what customers really want.

Source: Ries, E. 2011. *The Lean Startup: How Today's Entrepreneurs Use Continuous Innovation to Create Radically Successful Businesses*. (New York: Crown Business).

Ries and his team at IMVU presented their nascent product to their customers early and often to learn what they liked, what they didn't like, and what needed to change. The term used to refer to such a nascent product is the *minimal viable product*, or "MVP." The MVP is introduced to potential customers for their feedback, and then based on this feedback a decision is made either to *pivot* to something different or to *persevere* along the current product development pathway. This process is referred to as the Build-Measure-Learn Feedback Loop and is illustrated in Exhibit 2.1.

As the illustration shows, the product development process advocated in the lean startup is circular, iterative, and experimental. Ideas are built into minimally viable products and tested in the target market. As the process cycles on, the product is changed and refined until a sufficient number of customers find that the offering is acceptable. Contrast this approach with the more linear

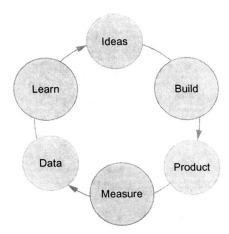

EXHIBIT 2.1
The build-measure-learn feedback loop.

stage-gate product development process used by most large organizations and shown in Exhibit 2.2.

The stage-gate process is characterized by specific "gates" following each stage of the product development process. These gates are managed by people inside the company who adhere to specific metrics to make decisions about whether developing products are allowed to proceed to the next stage. A negative review at any one of these gates results in a product being scrapped. Note in particular that products under development are measured at each gate using *internal* metrics, not direct *feedback* from customers. Unfortunately, too many technology startups still use this approach to product development. Basically, they are playing an all-or-nothing game where the product that makes it to launch had better be what customers want, because it's usually too late to pivot or turn back once the final product is launched.

In addition to the build-measure-learn feedback loop of iterative MVPs, Ries also identified a number of other principles embodied in a lean startup:

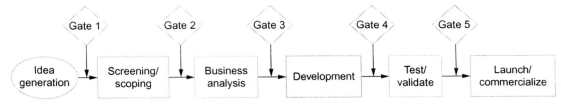

EXHIBIT 2.2
The stage-gate product development process.

Entrepreneurial management: A startup requires management techniques specifically geared to the context of extreme uncertainty. Rather than making complex plans based on "moving target" assumptions, management in a startup needs the discipline to adhere to the build-measure-learn feedback loop of iterative MVPs. This discipline enables entrepreneurs to make constant adjustments to the direction taken by the venture and to make the sometimes hard decisions to pivot or persevere. The startup should not focus on how closely work proceeds according to a preconceived plan, but rather on achieving validated learning milestones. This can sometimes be best achieved by organizing people into flexible cross-functional teams, rather than separate departments dedicated to particular functions.

Validated learning: The goal of a startup is to *learn* how to build a repeatable, scalable, business model. This requires entrepreneurs to run experiments that are rigorous and conclusive in order to test each element of their business model. Exposing customers early and often to a series of MVP releases and getting feedback from customers ensures that maximal learning is occurring at minimal expense.

Innovation accounting: Since the startup cannot measure its success by virtue of standard business measures such as revenue and profitability (at least in the early days), it needs to measure something else, for example, how much the venture is learning and its progress towards finding a scalable, repeatable business model.

The lean startup is a particularly compelling framework for technology entrepreneurs because they often have opportunities to learn customer needs and wants with less-than-perfect finished products. Of course, technology categories such as medical devices are not able to go to market with products that are flawed. Medical devices can't go to market at all until they pass through rigorous Federal Drug Administration (FDA) testing and approvals. Still, technology entrepreneurs of all stripes should consider how they can develop minimally viable products they can run through the build, measure, learn cycle to meet market needs.

2.4 PILLAR #3: CUSTOMER DISCOVERY AND VALIDATION

Customer discovery and validation should be the primary focus of technology entrepreneurs during the startup phase of their company. Blank and Dorf have reconceptualized the startup venture as a *temporary organization*, which means that the startup is fundamentally different from an established organization in that the startup must *discover* a scalable, repeatable business model. Once that has been achieved, via customer discovery and validation, then the company must transition from *searching* for a business model to *executing* a business model. In other words, it transitions from being a startup venture to an established organization. This is illustrated in Exhibit 2.3.

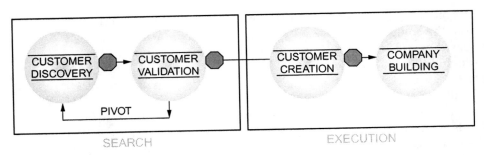

EXHIBIT 2.3
The customer development process.

As the left-hand side of this diagram shows, the focus for the startup should be customer discovery and validation as it searches for a scalable, repeatable business model. There is no mystery in this. Clearly one of the more common reasons that a startup fails is that it executes a preconceived business model that simply doesn't work with customers. A standard (and misleading) view of the technology startup is that it should develop a business plan, a product development roadmap, and a market requirements document before launching. This would suggest that the founding entrepreneurs have a prescient and accurate grasp of customers and their desires before they have tried to sell anything. But that is rarely the case.

A technology startup is a *temporary* organization that runs repeated experiments to gauge customer response to a series of MVP introductions. In this way, the startup can pivot from features and benefits that don't fit with customer needs and desires quickly and inexpensively. By way of contrast, if the standard product development pathway is followed prior to engaging customers, the startup might very well expend most of its capital (and also its goodwill with investors) and be unable to pivot or revise features because it has run out of cash. The Mini-Case below illustrates how Motorola made a lucrative pivot during World War II.

MINI-CASE
Motorola Pivots from Radio to Walkie-Talkies

Robert Galvin founded Motorola to build radios for automobiles. For several decades the business was good, but not great. After two decades, Galvin finally had an opportunity to take a vacation. Traveling in Germany in 1936, Galvin was convinced that Hitler was going to start a war. On return, he sent his assistant to an Army camp in Wisconsin to learn how the Army transmitted information in the field. He learned that the prevailing technique was to run a long wire from the front line to the back trenches. This was not only unreliable, but extremely hazardous for those in charge of running and maintaining the wire. Galvin reasoned, if the company could make radios for automobiles, couldn't it also develop a transmitter to enable two-way wireless communications? As a result of Galvin's hunch that Hitler was going to start a war, his company pivoted to a lucrative and strategically important business model that focused on supplying the U.S. military with the SCR 536, better known as the "walkie-talkie."

Sources: Csikszentmihalyi, M. 1996. *Creativity: Flow and the Psychology of Discovery and Invention.* (New York: HarperCollins Publishers.)

The fundamental purpose of the customer discovery and validation process is to turn guesses about markets, customers, marketing channels, and pricing into facts. Facts cannot be learned by writing a business plan. They can only be learned through direct contact with customers. Robert Galvin's direct contact with the Army led to the walkie-talkie. *Customer discovery* is defined as a process that captures the founders' vision and converts it into a series of hypotheses that can be tested with customers. *Customer validation* is the process of testing whether the evolving business model is repeatable and scalable.

Note that the customer development process illustrated above is circular and iterative. This is in stark contrast to the traditional product development model that moves in a straight line from idea, through design and development, to product launch and sales. Often, as a result of implementing this traditional, linear product development and sales process, startups move forward despite the obvious lack of customer interest. The customer development model, by way of contrast, encourages founders to go back and forth in the product development process based on feedback from customers. In fact, it encourages founders to repeatedly introduce their less-than-perfect product versions to customers and learn something new from each engagement.

Once customer discovery is completed to the founder's satisfaction, the customer validation process should begin. This process focuses on verifying that the emerging business model can be scaled to meet customer demand. During this phase, technology ventures release what are referred to as *high fidelity versions* of their products to test key features with customers. The use of these *test sales* helps the company identify important elements of the business model including:

- The key features that customers prefer
- The existence of a market large enough to be interesting
- The product's perceived value among customers
- Demand for the product
- The *economic buyer* of the product
- Pricing and marketing channel strategies
- The sales cycle and selling process

Each experiment in the customer validation process needs to be designed to address one or more of these key elements of the venture's business model. If technology entrepreneurs complete this process and define each of these elements, they have a better chance of raising the funds necessary to take the company to the next level of growth. In addition, because a repeatable and scalable business model has been verified with customers, the venture's valuation at its initial fundraising will be greater—preserving the founders' equity stakes.

2.5 PILLAR #4: THE BUSINESS MODEL CANVAS

The business model canvas is a new and powerful tool for technology entrepreneurs that was developed by Alexander Osterwalder.[7] A business model is defined as "the logic by which an enterprise sustains itself financially." Put more simply, a business model is "the way the business makes money." It's important for the technology entrepreneur to recognize that business models aren't declared and then executed at the launch of the venture, but instead have to be discovered through interaction with customers. In this way, a business model precedes development of the business plan. Technology entrepreneurs cannot develop a plan of execution until they have discovered a business model that they can execute. The key to an executable business model is that the technology entrepreneur has discovered a *repeatable* and *scalable* system that consistently delivers value to defined customers. By "repeatable" we simply mean that a system can be created that will produce and deliver value to customers on a consistent basis. By "scalable" we mean that the repeatable system can be made to handle enough volume to serve a growing customer base. A good way to begin thinking about the business model for any organization is by asking and answering two fundamental questions:

1. Who is the customer?
2. What does the customer need?

These two questions are really fundamental to either type of enterprise, for-profit or nonprofit. It's clear that these questions apply to profit-seeking ventures, but think about how nonprofits operate. Consider the American Red Cross, for example. The Red Cross has two kinds of customers—people in distress who need direct services, and donors who want to financially support the activities of the Red Cross. Each customer needs something a little different. Those in distress need relief in the way of food and shelter. Donors need to know that the Red Cross is using money wisely and effectively and that people in distress actually are getting the services they need.

To develop a viable business model, the technology entrepreneur—whether starting a for-profit or nonprofit enterprise—must find a way to bring in more cash than is used to create and deliver value to customers. The business model canvas is illustrated in Exhibit 2.4.

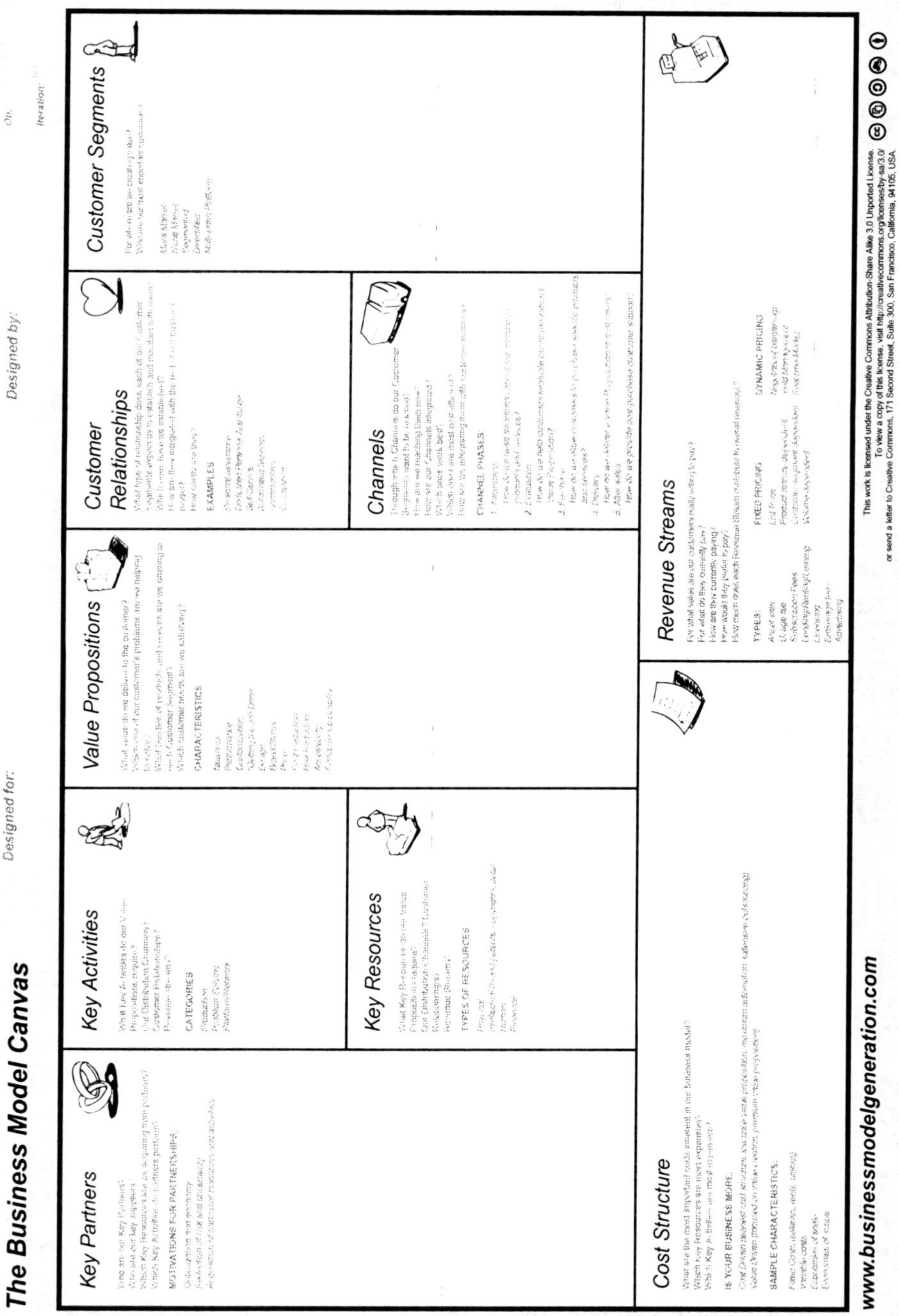

2.5 Pillar #4: The Business Model Canvas

EXHIBIT 2.4
The business model fst.

The business model canvas is divided into nine discrete segments. The technology entrepreneur must address each of the segments of the canvas in order to develop a scalable, repeatable business model, and the canvas provides a handy framework for its development by answering key questions:

Value proposition

- What value do we deliver to customers?
- Which one of our customers' problems are we helping to solve?
- What bundles of products and services are we offering to each customer segment?
- Which customer needs are we satisfying?

Customer relationships

- What type of relationship does each of our customer segments expect us to establish and maintain with them?
- Which relationships have we established?
- How are these relationships integrated with the rest of our business model?
- How costly are these relationships?

Customer segments

- For whom are we creating value?
- Who are our most important customers?

Channels

- Through which channels do our customer segments want to be reached?
- How are we reaching them now?
- How are our channels integrated?
- Which ones work best?
- Which ones are cost-efficient?
- How are we integrating them with customer routines?

Revenue streams

- For what value are our customers willing to pay?
- For what value do they currently pay?
- How are they currently paying?
- How would they prefer to pay?
- How much does each revenue stream contribute to overall revenues?

Cost structure

- What are the most important costs inherent in our business model?
- What key resources are most expensive?
- What key activities are most expensive?

Key partners

- Who are our key partners?
- Who are our key suppliers?
- What key resources are we acquiring from partners?
- What key activities do partners perform?

Key activities

- What key activities do our value propositions require?
- What key activities do our distribution channels require?
- What key activities do our customer relationships require?
- What key activities do our revenue streams require?

Key resources

- What key resources do our value propositions require?
- What key resources do our distribution channels require?
- What key resources do our customer relationships require?
- What key resources do our revenue streams require?

The business model canvas is used by startup technology entrepreneurs in an *iterative* way, meaning that while the startup team addresses each segment of the business model canvas, it may find that other segments need to be adjusted. This iterative process is usually managed in real time by flagging segments with notes using Post-it Notes, describing how that segment will be addressed. As the business model evolves, more flags are placed on top of the old ones. Thus, the top flags represent the latest thinking on how to address that segment of the business model. The notes stacked below the top note represent the historical record of the iterative process. By using this iterative process in each segment of the business model, the technology entrepreneur should eventually find a repeatable, scalable model to use for building and growing the venture.

2.6　PILLAR #5: THE ENTREPRENEURIAL METHOD

Scholars have suggested that expert technology entrepreneurs (and expert entrepreneurs in general) are skilled in what they refer to as the *entrepreneurial method*. This concept is a radical new understanding of entrepreneurship and builds on the concept of the *scientific method* practiced by expert scientists. Think about it for a moment. People can't become expert scientists without having learned how to implement the scientific method in their research, experimentation, and discovery processes. Perhaps it is equally true that people can't become expert technology entrepreneurs without having learned how to implement the entrepreneurial method in their customer discovery and validation processes.

Of course, the question that arises is: "What, exactly, is the scientific method and, incidentally, what is the entrepreneurial method?" Sparing you a long and philosophically obtuse explanation of the scientific method, suffice it to say that scholars have not pinned the definition down precisely, despite centuries of thinking and writing about the topic. In a nutshell, some leading scholars have determined that what makes expert scientists unique is that they have been trained to adhere to certain "principles."[8] Before you get concerned that this is getting too "theoretical," consider that the term "principles" translates to "consistent ways of thinking, behaving, and acting."

For example, one of the hallmarks of an expert scientist is the ability to construct experiments within the context of a specific discipline. Scientists are judged by how well they design and execute experiments, and then interpret and communicate the results of those experiments. That's why it is fairly easy to distinguish expert scientists from crackpots. The crackpot is no less enthusiastic about his or her "discoveries;" but what differentiates the crackpot from the expert is the ability to demonstrate adherence to professionally accepted practices—the principles of designing and executing sound experiments and of communicating results to other scientists.

If there are specific principles that comprise the scientific method, perhaps the same is true of the entrepreneurial method. In fact, there are four principles that have been identified as part of the entrepreneurial method:

2.6.1 Principle #1: Expert Technology Entrepreneurs Believe Value Creation is the Primary Purpose of their Business

Expert technology entrepreneurs have no illusions about the purpose of business. They don't have time to worry about whether the purpose of business is to maximize shareholder wealth, demonstrate "sustainability," or solve social problems. Startup entrepreneurs know that they must create value for other people to generate revenues and profits. And since the startup venture, unlike the large enterprise, must create value from extremely limited resources, the technology entrepreneur is more deeply involved in the actual value creation process than are most corporate managers and leaders. The concept of the "lean startup" aligns well with this principle. The lean startup counsels the entrepreneur to eschew business plan development and focus on bringing the MVP to the market as quickly as possible. According to this perspective, the value creation process is iterative and experimental, relying on customer feedback to discover what the venture should attempt to sell.

2.6.2 Principle #2: Expert Technology Entrepreneurs Rebound Personally and Professionally from Failure

There have been many studies on how entrepreneurs cope with and think about failure.[9] The expert entrepreneur is *resilient*, but that resilience is a result of a deep-seated belief that failure is acceptable and not a reflection of personal worth or capacity to launch and operate future ventures.[10] Expert entrepreneurs are competitive by nature and are prepared to compete in markets where there are clear winners and clear losers.[11] They are fully able and eager to embrace disruptive innovation that clears the way for new industries and new ventures.[12] And they believe that, as long as something useful was learned, rebounding personally and professionally from failure is a virtue.

2.6.3 Principle #3: Expert Technology Entrepreneurs Respect Private Property and Uphold Contractual Obligations

Expert technology entrepreneurs have respect for private property and contracts. To them, private property comprises the fundamental resources that they can leverage to create value for a market. Technology entrepreneurs understand that they are able to do what they please with private property to which they lay valid claim within the boundaries of common law and regulations pertinent to their industry. Expert entrepreneurs also realize that it is necessary to create contracts they are capable of fulfilling. Expert entrepreneurs act intuitively on the premise that private property needs to be respected and that contracts are to be honored. They regard it as morally virtuous to behave in this manner towards private property and contracts and as unseemly to behave otherwise.

2.6.4 Principle #4: Expert Technology Entrepreneurs Respect the Judgment of the Marketplace

Research into the entrepreneurial method is based in part on interviews conducted with expert technology entrepreneurs. One of the widely cited quotes from this research is "I don't believe in market research. I just go out and sell the stuff."[13] Of course, a single data point is insufficient for drawing general conclusions, but the point aligns with more robust research on entrepreneurial market making.[14] Scholars have investigated entrepreneurial opportunity recognition in great detail. Opportunity recognition is in part concerned with the identification of a market need. The concept of opportunity recognition is based on the assumption that a pre-existing market is somehow "out there" in the entrepreneur's environment waiting to be analyzed for "gaps" in the current offerings. However, it is clear from research on how expert technology entrepreneurs actually bring products to market that it is often the case that these entrepreneurs must act as often to *create* the market as to *exploit* it.[15] The act of *market creation* involves skills, techniques, and metrics that differ markedly from those that are necessary for traditional market analysis and new product launch. Market creation requires the entrepreneur to be skilled at listening, experimenting, gathering feedback, and rapid prototyping. Expert entrepreneurs regard it as a virtue to be willing to abandon deeply held beliefs about what customers want and to listen and respond to their feedback to refine and improve offerings. Expert entrepreneurs are willing to run numerous experiments with their product and service offerings and to "pivot," if necessary, based on market feedback. Expert entrepreneurs know that "building a better mousetrap" is not guaranteed to motivate customers to "beat a path to their door."

2.7 DELIBERATE PRACTICE

Becoming an expert technology entrepreneur requires practice. It is nice to believe that there is such a thing as a "born entrepreneur," but this is unlikely. In fact, we advocate that anyone can become an entrepreneur given the opportunity to practice the skills that we outline in this book. Some people seem to become entrepreneurs at a younger age than others—and many seem to do so without formally practicing the art of entrepreneurship. However, a closer look typically shows that they had been already exposed to the basic principles of entrepreneurship long before they achieved recognition as an entrepreneur.

Becoming an expert in anything requires practicing, learning, adjusting, and trying again and again. The scholarly literature on what it takes to become an expert has identified a technique called "deliberate practice." Individuals who engage in deliberate practice acquire superior knowledge structures and subsequently develop superior performance. Real expertise exhibits three characteristics: (1) it leads to performance that is consistently superior to that of the expert's peers; (2) it produces concrete results; and (3) it can be replicated and measured.[16] The principles of deliberate practice that you can use to develop your entrepreneurial expertise include:

- *Motivation*: Individuals should be motivated to undertake deliberate practice and develop expertise. Aspiring technology entrepreneurs should tap into whatever motivations are strongest for them, whether it is acquisition of wealth, solving a major social problem, or the sheer enjoyment of starting companies.
- *Understandability*: Rather than studying stories of successful entrepreneurs, which are mostly unique and nonrepeatable, your efforts to develop entrepreneurial expertise should focus on becoming skilled in the entrepreneurial method. The principle of deliberate practice means that you can only practice what you understand. The entrepreneurial method has broken down the tenets of expert entrepreneurship into meaningful and understandable principles. This means that, as an aspiring entrepreneur, you can learn to adhere to the principles to be applied across entrepreneurial domains and in each of the opportunities that you will encounter in your lifetime.
- *Feedback*: For deliberate practice to affect learning, immediate feedback on performance is important. This trial and feedback part of the learning

process is critical as you try new behaviors and modify them in the face of feedback and is used by experts as they continuously upgrade their skills. Look for opportunities to practice new behaviors and understandings, and gather immediate feedback on those performances.

- *Repetition*: Deliberate practice involves repeated performance of the same or similar tasks. The motivation required to repeatedly practice is one of the key distinctions between experts and people who merely have experience. Of course, practicing new behaviors initially can be awkward and uncomfortable. New behaviors usually do not become comfortable until you try them, practice them, and only then determine whether they are making things better for you and your venture.
- *Fit*: This component of deliberate practice asserts that the tasks being practiced must fit the individual and the contextual circumstances. For example, a person who aspires to be a computer game designer must not only have appropriate equipment, but must also be fortunately endowed with appropriate talent. If either is lacking, there would not be a fit between the practice and the goals. Aspiring technology entrepreneurs must learn that success is a function of talent, expertise, environment, and other factors. Expert technology entrepreneurs have either consciously or unconsciously responded effectively to the fundamental question of effectuation: "Given who you are, what you know, and whom you know, what types of economic and/or social artifacts can *you*, would *you* want to, and should *you* create?"[17]

Of course, it might be difficult to find opportunities to "practice" while you are striving mightily to build a company. Recognizing this, expertise scholars have developed a slightly different approach that you can use to develop your skills while you are striving to make your venture successful. This approach has been referred to as "*deliberate performance*."[18] Deliberate performance differs from deliberate practice in that deliberate performance occurs "on the job." Practice typically refers to undertaking simulated performance with the intention of improving performance in the "real world." But opportunities for aspiring entrepreneurs to practice are limited. Many are already to be in the "real world" and need to begin to create results. To remedy this, consider deliberate performance to be your practice routine as you build your venture and become an increasingly adept technology entrepreneur along the way. Many technology entrepreneurs are so-called serial entrepreneurs. That is, they are involved with multiple ventures over time. However, it is rare to find a serial entrepreneur who has not learned valuable lessons and applied them from one venture to the next. If such lessons are not learned, the "serial entrepreneur" had better consider getting a job, because investors aren't interested in entrepreneurs who repeat their mistakes.

2.7.1 When to Start Practicing

There are many opinions about when is the right time to become an entrepreneur. Some believe that entrepreneurship is only for the young. The many technology companies that have been launched by youthful entrepreneurs over the last several decades highlight this perspective. Of course, it is possible that the overweighting towards youth in the technology startups of the last two decades (especially Internet startups) is related partly to the newness of the technologies. Older entrepreneurs simply were not bred on the technologies and don't understand them as well as do youth who have been using and learning the new technologies from a young age. We are now seeing a trend toward second-generation technology entrepreneurs—those who had a successful startup when they were young and are now older and launching their second and third startups.

Being young is helpful when new technologies are rapidly emerging, but even the Internet is now two decades old, and more mature and savvy business models are taking over. Starting at a young age would also be helpful because of the tremendous levels of energy that are normally required to launch new ventures. On the other hand, greater levels of experience among older entrepreneurs can lead to greater efficiencies and less stress and strain.

In reality, there is no single time in life when it is better to launch your entrepreneurial career. In fact, the average age of the first-time entrepreneur is probably older than you think: research has indicated the average age of 39.[19]

Another factor that often concerns aspiring entrepreneurs is the amount of capital they have at their disposal. There is a common misunderstanding that launching a new venture requires large amounts of capital. In reality, the average amount of startup capital in new ventures is $25,000.[20] Expert entrepreneurs have learned that there are many significant ways to keep costs low when launching a new venture and that managing costs is paramount to long-term success. Of course, cost management means making difficult choices—such as forgoing fancy office space in favor of a larger marketing budget.

Another important lesson that expert entrepreneurs learn is that they must surround themselves with other effective people to become successful. A common mistake of aspiring entrepreneurs, by way of contrast, is to go into business with friends or family members *because* they are friends or family members. This more often than not leads to strains in the business and strains in the relationship. We are not saying that you shouldn't go into business with friends or family members, but be certain that you are doing so for the right reason. Anyone—including a friend or family member—whom you decide to align with as a business partner should have the necessary personal capabilities and skills to help the business succeed. Deciding when to launch into your first venture depends in part on the people you are able to recruit to your vision. Technology entrepreneurship is almost always a team sport. You are well advised to wait until you have recruited the necessary talent and experience to your venture before launching.

Finally, we will stress once again the need to find a good fit between your unique talents and the type of business that you launch. With entrepreneurship on the rise around the world it is becoming increasingly difficult for new entrepreneurs to find and fill a niche. You are far more likely to be successful if you build your business on talents that you possess to a greater extent than the average person. If you are a talented programmer, find a way to build a business around that talent. If you are a talented game designer, build a business around that talent. All of us possess some unique talent that can be leveraged in myriad ways to become successful. Do not launch your venture until you are sure that you are leveraging your greatest talents and strengths. Normally, as people age and mature they develop a keener sense of their unique talents, effectively narrowing the range of potential ventures that they might decide to launch.

2.8 CHAPTER SUMMARY

This chapter introduced you to what we call the "five pillars of technology entrepreneurship." Technology entrepreneurship differs from other types of entrepreneurship because it is generally easier to experiment with and rapidly iterate features and benefits of technologies than it is to do the same with features and benefits of, say, a restaurant. A restaurant entrepreneur generally opens the establishment with his or her full vision already in place. It would be difficult for the restaurant entrepreneur to mimic the customer discovery and validation processes that the technology entrepreneur should embrace. It would also be difficult for the restaurant entrepreneur to follow the precepts of the lean startup. What, for example, would constitute the "minimally viable product" of a restaurant? Certainly a restaurant entrepreneur can experiment with menus, ambience, and other features. But these are usually not introduced and discarded as quickly as features associated with a new technology could be.

As such, it's important for technology entrepreneurs to absorb the lessons of the five pillars discussed in this chapter. This is especially true for technology entrepreneurs who come from nontechnology entrepreneurial backgrounds. We've discussed that following a traditional, linear product development path from idea through sales is fraught with potential dangers for technology entrepreneurs. The dangers lie in the possibility that customers may not be interested in the finished product, and reversing course after a massive investment in product development may not be an option.

In addition to the lean startup and the customer discovery and validation processes, you also learned the importance of value creation and establishing a suitable value proposition. Creating a compelling value proposition is one of the segments of the business model canvas. The canvas is also an important tool for technology entrepreneurs because the goal of the startup (defined as a "temporary organization") is to discover a scalable, repeatable business model.

Finally, you were introduced to the concept of the "entrepreneurial method." To the extent that entrepreneurial expertise, like scientific expertise, depends on application of a consistent method, it is important to understand what you need to practice to develop such expertise. We've reduced the concept of the entrepreneurial method to a set of specific "principles" that you can practice while honing your own entrepreneurial skills. The principles center on value creation, respect for private property, resilience in the event of failure, and respect for the judgment of the marketplace. You can practice these via the processes of deliberate practice or deliberate performance, as we elaborated.

There is no time like the present to begin to develop the principles of the entrepreneurial method on your journey to becoming an expert technology entrepreneur. The challenge is to begin; the journey itself will be your reward.

KEYTERMS

Deliberate practice A technique for becoming expert in nearly any field.
Value proposition What a venture tells its customers about the value it provides to them.
Minimal viable product A nascent product that is presented to customers to gather feedback.
Pivot or persevere A move away from the current business model to a revised one; persevering connotes continuation with the current business model.
Temporary organization A startup with the goal to discover a business model; once that is done, execution and growth are paramount.
Customer discovery The process of converting the founders' vision into hypotheses that can be tested with potential customers.
Customer validation After customer discovery, the process of validating a scalable, repeatable business model.
High fidelity versions Products that have been refined iteratively through customer discovery and are now used in the customer validation process.
Test sales The activity of introducing the high-fidelity version of the product to customers to gauge their responses to the venture's sales approach and value proposition.
Economic buyer The person who finally decides whether to purchase a product.
Scalable business model A business model that can be grown to meet increasing customer volume.
Repeatable business model A business model around which a system can be built that delivers consistent value to customers.
Iterative The need to repeatedly address a problem until a suitable solution has been developed.
Entrepreneurial method It is thought that expert entrepreneurs practice a method similar to the way scientists practice the scientific method.
Market creation The entrepreneur who has a core competency, or develops a product, and then looks for applications of or creates a need or want in the marketplace.
Deliberate performance Practicing the techniques of deliberate practice while "on the job."

ADDITIONAL READING

Osterwalder, A., Pigneur, Y., 2010. Business Model Generation: A Handbook for Visionaries, Game Changers, and Challengers. John Wiley & Sons, Hoboken, NJ.

Anders Ericcson, K., 1996. The Road to Excellence: The Acquisition of Expert Performance in the Arts and Sciences, Sports and Games. Lawrence Erlbaum Associates, New York.

Ries E. 2011. The Lean Startup: How Today's Entrepreneurs Use Continuous Innovation to Create Radically Successful Businesses. Crown Business, New York.

Blank, S., Dorf B., 2012. The Startup Owner's Manual: The Step-by-Step Guide for Building a Great Company. K&S Ranch Press, San Mateo, CA.

WEB RESOURCES

http://www.businessmodelgeneration.com/: This is the website for the business model canvas and contains other useful tools for technology entrepreneurs.

http://ecorner.stanford.edu/authorMaterialInfo.html?mid=1465: This link is to Stanford's "ecorner," which hosts a wide range of resources for technology entrepreneurs. This particular link takes you to some helpful videos on the process of value creation.

http://www.iwanamaker.com/: This is the website of the company discussed in the short vignette at the beginning of this chapter. You may want to check in to see how the venture is faring.

ENDNOTES

1. Ericcson, K.A., M.J. Prietula, and E.T. Cokely. (2007). The Making of an Expert. *Harvard Business Review*, 85(7/8): 114–121.
2. Farzad R. 2006. "A Deal that Paid for Itself," *Business Week*, October 30, p. 38.
3. Tom Cheshire. 2011. How Rovio Made Angry Birds a Winner (and What's Next). *Wired*, 19(4).
4. See www.pg.com.
5. Stringer, S. 2008. "Connecting Business Needs with Basic Science," *Research Technology Management*, January/February, 9–14.
6. Ries, E. 2011. The Lean Startup: How Today's Entrepreneurs Use Continuous Innovation to Create Radically Successful Businesses. (New York: Crown Business).
7. Osterwalder, A. and Y. Pigneur. 2010. Business Model Generation: A Handbook for Visionaries, Game Changers, and Challengers. (Hoboken, NJ: John Wiley & Sons.)
8. Duening, T.N. and M.M. Metzger. (2013). The Entrepreneurial Method: Moral Virtues as the Foundation of Entrepreneurial Expertise. *American Journal of Entrepreneurship*
9. Cardon, M.S., C.E. Stevens, and D.R. Potter. 2009. Misfortunes or Mistakes? Cultural Sense-Making of Entrepreneurial Failure. *Journal of Business Venturing*, 26(1): 79–92.
10. Hayward, M.L.A., W.R. Forster, S.D. Sarasvathy, and B.L. Fredrickson. 2009. Beyond Hubris: How Highly Confident Entrepreneurs Rebound to Venture Again. *Journal of Business Venturing*, 25(6): 569–578.
11. Isenberg, D. 2013. Worthless, Impossible and Stupid: How Contrarian Entrepreneurs Create and Capture Extraordinary Value. (Cambridge, MA: Harvard Business Review Press.)
12. Christensen, C.M. 1997. The Innovator's Dilemma: When New Technologies Cause Great Firms to Fail. (Cambridge, MA: Harvard Business School Press.)

13. Read, S. and S. Sarasvathy. 2005. Knowing What to Do and Doing What You Know: Effectuation as a Form of Entrepreneurial Expertise. *Journal of Private Equity*, 9(1): 45–62.

14. Read, S., N. Dew, S.D. Sarasvathy, M. Song, and R. Wiltbank. 2009. Marketing under uncertainty: The logic of an effectual approach. *Journal of Marketing*, 73(3): 1–18.

15. Sarasvathy, S.D. 2001. Causation and Effectuation: Toward a Theoretical Shift from Economic Inevitability to Entrepreneurial Contingency. *Academy of Management Review*, 26(2): 243–263.

16. Ericsson, K.A., M.J. Prietula, and E.T. Cokely. 2007. "The Making of an Expert." *Harvard Business Review*, 85(7/8): 114–128.

17. Sarasvathy, S.D. 2001. op. cit.

18. Fadde, P.J. and G.A. Klein. 2010. "Deliberate Performance: Accelerating Expertise in Natural Settings." *Performance Improvement*, 49(9): 5–14.

19. Shane, S. 2007. The Illusions of Entrepreneurship: The Costly Myths That Entrepreneurs, Investors, and Policy Makers Live By. (New Haven: Yale University Press).

20. Shane, S. 2007. Ibid.

CHAPTER 3

Technology Venture Idea Generation

NASTY GAL HITS IT BIG IN COMPETITIVE FASHION INDUSTRY

Sophia Amoruso is not your typical successful technology entrepreneur. She is a photography school dropout who has a taste for vintage clothing. Amoruso started her business in her ex-boyfriend's San Francisco apartment by selling her favorite vintage clothing "finds" on eBay and via her Nasty Gal Myspace page. She used her own models and took her own photographs. For example, Amoruso bought a Chanel leather jacket at a local Salvation Army store for $18. Photographed on an attractive model, the jacket sold for more than $1000. "In the beginning, I was basically paying the models with hamburgers," Amoruso said.

Amoruso left home at the age of 17 and traveled to Sacramento, California where she found herself dumpster diving, working in record stores, and generally drifting up and down America's west coast. She launched Nasty Gal at the age of 22 and quickly developed a loyal following. When the demand from Nasty Gal's 60,000 plus users exceeded her supply, Amoruso approached some fashion design shops to help her create her own brand of vintage clothing. Today, her company sells its own racy vintage-looking clothes at affordable prices, as well as some actual vintage clothing. By 2010 Amoruso was entertaining offers from a host of venture capital firms. Mostly, she turned them down. However, when she expanded in 2012 she accepted a deal with Index Ventures for a $50 million equity investment. In 2013, Amoruso was 29 years old and in charge of a $240 million technology based fashion retail empire.

Sources: Adapted from John Ortved, "Sophia Amoruso Expands Nasty Gal," *The Wall Street Journal*, August 22, 2013; Victoria Barrett, "Nasty Gal's Sophia Amoruso: Fashion's New Phenom", Forbes, June 28, 2012; Nicole Perloth, "Nasty in Name Only," *The New York Times*, March 24, 2013.

3.1 INTRODUCTION

The question that all aspiring technology entrepreneurs need to answer in their own unique way is: What business should I start? Aspiring entrepreneurs are often confounded by the challenge of developing a credible idea for a new product or service. Consulting friends and family can be a useful start in developing and refining an idea for a new venture, but that may not be the best approach. Those friends and family members might be tempted to tell you what you want to hear, rather than providing the critical feedback that will help you refine your idea.

A good place to start generating an idea for a new technology product is to consider the type of venture you'd like to start. The range of possible ventures is as broad as the human imagination. Today, we are witnessing the first private space ventures launched by some of the leading entrepreneurs of the era. As we learned above in the Nasty Gal story, new forms of retail are launching nearly every day. Seemingly, there is no end to the proliferation of venture types. Fortunately, most technology ventures can be understood as one of only four fundamental types. We will explore these four types in detail in this chapter.

All technology ventures are based on the founders' vision for a new product or service. The challenge for the aspiring entrepreneur is to turn this germ of an idea into an operating venture. In this chapter, we introduce a four-step process to generate and crystallize ideas for a new technology product or service. This process will help any new entrepreneur identify a target market, develop and refine a product-service concept tailored to that market, test various business models for the chosen market, and explore how to acquire customers.

As we discuss in Chapter 2, developing ideas into thriving ventures is an iterative process.[1] That means most ideas aren't hatched fully formed, but rather change and develop over time as new information is gathered and results of experiments are analyzed. Expert entrepreneurs know that regardless of how well they understand a market, it will nearly always behave differently than they expect—sometimes radically so. The expert entrepreneur is determined to make his or her idea successful, but he or she has also learned to be flexible and adaptable. Adhering to an original idea out of stubbornness or pride will not make a bit of difference to customers. Ultimately, the goal of every technology entrepreneur is to *create value* for customers. And customers have many alternative choices today if a product does not fit their needs and desires.

Because product idea development is iterative, it is useful for aspiring entrepreneurs to use what is called an opportunity register. The opportunity register is a type of a personal journal that is dedicated to helping entrepreneurs keep track of their evolving product or venture ideas. An opportunity register can also be a means of recording intellectual property that is being created. In this chapter we will introduce the concept of an opportunity register and how to maintain a formal one.

Finally, to help generate exciting ideas for a new technology venture, we introduce some alternative sources of idea generation. The most likely source of idea generation is a person's own experiences, talents, and background. Still, on occasion, aspiring entrepreneurs are not able to generate a spark of insight or innovation on their own. They need an external stimulus to get their creative juices flowing.[2] Sometimes people can be inspired by reading something different from their normal fare, they can listen to inspiring people, and they can experiment with other techniques that we will recommend.[3] Let's begin this chapter by identifying the four fundamental venture types.

3.2 FUNDAMENTAL VENTURE TYPES

Entrepreneurs are among the most creative people on the planet. Imagine all the myriad ventures that have been created just in the past decade. Some of these ventures have had dramatic impacts on our economy, society, and ways of life. Facebook has more than 1 billion users worldwide and people are now staying connected to extended family and friends like never before. Twitter has revolutionized the art of the pithy comment. Tesla Motors has created the first all-electric sports car. In addition to these companies, here's just a short list of ventures that were founded by technology entrepreneurs and that have the potential to change the way we live and work[4]:

- Chegg
- Netflix
- Hulu
- Pinterest
- Instagram
- ZipCar
- LinkedIn
- AirBNB

Of course, these are all technology or Internet-related ventures, but there are many other nontech new ventures that have also changed our lives. Think of how FedEx has affected package delivery or how Southwest Airlines has changed passenger air travel. Think about beverage companies that have made bottled water—bottled water!!!—nearly ubiquitous.

Despite the many ways that entrepreneurs make millions through new ventures, in reality there are only six fundamental venture types that can be created. The six types are depicted in Exhibit 3.1.

As Exhibit 3.1 indicates, you can launch a venture that creates value for businesses, consumers, or the government. You can also create a venture that is primarily product-centric, or services-centric. That's really about your only choices, six fundamental venture types. This framework provides a handy

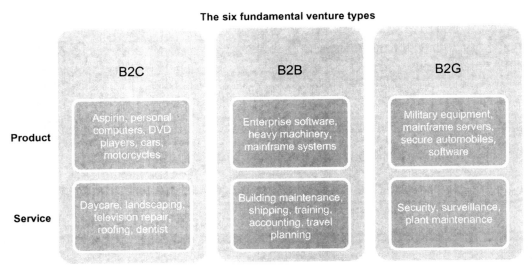

EXHIBIT 3.1
The six fundamental venture types.

starting point in your venture idea development process, but it is usually not as simple in reality. Many ventures offer both products and services to their clients. For example, new ventures that specialize in enterprise software often also have a service component to their businesses. Software providers might need to help install the new software for their clients, and many charge a fee for such services. Many also offer "maintenance plans" for a fee, including access to technical support in the event that problems arise with software usage.

Many technology ventures also serve both business and nonbusiness customers.[5] For example, Dell sells computers both to business customers and to retail customers. The company has separate web environments to service the different types of customer. Their corporate structure and reporting alignment separates these two business functions because the types of products, price points, marketing strategies, and service arrangements are vastly different for each customer type.

In general, aspiring entrepreneurs should attempt to limit their idea development—at least initially—to only those in one of the quadrants shown in Exhibit 3.1. It is difficult enough to create value for a single type of customer without having to worry about serving two vastly different customer types. A venture launched in one quadrant can always develop products or services in the other quadrants as its capabilities and resources expand. Let's explore each quadrant in more detail, beginning with business-to-business ventures.

3.2.1 Business-to-Business

Ventures that are launched to sell to other businesses are referred to as **"business-to-business"** or **"B2B"** companies. There are many good reasons to consider launching a B2B venture. One good reason is that businesses usually have more money than consumers and may be able to purchase better the venture's products and services. B2B ventures come in many varieties because the needs of business are many. For example, many companies do not manage their own data centers or server farms. Companies such as Amazon Web Services provide these services to other businesses—even to those that may directly compete with Amazon's other business units. Amazon Web Services is a significant part of Amazon's overall revenue, but most of us know Amazon primarily for its e-commerce business.

New B2B ventures don't have to confine their business models to providing services to other businesses. There is just as much need for innovative new products. If you are thinking about building a venture in a B2B quadrant, there are three primary ways to create value:

1. Help companies make more money
2. Help companies save money
3. Help companies comply with government regulations and/or community norms

B2B ventures have to create value in one of these three categories or they simply won't attract business customers. Businesses are not interested in "fashionable" or "whimsical" items in the way consumers often are. Most businesses are focused on making profits, and the only way to do that is through increasing sales, decreasing costs, or avoiding litigation and/or fines as a result of running afoul of regulations and norms.

Helping businesses increase sales is a popular focus for new ventures. For example, software designed to make a sales force more effective is a very crowded arena for venture development. Sales force automation, lead generation, and customer relationship management are just three types of enterprise software that target this space. Each of these three categories is crowded both with major incumbent firms, such as SalesForce.com, and a steady stream of new entries. Even though the space is crowded, significant new innovation over the existing products can still break through. However, any aspiring enterprise software entrepreneur should be aware that most companies are slow to change their existing ways of doing business, including shifting from one software type to another. There can be many costs associated with switching that prohibit consideration of a new product or service provider.

Effective cost-saving products or services for businesses almost always find a market. The focus for many of today's medium-sized and large companies is referred to as "lean." Lean is the catch-all term for constant attention to cost and cost-saving tactics. The lean enterprise is generally organized around specific programs, such as "Six Sigma," but there is plenty of opportunity for innovative new approaches to helping businesses reduce their costs.

One of the most potent new approaches to cost saving for companies is business process outsourcing, or BPO.[6] Businesses have learned that they can outsource their noncore functions to other companies that perform that function as their primary business activity. Entrepreneurs who saw this emerging trend were able to capture this booming business. Of course, many of the top outsourcing firms, such as InfoSys and Tata Group, are located in India. These ventures have been among the fastest growing companies in the world. For example, Infosys was founded by six Indian engineers in 1989, and the company reported 2012 revenue of nearly $7 billion.

Another category for B2B ventures that has emerged in recent years concerns the need for companies to comply with government regulations and/or demonstrate adherence to prevailing community norms. Few efforts to comply with government regulations or community norms are directly related to increasing sales or decreasing costs. Most are seen primarily as "preventative" in nature. For example, many companies today have implemented extensive, and often expensive, "sustainability programs."[7] Many such programs have little direct impact on a company's sales or assist the company in reducing costs. However, proponents of such programs argue that having a sustainability program promotes positive impressions of the company in the community and prevents possible negative goodwill. Many entrepreneurs have taken the initiative to develop products and services to serve this type of business demand. For example, Sustainable Minds is a venture that was formed to help companies analyze and manage the lifecycle impact of their products and services. The company was formed initially to provide companies with an analytic software tool. It has since grown to include a range of related products and services.

3.2.2 Business to Consumer

New ventures that target the consumer abound and are referred to as **"business to consumer,"** or "B2C," ventures. There are many ways that technology entrepreneurs can create value for consumers in both the service and product categories. People have a seemingly insatiable desire not only for functional products and services that help them meet their daily living needs, but also for fashionable, adventurous, whimsical, or entertaining products and services. For example, think about the enormous number of consumers who purchase games for their computers. Today, fully 58% of Americans play video games, and 51% of American households own at least one computer game console. As you can see from Exhibit 3.2, video game usage has displaced other forms of consumer products.

Consumer products and services are in some ways more difficult to develop because of the great variety of ways in which value can be created. We saw in the

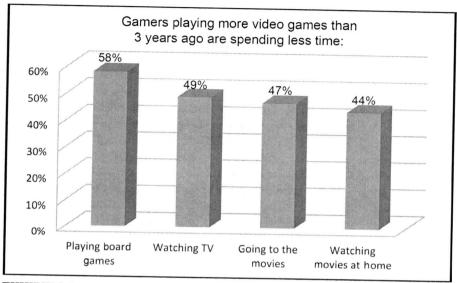

EXHIBIT 3.2
Video games have replaced other consumer products. *Source: The Entertainment Software Association, 2012. Essential Facts About the Computer and Video Game Industry. The Entertainment Software Association, Washington, DC.*

section above that there are really only three ways to create value for businesses: help them make more money, help them save them money, or help them comply with government regulations and social norms. Below is a noncomprehensive list of ways in which technology entrepreneurs can serve consumers:

1. Functional: 3D printing, home automation, and appliances
2. Entertainment: Video games, streaming media, and laser tag
3. Adventure: Virtual worlds, space tourism
4. Fashion: Google glasses, smart watches, and nano-fiber apparel
5. Transportation: Electric cars, electric bicycles, and hybrid vehicles
6. Health: Medical devices, fitness gear, and quantified self
7. Communication: Smartphones, online television, and e-mail
8. Shelter: Camping gear, smart homes, and emergency shelters
9. Food: Delivery services, energy drinks
10. Art: Electronic galleries, electronic photo frames
11. Literature: Blogs, tweets, and e-readers
12. Music: Streaming audio, electronic instruments

And the list could go on for likely hundreds of different categories, including some that have not even been invented yet.

Creating products and services for consumers is competitive, and most entrepreneurs that have been successful with consumer products counsel against being a **single-product venture**. It is fine to *launch* a venture around a single consumer product, but it is difficult to *grow* a company around a single product. The reason for this is well known to those who have competed in the consumer product space. If a product is successful, it can rapidly grow through multiple distribution channels, including the big retailers such as Costco and Wal-Mart. However, that very success can also prove to be a venture's undoing. Other companies that sell products through Wal-Mart and the other big retailers will likely not be far behind with an offering that directly competes. While an entrepreneur may have patents and other intellectual property rights to prevent outright copying, there are sometimes ways to get around these barriers. So while a single-product venture might enjoy success over the short-term, it is also likely to see its revenues decrease as the market becomes more competitive. The moral of the story is, if you desire to start your own consumer products company, you should have a number of potential product ideas that you can bring to market in succession and that are consistent with the brand image that you're developing. With such a strategy, you can build your company brand, and consumers will begin to look for that brand when deciding which new products to purchase.

3.2.3 Business to Government

The **business to government** (B2G) category is in many ways similar to the B2B category. The primary difference in doing business with the government over doing business with other companies is that the government often has highly restrictive rules that must be followed by its contractors. Usually, acquiring a government contract begins with bidding on a request for proposals (RFP) offered by a government entity. The RFP may specify that a single company will be awarded the contract, usually referred to as a sole-source contract. However, government projects often require that at least a portion of the work be done by a small business and/or a minority-owned enterprise. These so-called **set asides** ensure that the government is employing diverse businesses in its contracting.

In addition to rigorous stipulations that go with the bidding process, there also are rigorous standards that apply to those entities that win the bid and are contracted to conduct the work. For example, the government normally has very restrictive profit margins that it allows the performing ventures to make. That is, contracted ventures will be required to keep their profits within the allowable margins. Failure to comply with stipulations such as this exposes the venture to potential government auditing and, possibly, substantial penalties for noncompliance. In fact, for ventures that contract with the military, for example, their accounting reports will be scrutinized by the Defense Contract Audit Agency (DCAA). It is advised that technology entrepreneurs who seek to operate B2G ventures first learn about the accounting standards that apply to such ventures before bidding and executing on contracts.

Governments around the world purchase both products and services from private companies on a daily basis. The government can potentially be a lucrative client for a technology entrepreneur, but it is also advisable to avoid having government agencies as the only clients for the venture. This is because government contracts often are subject to funding cycles and political processes that could jeopardize continued operation. In the United States, for example, many government contracts with private business organizations were adversely affected by the so-called sequester. This was a congress-led response to perceived overspending and resulted in the cancelation or postponement of many contracts with private companies.

3.3 THE IDEA GENERATION PROCESS

As we discussed in Chapter 2, the product development process is iterative and is best when it involves direct customer engagement early and often. Of course, before you have a product that you can present to customers you must first have an idea for a product. We've developed a four-step process that you can use to develop your idea.

1. Identify a point of pain in a target market
2. Innovate a product or service
3. Test business models
4. Explore how to acquire customers

3.3.1 Step 1: Identify a Point of Pain

It is much easier to start a new venture if you have some idea of what your **target market** will be and what **point of pain** you intend to address.[8] Your target market doesn't determine the entire market opportunity for your venture's products or services, but it does give you an important starting point. Every technology venture has to start with the first customer, and it is far easier to get your first customer if you identify your target market, thoroughly study its pain points, needs, and idiosyncrasies, and tailor your offerings to that market.

We will discuss target market identification and analysis in far greater detail in Chapter 5. For now, simply be aware that you need to identify a point of pain in a target market, preferably one with which you have some familiarity and connections. Customer pain points can be identified using a number of techniques. One useful way of considering the typical customer experience is through a model known as the **consumption chain**. The consumption chain illustrates a typical set of steps customers must go through to acquire and use goods and services. Any one of the steps in the consumption chain may be a customer point of pain and ripe for a new solution. The consumption chain is illustrated in Exhibit 3.3.

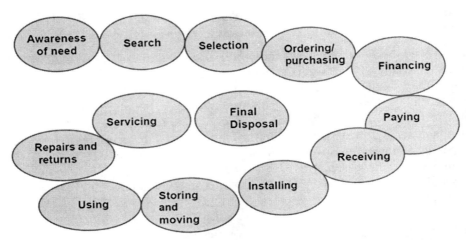

EXHIBIT 3.3
The consumption chain. *Source: McGrath, R.G., MacMillan, I., 2000. The Entrepreneurial Mindset. Harvard Business School Press, Cambridge, MA.*

This illustration is a handy way of finding pain points in nearly any market. Consider the Birchbox example (Mini-Case below). The founders realized that one of the pain points for consumers who comprise a target market interested in cosmetics is searching for the right new products. Driving to stores and trying new products can be time consuming and prohibitive for many shoppers. However, having the opportunity to try new products in the comfort of their

MINI-CASE
Birchbox Addresses Customer Pain Point

Birchbox is a pioneer of what is known as "subscription e-commerce." The business model is based on registered users receiving a monthly box of personal care products. After trying the products, Birchbox provides opportunities for users to purchase full versions of the products via its website. The company sends its 400,000 registered users one new box per month at a price of approximately $10/month. This part of the business model generates revenue, but not much profit. Birchbox founders knew that profitability would only result from converting the users into buyers of the products they were sampling. Birchbox co-founder Katia Beauchamp said "When customers find a product that they love from Birchbox, we try to make it as easy as possible for them to shop and then transact." Apparently, the model is beginning to work. One upscale cosmetics brand indicated that over 11% of Birchbox users who received a free sample of their product were converted into buyers. The company, founded in 2010, was continuing to grow. In 2013 it reported that its e-commerce sales were on track to triple, with 25% of sales generated through its e-commerce store and more than half of its users making purchases there.

Sources: Adapted from K. Mai-Cutler, "Birchbox Says Its Subscription Model Is Working, with 25% of Revenues Coming from Its E-Commerce Store," TechCrunch, August 15, 2013; D. Ransom, "Birchbox and Changing the Way We Shop," Entrepreneur.com, May 16, 2012; C. Rubin, "Hayley Barna and Katia Beauchamp: Founders of Birchbox," Inc.com, June 27, 2011.

own homes eliminates that problem and encourages consumers to purchase products they might otherwise have never discovered.

3.3.2 Step 2: Innovate a Product or Service

You may already have a rough idea of the product or service that you'd like to take to your target market. Be careful, however, not to be satisfied with your initial intuitions about what your customers want. To innovate a product or service idea is both an art and a science. There are a number of possible strategies that you can use. For example, the point of pain may be a common problem that you'd like to solve. The Mini-Case below highlights how one technology entrepreneur used his own experience and technical background to come up with a new product idea.

MINI-CASE
Coffee Joulies Keeps the Temperature Just Right

David Jackson wanted to start a business with his friend and fellow engineer Dave Petrillo based on their mutual interest in phase-change materials. One morning, during his usual routine, Jackson was making a latte for himself. He lamented that it would take 20 min to make the perfect latte, but once it was ready it would be either too hot or too cold to drink. When thinking about how to fix the problem, he realized that phase-change materials could be the solution to that problem. When he called his friend to talk about it, they realized that they were onto something. As a result of thinking about this common problem, Coffee Joulies were born. Coffee Joulies are metal beans about the size of ice cubes that are filled with phase-change material. When placed in coffee, the Joulies absorb excess heat to cool the coffee to 140 degrees. When the coffee cools below that temperature, the Joulies release the stored-up heat to keep the coffee at a stable temperature for up to 5 h. Jackson and Petrillo sought funding for their idea via Kickstarter and received more than $300,000 in pledges. Jackson and Petrillo combined their passion for working with phase-change materials with their everyday experiences to generate an idea for a new product. The product would not have much merit if Jackson were the only latte drinker in the world. But they both knew that latte drinkers comprise a large target market opportunity for their product idea.

Source: Westervelt, A. 2013. "A Morning Routine Sparked a Startup." *The Wall Street Journal*, August 19, p. R7.

In addition to the problems of everyday experiences as a source of product ideas, aspiring entrepreneurs have other potential sources of great new ideas. Below is a list of some techniques used by technology entrepreneurs to innovate new product ideas[9]:

1. **"Look at what is bugging you"** (David Cohen, Founder and CEO of TechStars): Startups are often based on a problem that needs to be solved. Identifying the problem usually happens while you're busy working on some project. The idea for Coffee Joulies came to David Jackson while he was working hard to make the perfect latte.
2. **"Be present in life"** (Angela Benton, Founder and CEO of NewME Accelerator): Many successful entrepreneurs do their brainstorming

around problems in which they are personally invested. Businesspeople tend to ignore their creative sides. Innovation often comes from maintaining a balance between technology and the arts.

3. **"Let your subconscious do the work"** (Ben Baldwin, Co-Founder and CEO of ClearFit): Creativity often happens when the mind is occupied with a monotonous task, leading to the coveted "eureka moment." Baldwin came up with the idea for ClearFit, a venture that makes it easy for companies to find employees that fit their culture, while he was driving 80 miles an hour and not thinking about work. He advises that aspiring entrepreneurs take a break, smell the flowers, and while they are doing that their subconscious mind may be devising a solution to a problem.

4. **"Attack practical problems"** (Brian Spaly, Founder and CEO of TrunkClub): Spaly advocates making a mental or actual note whenever you encounter a customer service or customer experience that is frustrating. Ask yourself whether you can come up with a solution to the problem you've encountered. Spaly's insight was that men's clothing stores were not helping men achieve their own unique style. His company, TrunkClub, is a men's clothing store that asks men to enter information about their style preferences, sizes, and stores in which they shop to match them with a personal stylist.

5. **"Think big"** (Kevin Colleran, Venture Partner, General Catalyst Partners): Colleran advises technology entrepreneurs to "go big or go home." Other advice he provides is to think about how you can make the world a better place. He says that the best technology entrepreneurs pursue ventures that simplify or improve the lives of many people.

You are likely to cycle through multiple revisions of your product or service concept before it is ready for the market. A good way to iterate and develop your product over time is to demonstrate it to selected members of your target market and ask for their reaction. This process is referred to as **customer validation.** For example, Jim Holley, an inventor and patent holder of baby products brought his first product to market after more than a year of iteration. His first product was a baby bottle called "U Mix." U Mix was a unique design that held the fluid in one part of the bottle and the dry formula powder in another. When the baby was ready to feed, the bottle was rotated to enable mixing of the fluid and dry formula. Holley intercepted mothers in grocery aisles with his product prototype as one technique to gather highly relevant feedback. He listened to their feedback and revised his product over and over until it was ready for mass distribution.[10] The customer validation process is illustrated in Exhibit 3.4.[11]

Customer discovery is defined as the process of identifying a target market. The customer validation process involves validating that the product(s) you intend to introduce to that market are actually desired. If there is little or no interest

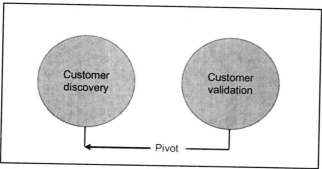

EXHIBIT 3.4
The customer validation process. Source: Blank, S., Dorf, B., 2012. The Startup Owner's Manual. K&S Ranch, San Jose, CA.

from your target market, it may be necessary to pivot to a different market, refine the product, or try a different business model.

The customer validation process does not need to end once you have entered the marketplace with your products. In fact, in order to keep a competitive edge in the modern economy, many analysts recommend continuous innovation as a core business strategy.[12] **Continuous innovation** means making ongoing improvements to the products you already have in the marketplace and inventing new ones to extend your brand and develop additional streams of revenue.

3.3.3 Step 3: Test Business Models

A **business model** is simply the way a business makes money.[13] Companies make money in a variety of ways, and often it takes a bit of creativity to determine which business model will be right for your venture. For example, many of the social media companies that are based on user-generated content (e.g., Twitter, Facebook, and others) don't have any obvious business model. Registration is free, usage of the respective platforms is free, and there are no charges for uploading content. So how do these social media ventures make money? In part, they collect private user data and sell that data to other companies. This is a controversial part of their business models.[14] Some of these companies also sell advertising, or they sell the rights to market directly to their members. Facebook, for example, reported operating results at the end of Q2 2013 that impressed investors because of the improvement in advertising revenue for users of their mobile platform. Its stock price, which had been languishing below its $38 IPO price for many months, rose above that level for the first time because investors realized the immense potential of its new business model.

Recall the Birchbox Mini-Case. When you think about it, Birchbox really doesn't have any novel innovation except for its business model. Birchbox doesn't have any proprietary products or services; it sells goods made by other companies. This is an important point. Aspiring entrepreneurs should recognize that unique new ventures are often launched based solely on innovative business models that create new value for customers. In the case of Birchbox, the founders realized that people enjoy getting monthly packages filled with new product samples and that they are willing to pay for this service. They also realized that the key to their eventual success would be to choose products that users would be influenced to purchase through their e-commerce store.

We will be returning to the concept of a business model throughout this text, but the following is a short, noncomprehensive list of some business models that technology entrepreneurs might use to monetize their products or services:

1. Pay as you go: This model is probably the most well known. It simply means that your products and services are offered at a price and your customers purchase them as they use them.
2. Freemium: This approach focuses on providing a base set of products and services for free and an enhanced set at a reasonable price. Many Internet companies, such as Dropbox, use this business model.
3. Billable hours: This business model is used by consulting firms and other professional service providers, such as software developers. Under this model, the consultant bills the client on an hourly basis for services rendered, often with a minimum fee or "retainer" as part of the business model.
4. Advertising based: This model is often used by Internet companies that do not charge for the content they offer. The goal is to aggregate users and sell advertising based on the number of daily and monthly visitors. Facebook is an example of a technology venture that is monetized entirely through advertising to its one billion plus registered users.
5. Subscription pricing: Subscription pricing involves payment of a fee to gain access to products (usually information products) and services for a fixed period of time. Many online content companies, such as *The Wall Street Journal*, charge subscriptions to users for access to their various publications.
6. Distributor or reseller: A distributor business model doesn't actually make anything, but, rather, aggregates products from companies that do, selling these products to end users. Many in-home medical technology companies use this model. They don't make the products they install in people's homes. Apria Healthcare is an example of such a company.
7. Retailer/Wholesaler: These are familiar business models where the goal of the entrepreneur is to aggregate and hold products for end users and make them available as needed. Online retailers abound, including such well-known names as Amazon, Fab, and Gilt. Online wholesale is also big

business, with companies such as Coffeeam and Houz Depot leading the way in their unique product categories.

8. **Landlord-Leaser-Licensor**: This business model is based on holding title to particular assets (such as office space or intellectual property) and permitting others to use the assets in return for some consideration, typically a fee. This is a popular business model for data centers and cloud-based technology "service providers." The term "service providers" is somewhat of a misnomer; while certain services are provided, the primary benefit provided the consumer is the use of the systems. Users of such services don't take title to the servers or software. They simply lease as much of the provider's assets as they need. Amazon's Web Services unit is an example of a lease-holder business model.

9. **Broker**: A broker doesn't buy or sell, but rather facilitates the transaction between buyer and seller and generally charges a percentage of the transaction. eBay is an example of a successful e-commerce venture that is based on the broker model.

This brief list is not comprehensive, but it should help you think about how your products/services actually will be sold in the marketplace and how they can attract sufficient customers to generate a profit.

Your business model will evolve over time as you learn more about your customers: their tastes and their ability to purchase your products. The business model canvas reviewed in Chapter 2 is a great tool to use to discover how you are going to deliver value to customers consistently and reliably. Remember, in designing your business model you ultimately are seeking a scalable and repeatable model around which your venture can serve a massive market opportunity. There is no point in devising a model that is not scalable because that will serve to frustrate customers who are forced to wait for their purchases when demand exceeds supply. There is no point in devising a business model that is not repeatable unless you are building a craft business that features one-of-a-kind products. There is nothing wrong with models that are not scalable or repeatable, per se, but they are not what technology entrepreneurs typically aspire to create. Artists and craftspeople may desire creating ventures that feature one-off products that are not repeated or intended for mass markets. Technology entrepreneurs generally set out to solve massive human problems and serve massive markets.

3.3.4 Step 4: Explore How to Acquire Customers

Customer acquisition is often more involved and expensive than many aspiring entrepreneurs think it is. It is involved because there are a vast number of unknowns that entrepreneurs will encounter regardless of how much prior market research they've conducted. Customers will, more often than not, respond in unexpected ways when presented with new products to evaluate or purchase. For example, a venture that launches a product might discover that it is wildly popular with customers. While this may seem like a great outcome, it can be damaging to the venture if it doesn't have adequate supply to meet the demand. Frustrated customers may be lost forever, and the problem often spreads as negative word-of-mouth.

Acquiring customers can also be a function of where they are located, geographically speaking, in relation to the venture. If the market is the local neighborhood (which is what it would be in a restaurant), it may be enough simply to put up a sign and announce that you are open for business. On the other hand, if a technology entrepreneur wants to acquire customers worldwide, there may be much more complexity and expense in announcing the venture's existence and value proposition. Many aspiring entrepreneurs have learned too late that only putting up a website is not enough to generate sufficient sales. A full-blown, thought-out strategy for acquiring, serving, and maintaining customers should be part of your new venture idea. This topic is discussed in greater detail in Chapter 10.

3.4 THE OPPORTUNITY REGISTER

Ideas are easy to come by, but unfortunately many aspiring technology entrepreneurs never make it past the idea stage. One of the major reasons why this happens is that the entrepreneurs don't develop their ideas sufficiently, and therefore they can't turn those ideas into ventures. The ideas are not developed because many aspiring entrepreneurs lock into a single idea and believe they must do all in their power to turn THAT idea into a venture. What the aspiring entrepreneur doesn't know that the expert knows all too well is that most ideas will fail as new ventures. To hedge against the risk of failure that attends a single idea, the expert entrepreneur knows that every good venture is based on a multitude of ideas that have gone through multiple and even continuous rounds of iteration, recombination, and improvement.

As an aspiring entrepreneur yourself, there is no time like the present to begin to keep track of your ideas as they evolve. A tool that many aspiring technology entrepreneurs use to track their flow of ideas is called the **opportunity register**.[15] An opportunity register is simply a notebook or computer file that you will return to again and again as new ideas come to you and as you modify the ideas that you've already recorded. Exhibit 3.5 provides an example of an opportunity register:

Using the analytic tools that you will be learning throughout this book you can continuously evaluate and refine your business ideas. You should take care to record your ideas in enough detail that you will be able to subject them to

Field	Your Input
Business concept: Short description	
Possible upside: What might the concept generate in revenue or profit terms?	
Related trends	
Relevant data: This is a place to jot down any material numbers or information that you may have.	
Obstacles and barriers: What might stop you from grasping the opportunity?	
Position: What competencies, skills, or resources might make this a good opportunity for you?	
Competition: Who is likely to contest this opportunity and what are they likely to do?	
Sources: Where did you get your information?	
Timing: How long is the window of opportunity? How quickly must you pursue it?	

EXHIBIT 3.5
Example of an opportunity register. *Source:* McGrath, R.G., MacMillan, I., 2000. The Entrepreneurial Mindset. Harvard Business School Press, Cambridge, MA.

thorough analysis. While, as will be discussed in Chapter 5, not the best evidentiary record of technology development, a loose-leaf folder is often best for an opportunity register in that you can add pages to each idea as needed. Each idea in your opportunity register should include:

1. The date on which the idea was initially entered. Each update to the idea should also include a date.
2. The context of the original idea and all future revisions. For example, if you are revising an idea, be sure to record the thoughts or events that led to the revision. It may be that you received some important feedback from a trusted source. Record the source and the nature of the feedback. As time goes on, you may need to recall the thought processes that led to key decisions as much as the decisions themselves.
3. As your idea matures over time, you should consider whether there is any intellectual property that you will need to protect. For example, new inventions evolve over time and eventually may result in some truly novel insights that can and should be patented or otherwise protected. This will be discussed in more detail in Chapter 5.

Its also important for aspiring entrepreneurs to work with **deliberate speed** in the development of their ideas.[16] It is exceedingly rare for potent new ideas to be generated and developed in a complete vacuum. More likely than not, any good idea that you have is already being developed by someone else.

We use the term "deliberate speed" because it is also important to recognize that many ventures fail because they don't develop the basic business concept and business model enough before going to market. Going to market too early can be very expensive, but getting to market too late can mean losing out to competitors. It is best to learn all you can about your markets and customers in the process of introducing your minimally viable product, and then continue to listen to and learn from customers after full release. Listening to your customers and building the feedback into subsequent releases of your products or refinement of your services better to meet customer needs is vital to your success.[17]

Of course, an opportunity register doesn't constitute a new venture. To launch a new venture, you'll eventually have to choose which of your various ideas gives you the best chance to succeed. It is possible that you'll be able to develop several of your ideas through your new venture, or build several ventures around several ideas. Still, you need to start with a single idea—preferably the one you feel most confident about developing into a successful venture.

Far too often, aspiring entrepreneurs have a lot of ideas that they'd like to build into a new venture, but they are unable to focus on a single idea to start. Expert entrepreneurs know well that developing a single idea into a going concern is a difficult challenge. Attempting to develop more than one idea during the venture launch process is almost certainly a recipe for failure. You increase your chances for success if you choose the idea that you think is the most likely to have a ready market and for which you have most of the necessary resources.

3.5 NONTRADITIONAL IDEA SOURCES

If your search for a venture idea is not going well, you may be able to find inspiration via some nontraditional sources. Remember, however, that we highly recommend (and research into expert entrepreneurs supports this point) that you should build a business that is a good fit with who you are. The Mini-Case below highlights how one technology entrepreneur leveraged his love of rap music to create a new venture.

MINI-CASE
Rap Genius Answers "What do those lyrics mean?"

In 2009 computer programmer and rap music aficionado Tom Lehman became curious about the line "Eighty holes in your shirt, there: your own Jamaican clothes" in rapper Cam'ron's song "Family Ties." Lehman asked his Yale University classmate, Mahbod Moghadam, what the lyric meant and was told that it referred to the tattered clothing worn by poor people in Jamaica. Lehman and Moghadam, along with a third founder Ilan Zechory, decided to start a blog called Rap Genius to explain the lyrics on the songs from Cam'ron's album "Purple Haze." The site's unique feature was that it allowed users to annotate the lyric explanations by highlighting a phrase or a word and entering an explanation of the meaning. Users could then click on the highlighted material to generate a pop-up box that would contain the explanation. The founders opened the site up to the public to enter rap lyric explanations in 2010. By mid-2011 the site had more than 1 million unique users per month, and users were actively reading and rating the annotations and adding comments of their own. Rap Genius received a $170,000 investment from Y-Combinator, and eventually it received a $15 million investment led by venture capital firm Andreesen Horowitz, and including such celebrities as Ashton Kutcher and rapper Nas. The founders of Rap Genius have subsequently launched two related sites called Poetry Genius and News Genius, and they have plans for other "Genius" sites.

Source: Adapted from Nicole Hong, "A Cryptic Lyric Inspired a Hot Startup," The Wall Street Journal, April 19, 2013, p. R3; Amanda Holpuch, "We're Trying to Make Rap Genius Into Everything Genius," The Guardian, July 23, 2013; Sean Ludwig, "Are Rap Genius's Founders Insane, or Is It Just a Gimmick?" VentureBeat, May 1, 2013.

While inspiration like that which motivated the Rap Genius founders is nice, it's not always the case that great ideas come to entrepreneurs in a sudden flash of insight. Some of us are less adept at the idea generation process than others and can benefit from inspiration from nontraditional sources. Below is a short list of some nontraditional sources of inspiration that you may want to investigate.

3.5.1 Read Alternative Literature

The long history of the sciences and the arts has proven that good ideas in one domain often can be translated and transferred to another domain. For example, in the computer sciences extensive work has been conducted on mimicking the human brain via software algorithms. One of the more promising lines of research in this area is referred to as "neural networks." These are networks of interconnected and parallel processing circuits that can do some of the things that the human brain can do. As it turns out, neural networks have an application to the world of high finance. Many leading investment houses today use neural network technologies to mimic the behavior of securities traders so that they can make predictions about buying and selling behaviors.[18]

In order for you to be able to transfer ideas from a domain outside your disciplinary expertise, you need to dive into the literature. If you are a business major steeped in the literature of finance, accounting, management, and marketing, you may find inspiration from reading evolutionary theory, cosmology, or art history. Steve Jobs was famous for the innovative and often beautiful products he created via his multiple ventures, Apple, Pixar, and Next. In his biography, Jobs stated that his ability to create one insight after another was because he was always seeking to remain on the intersection between technology and the liberal arts.[19]

3.5.2 Visit New Places and Experience New Things

Experience is a great teacher. There are some things that simply cannot be learned by reading about them. Imagine a book that explains to you how to ride a bike and how you will feel when you ride down hills with the wind in your hair. Do you think it is possible to learn to ride a bike through reading a book, or to understand the thrill of the wind in your hair by reading about how good it will feel?

In order to stretch your world and develop new ways of understanding how people live and work, you'll need to get outside of your normal comfort zone. We're not recommending that you take more risk than you're willing to take, but it is very unlikely that you'll find some unique niche in the wide world of human affairs without venturing beyond your routines. You may engage in some world travel to witness the types of things that people struggle with in foreign lands. You may want to investigate what they eat, how they live, and their sources of entertainment. More than one entrepreneur has discovered novel products in foreign lands that sold well when imported to the home market.[20]

We're also not advocating that you do these things without prior forethought about what you'd like to learn. It is one thing to take irresponsible trips and do irresponsible things with no forethought about learning valuable lessons, but it is quite another to take deliberate action intended to stretch your thinking and help you learn new and potentially valuable things. We are advocating that you use new experiences to refine the ideas in your opportunity register and that you build on this experience over time to continue to refine and improve your venture ideas.

3.5.3 Meet Thought Leaders

Thought leaders are people who are on the leading edge of ideas and innovations in a particular discipline. Where do you meet people like this? They are oftentimes featured on television, especially via cable networks like the Discovery Channel. Stephen Hawking, for example, is a well-known thought leader who is often featured talking about cosmic questions. Many thought leaders also appear at leading conferences. The Consumer Electronics Show is an annual draw for people to see and hear from leaders in consumer technologies. The South by Southwest Conference in Austin, Texas is also a major conference for thought leaders in technology and design. The Internet is also a rich source of access to thought leaders in nearly any category of human achievement. Below is a short list of websites that you may want to visit:

- TED (www.ted.com)
- The Edge (www.edge.org)
- Stanford University's eCorner (www.ecorner.stanford.edu)
- Fora TV (www.fora.tv)

Sometimes the ability to think differently about products and services is referred to as "thinking outside the box." What that means is that most people think in recurring patterns and often cannot think different without an external stimulus. This phenomenon is true for most people, including some of the world's most accomplished scientists. Research into how science evolves has demonstrated that most of us think about the problems we're trying to solve in terms of well-received "paradigms." Oftentimes in the history of science, the established paradigms actually blocked progress. Those who held fast to standard ways of thinking about problems were not the ones to create the revolutionary breakthroughs. In fact, the history of science is replete with "rebels" who dared to think differently, who oftentimes were vilified in their era, but who are today recognized as the giants of their discipline.[21]

In what ways is your way of thinking about the world stuck in old paradigms? How can you persuade yourself to think differently about the world and the people in it? We have given you a few starting points in this chapter, but it truly is up to you to follow through and create the next great entrepreneurial venture.

3.5.4 Team Up

If, after all is said and done, you are unable to generate an exciting idea for a new technology venture on your own, you still have an option—align yourself with someone that has that exciting idea. Remember, business is a team sport. There are a number of different elements necessary for the success of a startup venture, and the idea—intellectual property—is only one of those necessary elements. As will be discussed in Chapters 5 and 6, it is rare that any single individual can contribute all of the necessary elements to the venture. If you are unable to generate the seminal idea, you can still participate in the venture by making some other form of contribution—things like: resources, services, leadership, expertise, contacts and relationships, credibility, and capital.

3.6 CHAPTER SUMMARY

This chapter began by introducing you to the four types of venture: B2B product, B2B service, B2C product, and B2C service. We cautioned that it's really not this simple in practice, because many firms provide both products and services to their customers. Still, we asserted that startup ventures should choose only one quadrant to begin and then think of ways to expand into other quadrants after the initial effort begins to work.

Next we examined our four-step process for generating an idea for a product or service. The first step is to identify a point of pain in a particular target market.

Your target market is the first market that you will attempt to sell your products into. It is helpful if you have some familiarity with this market because it can be very difficult to please customers that you don't know anything about. The next step is to innovate a product or service that is designed to address the point of pain that you've identified. Remember, the product development process should be considered to be iterative, and it is helpful to get feedback from customers early and often—even before the product is ready for "prime time." The next step is to test various business models. A business model is defined as the way the business makes money. You'll need to test various models with prospective customers to find out what and how they are willing to pay for your product. Finally, you'll need to explore how your venture will acquire customers. It is no use to start a venture if you have no idea how you will acquire customers. Customer acquisition is often far more expensive than estimated and is normally far more difficult to achieve than initially thought.

We next discussed the notion of the "opportunity register." The opportunity register essentially is a type of journal that allows you to keep track of your product ideas, business model ideas, and customer acquisition ideas as you develop and refine them. We indicated that this is also a useful tool for recording your intellectual property developments. One of the ways in which patents are issued depends on the priority of the invention—meaning that if there is more than one party claiming similar patents, the party that invented it first will have priority. Good records in an opportunity register can assist with priority claims.

Finally, we explored some nontraditional ways for you to generate ideas for your new venture. These techniques include reading new things, visiting new places and meeting new people, and listening to thought leaders in a variety of disciplines that interest you.

KEYTERMS

Business model The way a business makes money.
Business to business (B2B) Ventures that sell to other companies.
Business to consumer (B2C) Ventures that sell to consumers.
Business to government (B2G) Ventures that sell products or services to government agencies.
Continuous innovation Making ongoing improvements to products you already have in the marketplace.
Customer validation The process of testing your product with customers and learning from their feedback.
Deliberate speed Moving quickly, especially technology entrepreneurs, to stay in front of potential competitors around the world.
Opportunity register A personal journal of you evolving venture ideas.
Point of Pain The problem you intend to solve for customers.
Single-product venture Ventures that sell only a single product to consumer markets.
Target market The set of customers you are targeting with your new technology product.

ADDITIONAL READING

Ulrich, K., Eppinger, S., 2011. Product Design and Development. McGraw-Hill, Irwin, New York.

Isenberg, D., 2013. Worthless, Impossible, and Stupid: How Contrarian Entrepreneurs Create and Capture Extraordinary Value. Harvard Business Review Press, Cambridge, MA, July 9.

McGrath, M., 2000. Product Strategy for High Technology Companies. McGraw-Hill, Irwin, New York, October 12.

WEB RESOURCES

www.edge.org: In case you need some help in generating new ideas for a venture, this website features some of the world's leading thinkers on a wide range of science and technology topics. The articles on this site are sure to stir your imagination and set your creative juices flowing.

www.killerstartups.com: This website is similar to TechCrunch. It highlights startups that are launching all over the world and provides insightful commentary and background on the ventures and the founders.

www.techcrunch.com: This website provides continuous updates on the world of technology ventures. It is a very useful site to visit often while you develop the ability to see the trends in new technology ventures and technology investor activity.

ENDNOTES

1. Fiet, J.O., and P.C. Patel. 2008. "Entrepreneurial Discovery as Constrained, Systematic Search." *Small Business Economics*, 30(3): 215-229.
2. Gimpl, M.L. 1978. "Obtaining Ideas for New Products and Ventures." *Journal of Small Business Management*, 16(4): 21-26.
3. Kutaula, P.S. 2008. "Funcastle: Creating a Business Opportunity from a Design Consultancy Assignment." *Design Management Review*, 19(3): 23-29.
4. Bienstock, C.C., Gillenson, M.L., and T.C. Sanders. 2002. "The Complete Taxonomy of Web Business Models." *The Quarterly Journal of Electronic Commerce*, 3(2): 173-182.
5. S. Burnaz and P. Bilgin. 2011. "Consumer Evaluations of Brand Extensions: B2B Brands Extened Into B2C Markets." *Journal of Product & Brand Management*, 20(4): 256-267.
6. Click, R. and T.N. Duening. 2004. Business Process Outsourcing: The Competitive Advantage. (Hoboken, NJ: John Wiley & Sons.)
7. O'Neil, G.D. Jr., J.C. Hershauer, and J.S. Golden. 2009. "The Cultural Context of Sustainability Entrepreneurship." *Greener Management International*, 55: 33-46.
8. Lehoczky, E. and D. Bortz. 2011. "Litmus Test Your Biz Idea." *Entrepreneur*, 40(9): 34.
9. Anonymous. 2013. "How Entrepreneurs Come Up with Great Ideas." *The Wall Street Journal*, April 29, p. R1.
10. Holley, J. 2011. The Startup Experience: Inventing and Bringing Products to Market. (Colorado Springs, CO: Businesses2Learn, LLC).
11. Blank, S. and B. Dorf. 2012. The Startup Owner's Manual. (San Jose, CA: K&S Ranch).
12. Ries, E. 2011. The Lean Startup: How Today's Entrepreneurs Use Continuous Innovation to Create Radically Successful Businesses. (New York: Crown Publishing Group.)
13. Weill, P., T.W. Malone, V.T. D'Urso, G. Herman, and S. Woerner. 2005. Do Some Business Models Perform Better than Others? A Study of the 1000 Largest U.S. Firms. MIT Sloan School

of Management Working Paper No. 226. Accessed at http://ccs.mit.edu/papers/pdf/wp226.pdf; on January 3, 2012.

14. Dwyer, C. 2011. "Privacy in the Age of Google and Facebook," *IEEE Technology & Society Magazine*, 30(3): 58-63.

15. McGrath, R.G. and I. MacMillan. 2000. The Entrepreneurial Mindset. (Boston: Harvard Business School Press.)

16. Cohen, D. and B. Feld. 2010. Do More Faster: TechStars Lessons to Accelerate Your Startup. (Hoboken, NJ: John Wiley & Sons.)

17. Cooper, B. and P. Vlaskovits. 2010. The Entrepreneur's Guide to Customer Development: A Cheat Sheet to the Four Steps to the Epiphany. (Cooper-Vlaskovits.)

18. Huang, W., K. Keung, Y. Nakamori, S. Wang, and L. Yu. 2007. "Neural Networks in Finance and Economics Forecasting," *Information Technology & Decision Making*, 6(1): 113-140.

19. Isaacson, W. 2011. Steve Jobs. (New York: Simon & Schuster.)

20. Albergotti, R. 2012. "The NFL's Top Secret Seed," *Wall Street Journal*, January 20, 2012.

21. Kuhn, T. 1970. The Structure of Scientific Revolutions. (Chicago: University of Chicago Press.)

Countdown to Launch

PART 2

CHAPTER 4

Markets and Product or Service Development

redBUS BUILT ON FIRST-HAND MARKET EXPERIENCE

During the holiday season of 2005, M. Phanindra Sama became frustrated with Bengaluru, India traffic as he waited long hours in traffic and could not make it to the ticket counter in time to buy a bus ticket to spend the holiday with his family. He realized at that moment that many people like him faced the same situation. That triggered the concept of redBus. Phani realized the inconvenience of not knowing the schedule and availability of buses, and the problem of not being able to buy a return ticket from another state until you reached your destination. Phani wanted to make the booking system and experience for customers similar to the online airline ticket reservation process.

With his college friends, he did market research and discussed the idea with several bus operators, consumers, and potential investors. They learned the gaps in the existing system and the value in streamlining the process. Phani explained that a travel agent in India might say that the last bus for Cochin was at 8:00 p.m. because that's the last bus of the operator he works with. But, that doesn't mean there aren't any buses available later from another operator. That was the common issue that many Indian passengers faced. Also, the return ticket could not be purchased until they reached their destination. redBus solved those problems by providing access to multiple bus operators' schedules and advanced booking options with seat availability in real-time inventory.

About 60,000 people ride buses in India every day, but the bus industry is very unorganized. There is no one easy and convenient platform for customers to buy tickets. redBus was the first company to recognize and implement a solution to make the bus reservation experience easier. It targets the middle and upper-class segments of society and is launching services like mobile payments and call centers in local languages. redBus faces huge competition from bus operators who may sometimes lower their cost and other online travel booking websites.

One of redBus' major tasks was to convince bus operators to allow online ticket sales as they were comfortable dealing with their traditional travel agents.

Phani submitted the idea with his two co-founders to TiE—The Indus Entrepreneurs—and received funding for their company. Initially, Phani started with only one office in Bengaluru and had 60 destinations on it schedule. As he received more angel funds, he opened four more offices in different cities.

Continued

> Pahni's concern is to keep pace with the organization's growth. The company is growing at such a fast rate that it is getting difficult for the founder to have time for his friends and family. He believes in human relations and does not want to be considered rude but he is spending less time with his family and friends.
>
> **Source:** http://www.redbus.in

4.1 INTRODUCTION

In the previous chapter we discuss the process of generating ideas for a product or service.

Just like what occurred in the case of redBus, a technology venture is often built around a new technological product or service that solves a significant, presently occurring, problem. In other cases, an idea, technology, or product may first be created and a market for that idea, technology, or product would then be thereafter developed. This is often the case with respect to particularly novel and disruptive technology.

In any event, once an idea is conceptualized, it needs to be screened—evaluated—through an appropriate systematic process. In previous chapters we discuss the "Lean Start" approach, where evaluation of the idea-technology-product is performed as an integral part of the build-measure-learn feedback loop involving presentation of successive minimally viable products to consumers. We will now discuss how screening can also be accomplished through, for example, a product planning and development process, an idea development process, and/or an opportunity assessment plan.

The screening process needs to consider the degree of newness that the new technological idea presents—the extent to which the idea involves new technology or a new marketplace—to the venture (including the entrepreneurs, the distribution system, and the consumer).

4.2 PRODUCT PLANNING AND DEVELOPMENT

A standard approach used by many technology entrepreneurs to evaluate new products and services is the product planning and development process. This process, indicated in Exhibit 4.1, consists of four stages, with an evaluation being done at the end of each stage—(1) idea stage, (2) concept stage, (3) product development stage, and (4) test marketing stage. To work effectively, evaluation criteria must be established and employed at each stage to either reject an innovation or allow it to proceed to the next stage. This is essentially the stage-gate product development process mentioned in Chapter 2 and shown in Exhibit 2.2.

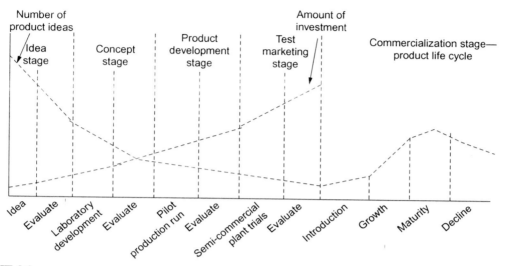

EXHIBIT 4.1
The product planning and development process. Source: Hisrich, R.D., Peters, M.P., Shepherd, D., 2013. Entrepreneurship 9E. McGraw-Hill/Irwin, New York.

The first stage, the idea stage,[1] is when the innovation is formulated and developed. There are many ways these innovative ideas occur. Sometimes the innovation comes from observing trends. There are several trends occurring today that will provide the opportunity for new innovations.[2] These include the organic food trend, the green trend, the health trend, the clean energy trend, and the social media trend. Just look at the increasing aging of the population around the world or the number of people tweeting, and you can see the trends and their increasing size. Other sources of innovative ideas include assessing the inventor's own research, evaluating existing products and services in the marketplace, listening to the consumers' complaints and suggestions, and observing new legal requirements to doing business in a country. Regardless of the innovation, it needs to be evaluated at this stage by applying the evaluation criteria established to determine if the product or service should move to the next stage.

After the innovation has passed the evaluation process in the idea stage, it moves to the *concept stage*.[3,4,5] In this stage, the modified innovation is tested to determine market reaction and the degree of acceptance. Although various methods can be employed, one of the easiest, cost-effective methods is to discuss the innovation with individuals in the defined market. This is called a "conversational interview." Various features, price, and promotional aspects of the innovation should be discussed in comparison to competitive products or services available in the market—particularly the ones presently being used to fulfill the market need. Again, the information needs to be analyzed using the evaluation criteria established before passing on to the next stage—the product development stage.

In the *product development stage*,[6] a final version of the product, based in part on the evaluations obtained in the concept stage, is subjected to consumer input. Again, consumer feedback is obtained, and the refined innovation is evaluated against the evaluation criteria established.

Following a successful product development evaluation, the innovation moves into the final stage for final evaluation and launch, which starts the product life cycle (see Exhibit 4.1). In this, the *test marketing stage*,[7,8] a market test of the innovation is done to help ensure successful commercialization. Due to the costs of test markets and the nature of some products or services, this stage is often bypassed with the innovation going from the product development stage direct to commercialization.

MINI-CASE
IDEO Specializes in Design Thinking

IDEO is a global design firm. The award winning consulting firm assists in growth and innovation of organizations in both the public and private sectors. In addition to management consulting, IDEO works with firms to create brands and design and launch new products and services. They use a "design thinking" approach to solve problems and construct innovative strategies. IDEO is ranked one of the most innovative companies in the world by Boston Consulting Group.

4.3 THE IDEA DEVELOPMENT PROCESS

The idea development process results in developing successful, sustainable, innovative new ideas.[9] It is composed of six stages as shown in Exhibit 4.2.

4.3.1 Describe the Idea and Competition

As indicated in Exhibit 4.2, the first stage is to describe the idea; evaluate any products or services that are filling the same need; and determine the advantages, benefits, and features that provide these benefits of your idea. Although this is further discussed in the Section 4.5, it is important to describe the details of the idea, the problem it solves, and the need it fills as succinctly and completely as possible. It is helpful to determine the country system code that best describes the product. For example, NAICS (North American Industry Classification System) is the country code in the United States, and SIC (Standard Industrial Classification) is the country code in China and Korea. This code can help you identify competitive products or services and their prices. This will allow the

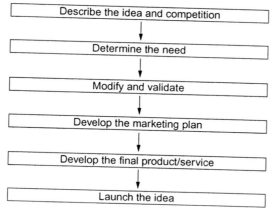

EXHIBIT 4.2
The idea development process.

determination of the unique features of your innovation (compared to the competitive products or services presently on the market filling the same need) and the features that deliver each of those benefits.

4.3.2 Determine the Need

The second stage is to determine the need for your innovation. This requires that the description developed in stage one is as in-depth as possible. At least three different groups of consumers or firms that might benefit from your innovation should be identified, and a customer profile for each group should be established. For business-to-consumer (B2C) groups, this profile should at least include age range, income range, and gender. Listing family size, education, occupation, and race are helpful, but not as critical. For business-to-business (B2B) groups, the profile should include industry, type of products or services, location, and size. A concept survey instrument should be developed that includes a brief description and, where applicable, both a diagram of your innovation and survey questions to elicit the desired feedback. The results of the concept survey of members of each of the three groups should be analyzed to indicate the important criteria that interest your potential customers.

4.3.3 Modify and Validate

Modify and validate, the third stage, involves developing a prototype template of your innovation and interviewing individuals concerning this prototype. The questions concerning the prototype should focus on features and benefits of your prototype, the important aspects of this particular product, and the buying process involved.

4.3.4 Develop a Marketing Plan

Stage four involves developing a detailed marketing plan. The marketing plan, which is further discussed in Chapter 7 when discussing the business plan focuses on pricing, distribution, and communication (promotion). The pricing data obtained in your competitive analysis (stage one) should be evaluated along with cost data to determine an initial price. A distribution plan laying down the geographic area where your innovation will be sold, how you will physically get it there, and distribution channel(s) should also be developed when applicable. Finally, all the possible marketing communication tools appropriate should be considered, including the importance of social media.

4.3.5 Develop the Final Product or Service

The development of the actual idea is the fifth stage. This involves the development of the features of the final idea, its marketing theme, naming, and packaging (if appropriate). The marketing theme needs to focus on key benefits or unique selling propositions that appeal to your target customers. The package, when appropriate, has many aspects, the most important of which is eye appeal when using a retail distribution.

4.3.6 Launch the Idea

The sixth and final stage is to launch the final product or service. This starts the product life cycle, which was previously discussed (see Exhibit 4.1).

MINI-CASE
Neostem Innovates from Its Research

Neostem is a firm in the emerging industry of cellular therapy. Cellular therapy is a new therapeutic technology that uses stem cells for tissue regeneration and disease prevention and treatment. Neostem develops proprietary cellular therapy products and provides contractual services to other firms in the regenerative medicine industry. By combining these revenue-generating services with their own research and development for new products, the firm is able generate cash flow and develop products in a more cost effective manner.

4.4 THE CONCEPT OF NEWNESS

One of the critical concepts affecting the successful launch and sustainability of an innovation is its *newness*—its novelty and disruptive nature. Even though, of course, newness or uniqueness is needed (and in fact is an essential part of an innovation), the degree of newness of the innovation affects its acceptability and the length of the adoption cycle for consumers, the individual investor and/or the organization, and the distribution system.

An example of a technology with a high degree of newness is regenerative medicine and, specifically, stem cell therapy. These therapies have the potential to empower a body to fight ailments and in some cases build entirely new parts such as new neurons or heart tissue. However, because of the newness of the technology, there is a resistance to its acceptance.

One area of medicine that holds much promise yet has all problems of newness is regenerative medicine and, specifically, stem cell therapy.[10] These therapies have the potential to empower a body to fight ailments and in some cases build entirely new parts, such as new neurons or heart tissue.

The science has been slowly advancing in bits and spurts in spite of religious outcries and political sanctions in some countries. Stem cell therapy, having been used successfully in animals and having successful preclinical trials in rodents, has resulted in healing such as having the paralyzed walk and making the impotent virile. Specifically, various forms of cancer, diabetes, heart disease, and Parkinson's disease have been eradicated in mice. When translated to humans, regenerative medicine and stem cell therapy will be a platform and breakthrough that rarely occurs in any field, and particularly medicine. Even though the effect will be substantial both economically and socially, such as what occurred with the Internet, transistors, and powered vehicles, all the problems with newness will need to be resolved. Particularly problematic are the long-time horizons, regulatory issues, money for the large research and development costs, public opinion, and political sensitivity.

MINI-CASE
Biogenomics Invents Biosimilars

Biogenomics is a young firm in the business of developing biopharmaceuticals and biotech products. Renowned scientists, business professionals, and a team of microbiologists, molecular biologists, and protein engineers comprise the firm. The firm has developed their own clones and cell lines to efficiently deliver cost-effective technologies and products. Biogenomics is particularly active in the manufacturing of biosimilars. Biosimilars are medical products that are created or derived from biological organisms using recombinant DNA techniques. This new technology holds the potential to provide effective medical products at a fraction of the cost of similar products on the market today.

4.4.1 Newness to the Consumer
Regardless of whether the innovation is for the consumer market (B2C) or the industrial market (B2B), if it is too far in advance of the present state of the market, problems can occur. Consider the newness of the innovation in terms of its disruption in the established consumption patterns or lifestyles of the target market. The least disruptive innovations—continuous innovations—have little impact or influence on the lifestyle of the purchaser and therefore usually do not take very long in the evaluation and adoption stage. The majority of innovations are in this category.

A single focus on a particular item in a restaurant chain is an example of a continuous innovation.[11] Even though it is important for the innovative item to have a broad enough appeal and versatility to be the main feature on the menu, the item itself can vary from eggs to soups to chicken to roast beef. The Another Broken Egg Café, an upmarket brunch restaurant, focuses on the preparation of eggs in various ways and provides about 150 items on the menu. Original Soup Man has 16 locations featuring between 6 and 12 varieties of soups daily, with staples such as lobster bisque and Mexican bean soup along with different sandwich selections. Kentucky Fried Chicken is an internationally oriented chain focusing on chicken, while Arby's focuses on roast beef sandwiches.

The concept needs to be broad enough and to have appeal as a main item, preferably for at least two meal times. This was not the case for PB Loco, a gourmet peanut butter and jelly chain that, in 2005, started franchising. Although initially heralded for its unique flavor combinations, the concept did not have appeal for breakfast and dinner and was considered an occasional novelty item, not a constant or regular eating experience.

On the other hand, it takes time for truly disruptive innovations to be accepted and adopted in the marketplace. Potential resistance to replacing an existing base of still operating old-technology products should be evaluated. Relatively recent examples of disruptive technology include digital video recorders and entertainment on demand, inexpensive video cameras and YouTube, and smartphones and wireless Internet access.

4.4.2 Newness to the Organization

The newness of the innovation to the sponsoring organization is also important to assess. The first time an individual invents, or the first time an innovation is in a new area of endeavor for a venture, there are more difficulties in developing and launching the innovation. This is reflected in the rule that most venture capitalists use: make sure someone on the management team, if not the entrepreneur, has experience in the industry of the new venture.

When an innovation has not only technological newness but also market newness for an organization, (referred to as *diversification*; see Exhibit 4.3), the highest level of problems (and even failures) is encountered by companies regardless of size. This is evidenced from the failure of the Gillette LCD watch, which was well outside the parameters of the typical products of the company.

4.4.3 Newness to the Distribution System

The final area of concern is newness to the distribution system.[12] Like consumers, individuals, and organizations, distribution systems have lifestyles—ways of doing things. An innovation outside its typical product category, size, shelf fit, or packaging will have a more difficult time gaining access. A new dog treat that was odorless to humans but loved by dogs could not access the retail stores

	Technology newness →		
Product objectives	No technological change	Improved technology	New technology
No market change		**Reformation** Change in formula or physical product to optimize costs and quality	**Replacement** Replace existing product with new one based on improved technology
Strengthened market	**Remerchandising** Increase sales to existing customers	**Improved product** Improve product's utility to customers	**Product life extension** Add new similar products to line; serve more customers based on new technology
New market	**New use** Add new segments that can use present products	**Market extension** Add new segments modifying present products	**Diversification** Add new markets with new products developed from new technology

(Market axis runs vertically downward on the left side)

EXHIBIT 4.3
New product classification system.

in the United States until a new package size and design was developed, allowing the dog treat to be displayed on the shelves set aside for this type of product.

4.5 OPPORTUNITY ASSESSMENT PLAN

Probably one of the best methods to use to ascertain the marketability of an innovation is the *opportunity assessment plan*.[13] The opportunity assessment plan is usually shorter than a business plan (discussed in Chapter 7); focused on the opportunity and market rather than the business; and has no financial, marketing, or organizational plan. It is used to determine if the innovation has at least three to five unique features (unique selling propositions) compared to the competitive product or service presently on the market and filling the same need. The opportunity assessment plan also determines if the product or service has a viable market that is large enough, growing, and accessible enough to warrant pursuing the innovation.

The opportunity assessment plan has four parts—two major and two minor parts. Part 1, a major part, focuses on the product or service idea and the competition. It requires obtaining the country system code of the country, such as NAICS for the United States and SIC for China and Korea. After defining the innovation as thoroughly as possible, the various aspects of the product or service filling the need are identified, thus indicating the innovation's unique selling propositions.

The second part—another major part—focuses on the market for the innovation. The size of the market over the past three to five years should be obtained in order to identify any trends. The growth rate of the market should also be obtained. An innovation has a much stronger chance of success in a large, growing market than one that has leveled or is declining.

Parts 3 and 4—minor parts—focus on the skills, experience, and background of the team, as well as the steps needed to launch the innovation (see Exhibit 4.4).

Exhibit 4.4 Opportunity Assessment Plan.

An opportunity assessment plan is NOT a business plan. Compared to a business plan, it should:
- Be shorter
- Focus on the opportunity, not the venture
- Have no computer-based spreadsheet
- Be the basis to make the decision on whether to act on an opportunity or wait until another, better opportunity comes along

Part 1
A description of the product or service
- What is the market need for the product or service?
- What are the specific aspects of the product or service (include any copyright, patent or trademark information)?
- What competitive products are available filling this need?
- What are the competitive companies in this product market space and their strengths and weaknesses?
- What are the country counting codes for this product or service?
- What are the unique selling propositions of this product or service?

Part 2
An assessment of the opportunity:
- What market need does it fill?
- What is the size and past trends of this market?
- What is the future growth and characteristics of this market?
- What are total industry sales over the past 5 years?
- What is anticipated growth in this industry?
- What is the profile of your typical customers?

Part 3
Entrepreneurial self-assessment and the entrepreneurial team:
- Why does this opportunity interest you?
- What are your reasons for going into business?
- How does it fit into your background and experience?
- What experience is needed to successfully launch the product or service?

Part 4
What needs to be done to translate this opportunity into a viable venture?
- Establish each critical step in order.
- Determine the time and money needed at each step.
- Determine the total amount of money needed and its source.

Part 4 is particularly important because it gives an indication of the time and money needed for successfully developing and launching the innovation.

4.6 DISRUPTIVE TECHNOLOGY

The newness of a product or service can cause difficulties to the consumer, organization, and distribution system (if applicable). When the technology is *disruptive*, as in the case of regenerative medicine and stem cell therapies, these difficulties are heightened. In disruptive technologies, multiple players are involved and need to be dealt with. Sometimes this requires the assembly of different downstream resources. One of the best cases is Thomas Edison. Not only did he invent the, but also (through Edison Electric Light & Co.) he developed the infrastructure needed to employ the light bulb, such as coal-fired generators, transmission lines, and the wiring system for streets and buildings.

Many disruptive technologies are not as clearly groundbreaking as the light bulb or the Internet. In some cases, which will be discussed in the next section, innovations making existing products or services better, cheaper, or more acceptable to the market can be disruptive. Amazon did not create online shopping, but it made online shopping easier and cheaper by bringing millions of products to one storefront. Similarly, Steve Jobs figured out how to make the existing expensive, custom-made computer mouse at 25% of the existing cost. Frends, a new company based in Encinitas, California, introduced its premium designer-style headphones for women in Apple and Best Buy stores in October 2012. They had sales of $1.6 million in 2012, which are expected to radically increase in 2013. Being wrapped in soft leather and accented with hand-polished antiqued metals, the headphones are less likely to cause hair snagging. The company combined fashion and electronics in a new disruptive way.

Frends,[14] Apple, and Microsoft were able to successfully bridge a gap which resulted in a sustainable technology (see Exhibit 4.5). This, of course, rarely occurs and particularly not as quickly as it occurred in these cases. Usually, the market does not embrace the new disruptive technology so quickly or even ever.

4.7 THE MARKET

Regardless of the nature of the technological idea and its degree of disruptiveness, it is essential for every idea to have a *market*. From an economic viewpoint, a market is a mechanism that bridges the gap between supply and demand and consists of a group that may buy the product. There are three types of markets for a technological product or service: *consumer markets*, *industrial markets*, and *government markets*. The consumer market, also known as the B2C market,

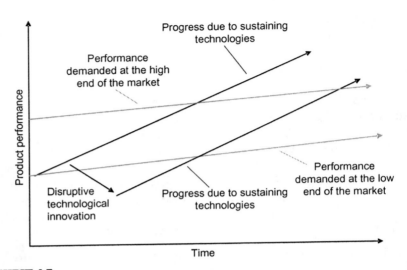

EXHIBIT 4.5
The impact of sustaining disruptive change.

consists of private individuals who purchase products or services for personal satisfaction and use. The industrial market, also known as the B2B market, consists of a variety of different entities involved in the purchase or resale of the product or service not for final consumption or use like the consumer market. The industrial market purchases for further process, use in operations, and resale and can be classified as retailers, wholesalers, institutional users, and manufacturers. The *government market* (B2G) is composed of purchasers in four broad categories: municipalities, county governments, state governments, and the federal government.

To meet the definition of a good market in any of these three markets, the following criteria must be present: measurability, accessibility, profitability, and stability. Measurability determines the degree, size, and other aspects of the market that can be determined. Some markets for technical products or services are more difficult to determine. Julie Uhrman had no idea how big the market was for her idea for her new game console, Ouya, with an open-source platform allowing developers to create games and incorporate accessories priced at $100. Unable to raise the needed capital from any other source, she was able to raise $2.4 million through Kickstarter, a crowd funding website that increased through capital to $8 million. She also ultimately received a $15 million VC funding round led by Kleiner Perkins Caulfield & Byers.

MINI-CASE
Kickstarter Helps Fund Disruptive Technologies

Kickstarter is a paradigm shifting way to fund projects. The website provides a platform where regular people can help fund projects in anything from films and music to design and technology. Since its inception in April 2009, almost $950 million has been pledged by over 5.5 million people, helping fund more than 55,000 projects. Kickstarter does not involve themselves in the projects, only the funding. Kickstarter is open to anyone with a project that fits their guidelines. The project creators make their own goals and deadlines for receiving funds. If the funding goal is met by the set deadline, then the full amount will be given to fund the project.

Accessibility, the second criteria, measures the capability of the company to effectively market and deliver the technical product or service to this defined market. Although the market identified may be of sufficient size, the venture may not be able to reach it profitably. Arnolite Pallet Corporation identified some potential large users for its new plastic-molded modular pallet, but could not effectively and profitably sell it due to high transportation costs.

The profitability criteria defines whether the market identified is large enough and easy enough to be worthwhile.[15] The size of the market identified needs to be large enough to justify the effort and expense of reaching and serving it. Many drugs that can cure diseases cure such a small number of patients that they are not worth marketing by large pharmaceutical companies. Some of these with smaller, but sufficient, size come to market as orphan drugs which are successfully marketed by smaller technology companies with a lower overhead burden.

The final criteria, the political stability of the market now and in the future, needs to be favorable. Many technology products that could be successful in developing economies may not be profitably marketed due to instability of the market or country.[16] The risk of building a plant which may soon be nationalized by a country precludes some products from being produced and marketed there.

4.7.1 Market Segmentation

Market segmentation, or further defining the market by some criteria, is important to focusing the efforts of the technology entrepreneur. Although a technology entrepreneur may think his or her technical product or service is ideal for every market identified, rarely if ever is that the case. A market segment, a smaller subset of the market, should be identified, and the appropriate marketing strategy (the right combination of product, price, distribution, and promotion, discussed in Chapter 10), can be developed and created. The segmentation techniques available by type of market are indicated in Exhibit 4.6

Exhibit 4.6 Market Segmentation by Type of Market

Segmentation Criteria	Basis for Type of Market		
	Consumer (Business to Consumer)	Industrial (Business to Business)	Government (Business to Government)
Demographic	Age, family size, education level, family life cycle, income, nationality, occupation, race, religion, residence, sex, and social class	Number of employees, size of sales, size of profit, and type of product lines	Type of agency, size of budget, and amount of autonomy
Geographic	Region of country, city size, market density, and climate	Region of country	Federal, state, and local
Psychological	Personality traits, motives, and lifestyle	Degree of industrial leadership	Degree of forward thinking
Benefits	Durability, dependability, economy, esteem enhancement, status from ownership, and handiness	Dependability, reliability of seller and support service, efficiency in operation or use, enhancement of firm's earning, and durability	Dependability, reliability of seller and support services
Volume of use	Heavy, medium, and light	Heavy, medium, and light	Heavy, medium, and light
Controllable marketing elements	Sales promotion, price, advertising, guarantee, warranty, retail store purchased service, product attributes, and reputation of seller	Price, service, warranty, and reputation of seller	Price, reputation of seller

for all three markets—consumer (B2C), industrial (B2B), and government (B2G). The basic segmentation criteria—demographic, geographic, psychological, benefits, volume of use, and controllable marketing elements—can be effectively used by the technical entrepreneur to define a target market for focus in the launch of the new technical product or service.

Demographic segmentation is one of the most widely used segmentation techniques for identifying potential individuals who are most likely to purchase and use the new technical product or service. This is in part because most published data on each of these markets is collected on some demographic basis. The most widely used segmentation criteria are age, gender, and income in the consumer (B2C) market, type of product, sales, and size of profit in the industrial (B2B) market, and size of budget and/or number of employees in the government (B2G) market.

Most all demographic data regardless of market is collected on a geographic basis—the next segmentation criteria. These geographic clusters look at the demographic information based on country, region, state, province, or municipality (Standard Metropolitan Statistical Area—SMSA). This allows for a focused launch because only a certain geographic area will be approached in the first year. This selected area is the basis for the rollout of the technical product or service in future years. The Internet allows for successfully marketing to multiple disparate geographic markets at the same time. Even though Frends targeted women in the New York SMSA to launch the company's new headphone, sales were achieved outside this targeted geographic area with online sales through the company's website.

Psychological segmentation is not frequently used by technical entrepreneurs, particularly in the heterogeneous consumer market due to the absence of published data and the high cost of collecting new original data on such things as personality traits motives or lifestyles of the market.[17] If original data can be collected, it is beneficial for the technical entrepreneur to particularly evaluate the technical product idea with those already on the market through psychological segmentation. Positioning the new idea in terms of market perception of it and existing leads provides a significant advantage to the technical entrepreneur.

Benefit segmentation, the fourth segmentation criteria, can be one of the most effective segmentations for the technical entrepreneur. This is particularly true when very specific benefit segmentation criteria can be identified in the consumer (B2C), industrial (B2B), and government (B2G) markets. Although only very general benefit criteria are indicated in Exhibit 4.6, very technical product or service criteria need to be identified for this technique to achieve the best results. An extremely successful new product or service launch can be achieved when the technical product or service attributes or its unique selling propositions, previously discussed in this chapter, match the specific wants or needs, or both, in the market. This occurred for a unique innovative delivered food concept—Gourmet to Go—when launched. It delivers entire nutritional meals for a family, accompanied by a foolproof recipe (cooking for dummies) that allows the dinner to be prepared in no more than 20 min. The benefits desired by the upper income market segment targeted are ease of cooking, desire to be involved in the preparation, time savings, and nutrition—a great match!

The volume of use segmentation criteria clarifies the market based on size. Because this occurs based on some demographic dimension anyway, it is not a widely used segmentation criteria.

This is similar for controllable marketing elements criteria; the specific aspects of product price, distribution, and promotion are selected in the marketing plan used to launch the new technical product or service.

4.7.2 Target Market and Positioning

The smaller subset of the market—selected through market segmentation—is the target market, all or part of which is the focus of the launch of the new technical product or service. Specific market positioning concepts for this target market need to be identified and the appropriate marketing mix needs to be developed for the launch, as discussed in Chapters 9 and 10.

4.8 CHAPTER SUMMARY

This chapter focused on developing the technical product or service and its market. Three techniques (the product planning and development process, the idea development process, and the opportunity assessment plan) for evaluating the new technological idea were discussed. Regardless of the technique(s) employed, the degree of newness to the consumer, organization, and distribution system needs to be taken into account. The chapter concluded with a discussion of the three types of markets—consumer (B2C), industrial (B2B), and government (B2G); market segmentation and its techniques (demographic, geographic, psychological, benefit, volume of use, and controllable marketing variables); and the target market.

KEYTERMS

Concept stage The stage where the modified innovation is tested to determine market reaction and the degree of acceptance.
Consumer market The purchase of products or services for personal satisfaction/use.
Demographic segmentation Characteristics of the target market.
Disruptive innovation A novel innovation that disrupts an existing market and ultimately displaces an existing technology.
Diversification When a product is not only new technology, but also in a new market for a venture.
Government market Consists of purchasers in four broad categories: municipalities, county governments, state governments, and the federal government.
Idea stage When the innovation is formulated and developed.
Industrial market Consists of a variety of different entities involved in the purchase or resale of the product or service not for final consumption or use.
Market A mechanism that bridges the gap between supply and demand and consists of a group that may buy the product.
Market segmentation Defining the market by certain criteria.
Newness A product or service that is unique or new to the market.
Opportunity assessment plan Is shorter than a business plan, is focused on the opportunity and market and not on a business, and has no financial plan, marketing plan, or organizational plan.
Product development stage The stage when a final version of the product, based in part on the evaluations obtained in the concept stage, is subjected to consumer input.
Test marketing stage The stage when a market test of the innovation is done to help ensure successful commercialization.

ADDITIONAL READINGS

Beard, C., 2012. Insights into US tech success. NZ Business 26(6), 68, The article gives advice on how technology startup companies should approach the U.S. market. Even though the article focuses on New Zealand startups, it is for all technology startups that want to enter the U.S. technology market.

Blank, S., Dorf, B., 2012. The Startup Owner's Manual: The Step-by-Step Guide for Building a Great Company Volume 1. K&S Ranch Press, Pescadero, California, The book is a guide to the customer development process for startups. Among other topics, it describes how to gain, maintain, and enlarge a customer base and how to fit your product to the market.

Engelen, A., et al. 2011. Market Orientation and Inter Firm Knowledge in Inter Organizational Relationships: Market-Driving Behavior: Performance Consequences in High-Tech Start-Ups. In: AMA Summer Educators' Conference Proceedings, August 2011, 22. p. 520, The article describes orienting a business to influence and shape the market, rather than orienting a business around a set of existing consumer needs. The study details how market-driving leads to greater business performance.

WEB RESOURCES

http://www.entrepreneur.com/businessideas/interest/40.html: This site is a business idea center at entrepreneur.com.

http://ideamensch.com/50-tech-entrepreneurs/: This site contains examples of tech entrepreneurs.

https://plus.google.com/102279327257897906655/posts: This is a free online learning site.

http://www.ugcs.caltech.edu/ kel/FEI/: This site contains free entrepreneurial ideas.

ENDNOTES

1. Grégoire, D.A. and D.A. Shepherd. 2012. "Technology-Market Combinations and the Identification of Entrepreneurial Opportunities: An Investigation of the Opportunity-Individual Nexus." *Academy Of Management Journal*, 55(4): 753-785. DOI:10.5465/amj.2011.0126.
2. Bluestein, A. 2012. "How and Where to Make Money in 2013 (and beyond)." *Inc*, 34(10): 58-65.
3. Wouters, M. 2010. "Customer Value Propositions in the Context of Technology Commercialization." *International Journal of Innovation Management*, 14(6): 1099-1127.
4. Lorinc, J. 2013. "The Inventor's Playground." *Profit*, 32(4): 9-10.
5. Katzy, B.R., G.H. Baltes, and J. Gard. 2012. "Concurrent Process Coordination of New Product Development by Living Labs - An Exploratory Case Study. *International Journal of Product Development*, 17(1/2): 23-42.
6. Song, L.Z., M. Song, and M.E. Parry. 2010. "Perspective: Economic Conditions, Entrepreneurship, First-Product Development, and New Venture Success." *Journal of Product Innovation Management*, 27(1): 130-135. DOI:10.1111/j.1540-5885.2009.00704.x.
7. Srivastava, P., J. Yoo, G.L. Frankwick, and K.E. Voss. 2013. "Evaluating the Relationship of Firm Strategic Orientations and New Product Development Program Performance." *Journal of Marketing Theory & Practice*, 21(4): 429-440. DOI:10.2753/MTP1069-6679210406.
8. Hong, J., T. Song, and S. Yoo. 2013. "Paths to Success: How Do Market Orientation and Entrepreneurship Orientation Produce New Product Success?" *Journal of Product Innovation Management*, 30(1): 44-55. DOI:10.1111/j.1540-5885.2012.00985.x.
9. Willoughby, M., J. Talon-Renuncio, J. Millet-Roig, and C. Ayats-Salt. 2013. "University Services for Fostering Creativity in High-technology Firms." *Service Industries Journal*, 33(11): 1103-1116. DOI:10.1080/02642069.2011.623777.
10. O'Brien, J.M. 2012. "The Great Stem Cell Dilemma." *Fortune*, 166(6): 186-195.
11. Food. 2011. *Hospitality Design*, 33(1): 74-75.

12. Neel, K.C. 2010. "Discovery's Dangar Finds Challenge in Distribution Game." *Multichannel News*, 31(4): 4.
13. Tibergien, M. 2013. "Vision, Mission, Goals, Objectives." *Investment Advisor*, 33(9): 59–60.
14. Wang, J. 2013. "The Disrupters." *Entrepreneur*, 41(7): 50–53.
15. Arora, A. and A. Nandkumar. 2012. "Insecure Advantage? Markets for Technology and the Value of Resources for Entrepreneurial Ventures." *Strategic Management Journal*, 33(3): 231–251. DOI:10.1002/smj.953.
16. Jain, V. 2011. "Indian Entrepreneurship and the Challenges to India's Growth." *Ivey Business Journal*, 75(5): 9–11.
17. Gabay, G., L. Flores, H. Moskowitz, and A. Maier. 2010. "Creating Technology-based Merchandising Ideas for Hair Coloring through Weak Signals, Concept Optimization and Mind-Set Segmentation." *Journal of Consumer Marketing*, 27(3): 211–223. DOI:10.1108/07363761011038284.

CHAPTER 5

Protecting Your Intellectual Property

VisiCalc PAYS FOR DECISION NOT TO PATENT FIRST SPREADSHEET PROGRAM

Many people believe that one of the key contributors to the growth of personal computers was a spreadsheet program called VisiCalc. VisiCalc was the brainchild of Dan Bricklin while he was studying for his MBA at Harvard Business School in 1978. He observed that an error in a single cell of a common (paper) spreadsheet required that the value in every other cell be changed. As a way to solve this problem, Bricklin envisioned electronic spreadsheets where a cell entry could be changed, and all the other cells would change automatically according to specified formulae.

To implement this new idea, Bricklin enlisted the expertise of a software programmer named Bob Frankston. Their partnership led to the creation of Software Arts, Inc. Soon, Bricklin and Frankston's fledgling company garnered the attention of Personal Software, a software publisher that loaned Software Arts an Apple II computer on which to perform a demo of VisiCalc.

Personal Software eventually funded the development of VisiCalc by licensing it from Software Arts and becoming its sole distributor. What once began as an innovative idea in Bricklin's apartment building in mid-1978 went on to become, in 1979, the first spreadsheet program ever released on the market. VisiCalc was considered revolutionary in many circles.

However, VisiCalc's story does not end there. No sooner had the company completed its initial investment and established a firm market for its product, other companies began to place conceptually identical spreadsheet programs on the marketplace. These newcomers not only profited from publicly available information of the original spreadsheet program, including its established market acceptance, but arguably they were able to muster greater resources to overcome VisiCalc's market leadership.

According to Bricklin, a conscious decision was taken not to pursue patent protection of VisiCalc. He stated that Personal Software (later renamed VisiCorp) had retained a patent attorney. The patent attorney explained that there were many obstacles involved in obtaining a patent on the software, estimating only a 10% chance of success. Based on this advice, and the potentially high costs involved, Bricklin decided not to pursue a patent. Instead, copyright and trademark protection were vigorously pursued to keep others from replicating their work.

Continued

"The enormous importance and value of the spreadsheet, and of protections in addition to copyright to keep others from copying our work, did not become apparent for at least two years, too late to file for patent protection," said Bricklin. "If I invented the spreadsheet today, of course, I would file for a patent," he added.

Source: Hormby, VisiCalc and the Rise of the Apple II, http://lowendmac.com/orchard/06/visicalc-origin-bricklin.html. U.S. Patent and Trademark Office; Dan Bricklin, Patenting VisiCalc, http://www.bricklin.com/patenting.htm.

5.1 INTRODUCTION

Intellectual property (IP), the intangible asset that comprises or results from creativity, innovation, *invention*, *know-how*, and reputation (and encompasses all rights to technology), is becoming an increasingly important aspect of any business. This is particularly true with start-up and emerging technology ventures. IP is typically the basis for the competitive advantage necessary for a technology venture to succeed and has become the primary asset of modern business enterprises. As Dan Bricklin found out to his chagrin, having the greatest technology in the marketplace will not sustain success if you do not take the appropriate steps to prevent your competitors from legitimately appropriating it.

In this chapter, we examine the various types of IP and how to protect them. We discuss the significance of IP, how to recognize IP assets, and the characteristics of the individual types of IP and protection mechanisms.

5.2 IP AND TECHNOLOGY VENTURES

The importance of IP to modern business ventures is clear. Only a few decades ago, the bulk of U.S. corporate assets were tangible in nature, which is depicted in Exhibit 5.1. Intangibles such as IP comprised ~20% of corporate assets. However, by 2005, the ratio of intangible to tangible corporate assets had essentially reversed; the market value of the S&P 500 was ~80% intangible assets.

According to the U.S. Department of Commerce, "[t]he entire U.S. economy relies on some form of IP, because virtually every industry either produces or uses it."[1] In 2010, about 28% of the jobs and 35% of the GDP in the United States are attributable to IP intensive industries. The emphasis on IP assets is not confined to the United States; as shown in Exhibit 5.2, intangible assets also represent at least approximately a third, and typically the majority, of value in significant non-U.S. markets. As of September 2013, about 35% of the jobs in the European Union and 39% of the GDP come from sectors that are dependent on patents and other IP to function.[2]

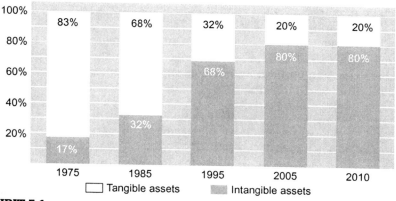

EXHIBIT 5.1
Components of the S&P 500. *Source: Ocean Tomo: http://www.oceantomo.com/media/newsreleases/intangible_asset_market_value_2010.*

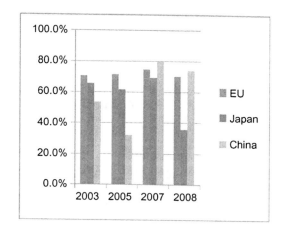

EXHIBIT 5.2
Intangible value as percent of market value for non–U.S. markets. *Source: Ocean Tomo: http://www.oceantomo.com/media/newsreleases/intangible_asset_market_value_2010.*

Consider the size and nature of the transactions driven by IP. For example, Microsoft paid $8.5 billion to acquire Skype and $6 billion to acquire a Quantive.[3] SanDisk paid $327 million to acquire Pliant Technology,[4] and Nuance acquired Equitrac for a price of $150 million.[5] Apple, RIM, and Microsoft teamed up to outbid Google to buy a bundle of Nortel patents relating to mobile phones and tablet computers for $4.5 billion.[6] Apple and Google teamed up to purchase Kodak's patent portfolio relating to the capture, manipulation, and sharing of digital images for $527 million.[7] Microsoft sold a package of its patents related to mobile, Web, and instant messaging technology to Facebook for $550 million and purchased AOL's patents covering Internet technology for $1.05 billion. InterDigital sold a bundle of patents relating to 3G/LTE technology to Intel for $375 million.

IP is particularly important to startup and emerging ventures. For a venture to succeed it must have a competitive advantage; some aspect of the venture—its operations, product, or services—must be (or at least must be perceived in the market to be) unique, better, or distinctive in comparison to that of the competition. This is particularly true for a start-up venture in a market with already established (and well-financed) competitors. Sustained success requires a sustained competitive advantage. Sustainable competitive advantage for technology companies almost always derives from some form of IP. Without the appropriate legal foundation to protect IP rights, however, competitors will be able to legitimately appropriate or copy the feature, and the competitive advantage will be lost.

In addition to providing competitive advantage, IP can contribute to the success of a venture in a number of ways:

- Demonstrable rights to IP are perceived as an indicator of the likelihood of success of a technology venture. A study of venture capital funding between 1995 and 2002 showed that businesses that had IP holdings (patents) were 34% more likely to be successful in obtaining subsequent rounds of venture capital funding.[8]
- Rights to IP are assets of the venture, often one of the most significant factors in the valuation of the venture.
- Patents are, in effect, a third-party certification of expertise.
- IP (most commonly patents) can be used as collateral for a loan.[9]
- IP can be leveraged (through strategic licensing, granting profit interests,[10] securitization,[11] etc.) to provide capital to the venture.

IP is also "currency" between participants in the formation of a technology start-up venture. (We discuss the various legal structures available for your start-up in the next chapter.) Exhibit 5.3 shows that each participant (co-venturer) in the start-up venture makes a *contribution* to the venture and in return for that is given some form of *consideration*.

A participant's contribution to the venture can be capital (cash or credit). However, in a technology venture, at least one participant makes a contribution in the form of IP. The IP contribution can be in lieu of capital. (Contributions to the venture can relate to any of the elements necessary for the success of the

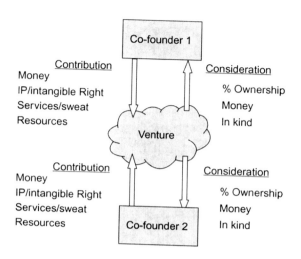

EXHIBIT 5.3

Transactional analysis of formation of a business. *Source: Michael Lechter, A., 2010. OPM: Other People's Money, The Ultimate Leverage, second ed. TechPress, Phoenix, AZ.*

venture, such as resources, leadership, "sweat," and connections. Initial contributions by principals are typically made in return for, at least in part, an equity interest (otherwise they would not be principals).

Recently, so-called *open* models of doing business have gained notoriety. Examples include various consortia[12] and *open source licensing* of software.[13] Even these business models for the most part rely on IP protection.[14]

A savvy technology entrepreneur will secure exclusive rights to the venture's IP to the greatest extent possible. However, the laws relating to IP are anything but intuitive. Valuable rights in IP can be unwittingly lost by seemingly innocent courses of action, and failing to consider third-party rights can lead to disaster.

5.2.1 IP Protection

IP can take many forms. As will be discussed, the terms trade secret, utility patent, design patent, copyright, mask work, and trademark have come to denote both an IP asset and legal mechanism for protecting the underlying assets. These mechanisms provide a framework for establishing and maintaining rights in IP. A strategy employing combinations of the various legal mechanisms should be developed to maximize protection for facilitating the venture's goals. Exhibit 5.4 highlights some of the characteristics of the respective legal mechanisms.

Exhibit 5.4 Comparison of U.S. Intellectual Property Protection Mechanisms

Protection mechanism	Term	Source of Law	Protectable Subject Matter	Scope of Protection Precludes Unauthorized	Compatible Concurrent Forms of Protection
Trade secret	Potential infinite (as long as kept secret)	State law (Uniform Trade Secret Act)	Anything that can be kept secret	Copying or use. No protection against independent development	Trademark
Trademark	Potentially infinite (as long as in use and capable of identifying source) registrations renewable 10-year terms	State common law; State statutes; Federal statute (15 USC §1051 et seq)	Anything non-utilitarian that is capable of identifying the source of goods or services	Use of trademark in context that creates any likelihood of confusion as to source, sponsorship, or affiliation	Trade secret Design patent Utility patent Copyright
Copyright	70 years after death of the last surviving author, or if work for hire, 95 years from publication	Federal statute (17 USC §101 et seq)	Non-utilitarian works of authorship	Copying of copyrighted aspects of a work of authorship	Trademark Design patent Utility patent
Maskwork	Until end of 10th calendar year after earlier of registration or first commercial exploitation	Federal statute (17 USC §901 et seq)	Maskworks (images, representing three-dimensional patterns in the layers of a semiconductor chip)	Use of reproductions of masks introduction of competing chips	Trade secret Trademark Utility Patent
Design patent	14 years from issue date	Federal statute (35 USC §§1 et seq, §171)	Nonfunctional aspects of ornamental designs	Application of patented design to any article of manufacture, or sale or offering for sale such article of manufacture	Utility patent Copyright Trademark
Utility Patent	20 years from filing date of application	Federal statute (35 USC §1 et seq)	New and useful process machine, manufacture, or composition of matter, or new and useful improvement thereof	Making, using, offering to sell, selling or importing any patented invention	Copyright Design patent Maskwork Trademark

5.3 RECOGNIZING IP

A starting point for developing IP assets is, simply, recognizing their existence. Identifying IP assets is easy when a venture is built around a particular product or idea. For example, when technology is developed specifically to fulfill a want or need in the marketplace or to solve a problem generally encountered in the marketplace, it is clear that IP rights in that technology should be pursued. However, valuable IP also may reside in other aspects of the venture.

A potential IP asset is created every time a problem is solved—even when the problem relates to internal operations. When a problem is encountered in connection with a business, it is likely that its competitors will encounter the same problem. If the business develops a solution to the problem that is better than that of its competitors, obtaining the exclusive rights to use that solution can give the business an advantage—a great advantage if the problem is significant and the solution is notably better than that of the competition's.

Technology ventures should: continuously and systematically analyze their products and operations to identify potential IP; affirmatively seek opportunities to develop protectable IP assets; obtain and maintain exclusive rights to the IP to the maximum extent available; and exploit the assets as part of an overall strategy. Everything that gives the business an advantage should be identified. For example, this may involve:

- Identifying the reasons that customers are attracted to the business instead of to its competitors.
- Dissecting the operations and systems, products, services, and communications and analyzing each component and feature to determine whether there is anything about it that is unique, better, or distinctive.
- Developing a strategy to exploit the IP by securing exclusive rights through application of the appropriate legal mechanisms and/or promoting the use by others to drive the company's sales of other products or services.

Technologies or processes that are not used by the venture might still be of use to others. For example, you may develop alternative technological solutions to a problem and choose to actually employ only one of those alternatives. However, the alternatives that were not chosen may still have value. Rights to those developments may be used to broaden a competitive advantage or to generate income through licensing.

Any noncompetitive markets where the venture's technology can be utilized should be identified. Income can be generated, without affecting competitive advantage in your marketplace, through strategic field of use *license agreements*—permitting (for some consideration) a *licensee* to use the technology only in the noncompetitive market.

By properly identifying and protecting as many IP assets as possible, a venture will be more attractive to investors. Ventures with IP holdings are more likely to be successful in obtaining multiple rounds of venture capital funding. In analyzing a company, investors evaluate not only its IP, but also the systems and processes the venture has in place to protect IP assets.

5.4 RECORD KEEPING

The basic foundation for IP protection and for protecting against third-party charges of infringement is often a complete and accurate evidentiary record. It is sometimes necessary to prove the specific nature of technical innovations, the date they were made, and the project with which the innovations are associated.

> **KEY POINT**
>
> *Making an Invention Under U.S. Law: Two Distinct Steps*
>
> To *make* an invention under U.S. law you must *conceive* the invention and then *reduce it to practice*. Conception is the mental portion of the inventive act. Reducing the invention to practice is building the invention and proving that it works for its intended purpose. The filing of a patent application is a *constructive reduction to practice*, equivalent to actually reducing the invention to practice. The *diligence* with which the inventor reduces the invention to practice after conception can also be a factor in some circumstances.
>
> This two-step process becomes important in, for example, interference proceedings under the old law and in the application of contract provisions keyed to when an invention is *made, first made,* or *first actually reduced to practice.*
>
> Your contemporaneous evidentiary record should cover each of the elements of making an invention.

Typical situations that require an evidentiary record of development include:

- Disputes regarding ownership of or rights to use technology—whether a certain technology was first made under a particular development contract or government contract, is covered by a particular license agreement, or is subject to a confidentiality or nonuse agreement
- Proving that IP was independently developed and not derived from another's proprietary material (e.g., as a defense to trade secret, copyright, or mask work infringement, in patent *derivation proceeding* or *interference proceeding*)
- Proving that an invention was previously developed, not abandoned, suppressed, or concealed (as a defense to patent infringement with respect to patents applications filed prior to March 16, 2013)

063

P10F70MV8

GEORGIAN COLLEGE LIBRARY COMMONS

9

ISBN	Qty	Sales Order
9780124201750	1	F 19276983 1

Customer P/O No: 2016/02/16

Title: Technology entrepreneurship : taking innovation to the marketplace /

Format: P (Paperback)
Author: Duening, Thomas N.
Publisher: Elsevier Academic Press
Fund: BABUSBENTBK
Location: GEOB – BK
Loan Type: 1396A01
Coutts CN: 28338796

Order Specific Instructions

Ship To: JT 1396A01 F
GEORGIAN COLLEGE LIBRARY
C/O OCLC CANADA
LIBRARY TECHNICAL SERVICE
1465 ST. JAMES STREET
WINNIPEG MB R3H 0W9

Volume:
Edition: 2nd ed.
Year: 2015.
Pagination: xxiv, 370 pages
Size: 24 cm

Routing 1
SORTING
F02A03X
Shipping

Available:
MyiLibrary, EBSCO,
EBL, EBRARY,
ProQuest Ebook
Central

063578876 ustwlg11 RC2

PROQUEST LLC:

> **KEY POINT**
>
> *Change in U.S. Patent Law*
>
> In 2013 U.S. patent law changed. However, the old law still governs those applications filed prior to the date that the new law went into effect and patents issuing from those applications. This means that the different law applies to U.S. patents depending upon their effective filing dates. This will be a consideration until all of the patents with effective filing dates prior to March 16, 2013 have expired—at least until March 16, 2033.
>
> Among other things, the law changed from a "first to invent" to a "first to file" system. It also expanded the criteria for prior art against which patentability is measured; expanded a prior use defense; and changed or added various postgrant and inter-partes proceedings.

- Proving that an invention was in public use, on sale, or otherwise available to the public prior to the effective filing date of the patent (as a defense to patent infringement with respect to patents applications filed on or after March 16, 2013)
- Proving that an invention was in nonpublic commercial use before a certain date (as a personal defense to patent infringement with respect to patents applications filed on or after March 16, 2013)
- Holding interference proceedings before the United States Patent and Trademark Office or courts to determine priority of invention (with respect to patents applications filed prior to March 16, 2013)

Significantly, in most proceedings, the mere word of the inventor or inventors with respect to the elements of making an invention *is not considered competent evidence*; documentation and corroboration by a noninvolved witness are typically required. Accordingly, development should be contemporaneously documented in tangible media form and witnessed by, or placed in the possession of, a person or entity that was not involved in the development effort.

5.4.1 Record Keeping Procedures

It is best practice to establish procedures that facilitate complete and accurate records of the precise nature and time frame of development activities, keeping in mind the requirement for independent corroboration.

Many companies employ some form of invention disclosure forms to establish a conception date in the United States. These are not a substitute for contemporaneous records. The forms typically (1) are at least days after the fact and (2) do not contain the details necessary to prove all of the elements of making an invention, let alone diligence.

The evidentiary value of a record entry is directly proportional to the specificity of the entry and the care taken to establish the date and authenticity of the entry. For a physical record, this is accomplished simply by signing and dating each entry and having each entry read, signed, and dated by a witness. Electronic records (which can easily be modified), require a more elaborate procedure—typically involving placing the records into the hands of an independent "trusted third party" (escrow agent) who can testify that the records were deposited into escrow as of a certain date and have not been modified.

> **KEY POINT**
>
> *Technology Escrow*
>
> Technology escrow is a process where information regarding technology is placed into the hands of a "trusted third party" (escrow agent) who delivers the information to a specified entity upon the occurrence of a specified condition or event. In the context of establishing evidentiary records of development, the escrow agent holds the records and provides evidence (testimony) that the information is precisely that delivered to them on a certain date.
>
> In the context of trade secret licensing (e.g., software licenses) trade secret information regarding the licensed technology (e.g., source code) is delivered to the escrow agent by the licensor and delivered to the licensee upon occurrence of, for example, a breach of the license agreement or inability of the licensor to maintain the software. In some instances, the escrow agent will verify the completeness and sufficiency of the escrowed materials.
>
> When referring to the sale of trade secrets, a technology escrow is often used in precisely the same manner as what is used in the escrow process in the sale of real property.

The context of an entry within a set of records can sometimes be used to prove a date. For example, if an entry showing conception is found in a *bound* notebook between entries dated January 3 and January 5, then that is relevant proof the invention was conceived sometime between January 3 and January 5. It would not be as relevant, however, if a loose-leaf binder had been used instead of a bound notebook because loose-leaf binders allow for inserting pages.

5.4.2 Guidelines for Record Keeping

The goal of keeping records is to create an irrefutable evidentiary record of development and invention. It is imperative that each record entry identifies the subject of the work with particularity and contains all relevant details. An entry such as "worked on new sharpener" sheds little light on whether the "new sharpener" included a specific feature. Every entry should indicate the particular project with which the entry is associated, and, if possible, be signed and dated by (or "escrowed" with) someone who is not participating in the project.

All computations, diagrams, and test results should be contemporaneously entered into the record. Keep in mind the evidentiary value of the record is at least in part determined by how easily it can be falsified. It is as easy to do "pencil" calculations in ink in a bound notebook as it is to do them on scratch paper. If the entry is legible, there are no particular format or neatness requirements. It is also easy enough to scan documents into a computer to become part of an electronic record. However, if electronic records are used, they may be of little evidentiary value unless escrowed with—provided (preferably on a daily basis) to—an entity (preferably independent) who was not involved in the development effort and can verify that the records were provided to him or her on a certain date, and unless they are unchanged from the original provided. All persons involved in the work should be identified in the corresponding entries. Unless participants are identified, it is often difficult to establish, long after the fact, those involved in or who witnessed particular activities.

It is important that all loose papers, such as blueprints, schematics, flowcharts, oscillographs, photographs of models, and so on, be signed and dated, cross-referenced to a particular entry, and, preferably, mounted in or scanned into the body of the appropriate entry. Similarly, physical results of tests, for example, samples, models, and prototypes, should be carefully labeled with the date, cross-referenced to record entries, and retained.

Records should be maintained with the idea of proving not only the dates of conception and reduction to practice, but also reasonable diligence in reducing to practice. To this end, it is important to have the documentary evidence and dated record entries describe *all* testing performed (both good results and bad), materials and equipment ordered, the particular types of equipment used, and the results of the testing.

Procedures should be implemented to ensure that significant tests or demonstrations showing or comprising reduction to practice are witnessed by noninventors or that such tests or demonstrations are repeated before noninventor witnesses.

Audio and/or video recordings of significant tests showing and identifying all participants, preferably including noninventor witnesses, should be considered.

In all, a documentary record should be maintained that is capable of establishing the dates and activities comprising each of the elements of making an invention, identifying individuals involved in the work who can provide testimony, and identifying the particular project with which technical work is associated.

MINI-CASE
Invention Must Be Proven by Competent Evidence

Was Alexander Graham Bell entitled to the patents on the telephone? Was he the first to reduce a telephone to practice? The courts of his time ultimately held that he was, but primarily because of a lack of documentation evidencing the allegedly prior work of others.

During the 1860s and 1870s, there was considerable parallel experimentation on the transmission of the human voice over electrical wires. Notably, the experimenters included not only Alexander Graham Bell, but serial inventors Thomas Edison, Elisha Gray of Chicago, and Prof. Amos Dolbear. Bell was the first to file patent applications on the key components of the telephone (e.g., the microphone); he filed an application entitled "improvement in telegraphy" on February 14, 1876. Over the next five years, Bell filed applications on mechanisms for establishing a telephone connection and on twisted pair wire.

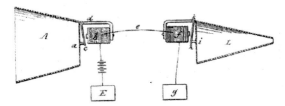

5. The method of, and apparatus for, transmitting vocal or other sounds telegraphically, as herein described, by causing electrical undulations, similar in form to the vibrations of the air accompanying the said vocal or other sound, substantially as set forth.

Over the next two decades, however, there were over 600 proceedings involving the Bell patents. For example, Western Union (which then dominated the U.S. telegraph system), purportedly acquired technology from Edison, Gray, and Dolbear and entered the telephone market in 1878. Bell sued for patent infringement, resulting in Western Union agreeing to pull out of the telephone market in 1879. Around that time, "People's Telephone Company" entered the market with technology acquired from one Daniel Drawbaugh. Bell sued for patent infringement. People's Telephone argued that Bell's patent was invalid because of prior invention by Drawbaugh; Drawbaugh claimed to have invented the microphone element of the telephone approximately a decade prior to Bell, but had not applied for patent or commercialized the technology due to lack of funds. In support of its "prior invention" defense People's Telephone presented the testimony of 300–400 witnesses (resulting in some 1200 pages of transcript). Notwithstanding all of the testimony, since nothing whatsoever "in print or in writing," "not even a memorandum or a drawing of any kind," was produced in support of Drawbaugh's claim, the majority of the court held that Drawbaugh did not make a telephone microphone prior to Bell. A number of judges, however, dissented. We will never know whether Drawbaugh's claims to have made a telephone microphone before Bell were factual. If they were, contemporaneous documentation would have made all the difference.

5.5 TRADE SECRETS

Secrecy is probably the most ancient form of IP protection. Any proprietary information that can be kept secret can be a trade secret. If the competition doesn't know it, they can't copy it—and will have to expend the time and effort to develop it on their own (assuming that they can). Trade secrecy is the primary protection mechanism for information, data, know-how, and expertise, but it can also be used to protect inventions.

Trade secret protection has a potentially infinite duration. The information underlying the trade secret is protected as long as it is not accessible to competitors. However, a trade secret can be very fragile. Once a trade secret becomes generally known, irrespective of how it becomes known, the protection is lost. A trade secret also provides no protection against independent development of the technology by others.

In the United States, trade secret rights that are enforceable against others are provided under the laws of the various individual states. In order to qualify for protection, the subject matter of the trade secret right must meet certain prerequisites. It must not be generally known, and must: (1) derive some value from being kept secret, and (2) be subject to reasonable efforts to maintain its secrecy. As will be discussed, reasonable efforts typically involve restricting access and establishing contractual obligations of confidentiality.

Some types of technology are simply unsuited for trade secret protection. For example, any technology that is evident from marketing materials, or that can be reverse engineered from a product that is sold to the public, cannot be maintained as a trade secret. **In general, in the absence of an express or implied contractual obligation, any unpatented technology that comes into an entity's possession legally can be freely used and copied.**

Information that is generally known in an industry and basic skills or practices employed in an industry, even though unquestionably valuable to a business, do not qualify for trade secret protection. Typical examples of such *nonproprietary know-how* are the skills and knowledge acquired by an employee who is trained in the operation of a commercially available machine. A venture's interest in nonproprietary know-how is best served by ensuring that the know-how is possessed by a number of people within the business and by ensuring that the know-how is well documented or recorded.

5.5.1 Procedures for Trade Secret Protection

In theory, the procedure for maintaining a trade secret is simple. Implementation of the procedure, however, requires discipline. To maintain trade secret status, you must be able to show that your procedures for maintaining secrecy meet the standard of "reasonable under the circumstances." All technology considered proprietary should be clearly treated and marked as such. For example, all printouts, flowcharts, schematics, layouts, blueprints, technical data, and test results that contain confidential information should be marked confidential. However, merely marking something "confidential" does not make it so. Providing someone with a document marked "confidential" does not in and of itself create any obligation to keep the secret. Conversely, indiscriminate use of proprietary markings can dilute the significance of the marking when used on things that are, in fact, proprietary.

KEY POINT

Inappropriate Confidentiality Markings Can Hurt You

A large aerospace company was sued for infringement of a patent with an effective filing date over two decades earlier. Finding witnesses that could testify with respect to the events occurring so long ago presented a problem. However, it happened that the claimed invention was described in a maintenance manual for an earlier model of the accused device. The manual was provided, without any confidentiality agreement, to all purchasers of the earlier model of the accused device, as well as to all organizations servicing aircraft and to anyone else that requested a copy. The dates were such that the manual, if a publication, would invalidate the claims that were being asserted. However, though publicly available, the manual was marked "confidential." The inappropriate confidentiality markings complicated the company's patent infringement defense; they had to prove that the manual was in fact publicly available notwithstanding the confidentiality markings.

Tight security should be maintained and access restricted to the area in which a trade secret is practiced or kept. Access to and knowledge of a trade secret should be permitted only on a need-to-know basis. Records should be kept identifying all persons given access to any portion of the trade secret. All copies of trade secret documents should be accounted for.

It is imperative that *confidentiality agreements* be executed with every entity that is given access to any part of the trade secret technology. Confidentiality agreements are the primary mechanism for creating the obligation of secrecy necessary for trade secret status. As a general rule, permitting any entity to have access to the trade secret information without first imposing an obligation of confidentiality will impair or destroy trade secret status. An illustrative generic confidentiality agreement is provided in Appendix I.

Although employees are obligated by law in most jurisdictions to keep their employer's trade secrets confidential, there is great room for dispute regarding the scope of that obligation and precisely what should be considered a trade secret. Accordingly, each principal and/or employee of a company who has access to company technology or sensitive business information should sign a confidentiality agreement that defines and creates presumptions with respect to things deemed confidential. Generally, such agreements should be executed at or prior to the time of hiring the employee.

When employees who have access to trade secret information leave the company to take a similar position with a competitor, disclosure or use of the former employer's confidential information is sometimes inevitable, notwithstanding any agreement to the contrary. In those circumstances, a *noncompetition agreement* is often used with the employee to prevent the employee from taking a position that would inevitably cause the trade secret to be disclosed. However, such agreements are disfavored and strictly construed by the courts.[15] Under any circumstances, the noncompetition agreement should be carefully crafted so that geographic scope, duration, and scope of prohibited employment are restricted to the minimum necessary to protect the proprietary rights of the former employer. If any of those aspects of the agreement are deemed to be overly broad or overreaching, a court is likely to find the agreement unenforceable.[16]

5.6 PATENTS

The basic premise of the patent system is to provide inventors a limited-term exclusive right to their inventions as an incentive to bring their inventions to the public, as opposed to keeping the inventions secret. Generally, the term of a patent on an invention is 20 years from the date on which the application for the patent was filed.

Patents are territorial, effective only within the jurisdiction in which it is granted. The patent law of the jurisdiction specifies the general fields of subject matter that can be patented and the conditions under which a patent may be obtained. As will be discussed in greater detail, a patent is divided into two major sections: the written description (typically incorporating a drawing), which discloses the invention to the public; and the claims, which define the particular IP to which the inventor obtains rights.

The significance of the patent system in the United States is reflected by its origin in the U.S. Constitution. Article I, section 8 of the U.S. Constitution reads: "Congress shall have power to promote the progress of science and useful arts, by securing for limited times to authors and inventors the exclusive right to their respective writings and discoveries." Under this power, Congress has enacted various laws relating to patents. The first patent law was enacted in 1790, and was revised in 1952, in 1999, and then again in 2011 (effective March 16, 2013). The 2011 revisions, referred to as the "American Invents Act" (AIA), significantly changed the U.S. patent law with respect to patents with effective filing dates of March 16, 2013 and later. There are four basic patent types (three in the United States):

Utility Patent ("patents for invention"): These are intended to protect utilitarian, functional items and relationships, physical embodiments of innovative ideas, and discoveries. For example, in the United States, a utility patent is issued, subject to the conditions and requirements of the patent law, to "any person who invents or discovers any new and useful process, machine, manufacture, or composition of matter, or a new and useful improvement thereof." The word "process" is defined as a process, act, or method and primarily includes industrial or technical processes. The term "manufacture" refers to articles that are made and includes all manufactured articles. The term "composition of matter" relates to chemical compositions and may include mixtures of ingredients as well as new chemical compounds. These classes of subject matter taken together include practically everything that is made by man and the processes for making the products. The utility patent generally permits its owner to exclude others from making, using, or selling the invention for a period of up to 20 years from the date of patent application filing. Generally, when someone refers to a "patent," they are referring to a utility patent. Approximately 90% of the patents granted by the USPTO in recent years have been utility patents. (See Exhibits 5.5 and 5.6.)

Petty Patent/Utility Model (not available in the United States): This type is intended to protect utilitarian/functional items and relationships that are worthy of an incentive but don't quite meet the threshold standard requirements for a utility patent. A utility model typically affords a lesser protection than a utility patent.

Design Patent/Industrial Design: This type is intended to protect nonutilitarian/nonfunctional appearance and designs. In the United States, they are issued for a "new, original, and ornamental design for an article of manufacture." The design patent permits its owner to exclude others from making, using, or selling the items embodying the design for a period of 14 years from the date of patent grant.

Plant Patent: This patent is issued for a new and distinct, invented or discovered asexually reproduced plant, including cultivated sprouts, mutants, hybrids,

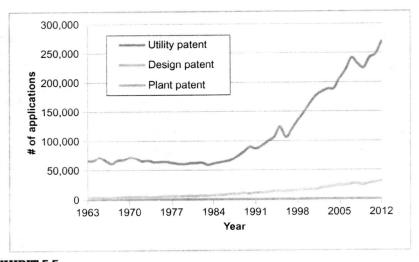

EXHIBIT 5.5
U.S. patent applications.

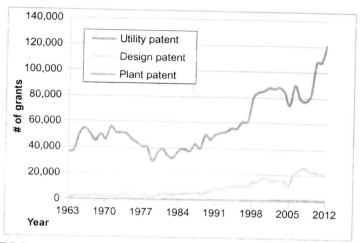

EXHIBIT 5.6
U.S. patents granted.

and newly found seedlings, other than a tuber-propagated plant or a plant found in an uncultivated state. The plant patent permits its owner to exclude others from making, using, or selling the plant for a period of up to 20 years from the date of patent application filing.

5.6.1 The Patent Application

A *patent application* is, in effect, a proposed patent, submitted to the appropriate government agency (patent office) for its approval. The requirements for the application are typically provided in the applicable statute. In the United States, a nonprovisional application for a patent must include a written description of the invention, a drawing when necessary for understanding of the invention, at least one claim (defining the patented invention), and an oath or declaration by the applicant.[17]

Some countries (e.g., the United Kingdom and the United States) provide for what is referred to as a *provisional application*, which does not require the formalities of a regular application. A provisional application, however, is not intended to provide any enforceable rights. It is a tool to prevent subsequent events from becoming prior art against which patentability is measured (which will be discussed) with respect to related nonprovisional applications. A provisional application is not examined and is automatically abandoned 12 months after filing. It does not ever mature into a patent. For a patent to issue on the subject matter described in the provisional, a regular nonprovisional application *claiming priority* on the provisional application must be filed within a year of the provisional.

In some cases, examination of an application by the patent office does not result in the allowance of all of the claims that are desirable with respect to the subject matter described in the application. In those cases, a *continuation application*, substantively identical (except perhaps in the case of claims) to (and containing a specific reference to) the original or "parent application," to which it specifically refers, may be filed while the parent application is still pending to pursue the claims that were not allowed in the parent application.[18] The continuation is treated as if it was filed on the date that the parent application was originally filed. However, the term of the patent issuing from the continuation application is measured against the filing date of the parent application.[19]

Once a formal application has been filed, no *new matter* can be added to the application. In this context, new matter refers to new embodiments or details not described or shown in the originally filed application. However, those further details, or embodiments, of the invention can be covered by filing a *continuation-in-part* (CIP) application. The claims relating to the material described in the parent application are given the benefit of the original filing date. The claims covering the additional details are given the filing date of the CIP application. However, the term of the patent issuing from the CIP application is measured against the filing date of the parent application, even with respect to the claims relating to the new matter. A CIP application can be filed at any time before the original application issues as a patent or is abandoned.

5.6.2 Written Description

A patent's *written description* (with the drawing) is the primary vehicle for making the invention known to the public. Each element of the invention is shown in the drawing, designated by a numeral. The written description describes the elements of the invention and how they work together, with specific reference to the numeric designations used in the drawing.

By way of illustration, the first page of a patent is shown in Exhibit 5.7. This is Patent No. 7,122,923, entitled Compact High Power Alternator, invented by Messrs. Lafontaine and Scott and owned by Magnetic Applications, Inc. It issued in 2006 from an application filed in 2004. Note how each element in the drawing is numbered. Note also all of the valuable information provided on the front page of the patent regarding, e.g., other potentially relevant patents.

The level of detail required in the description tends to vary from jurisdiction to jurisdiction. The United States requires a greater level of detail than most other countries; the U.S. law requires that the description be in sufficient detail to enable a *person of ordinary skill in the art* (hypothetical typical practitioner in the field of the invention) to make and use the invention. Basically, it is desirable to include as much detail as possible in the description and to be exceedingly careful to fully describe each and every feature that is to be protected. In addition, at

First page of a U.S. Patent

(12) **United States Patent**
Lafontaine et al.

(10) Patent No.: **US 7,122,923 B2**
(45) Date of Patent: **Oct. 17, 2006**

(54) COMPACT HIGH POWER ALTERNATOR

(75) Inventors: Charles Y. Lafontaine, Berthoud, CO (US); Harold C. Scott, Lafayette, CO (US)

(73) Assignee: Magnetic Applications Inc., Lafayette, CO (US)

(*) Notice: Subject to any disclaimer, the term of this patent is extended or adjusted under 35 U.S.C. 154(b) by 0 days.

(21) Appl. No.: **10/889,980**

(22) Filed: **Jul. 12, 2004**

(65) Prior Publication Data
US 2005/0035673 A1   Feb. 17, 2005

Related U.S. Application Data

(60) Provisional application No. 60/486,831, filed on Jul. 10, 2003.

(51) Int. Cl.
H02K 9/00 (2006.01)

(52) U.S. Cl. .... 310/58; 310/61; 310/60 A

(58) Field of Classification Search .... 310/52–64, 310/153, 112, 113; 290/1 B
See application file for complete search history.

(56) References Cited

U.S. PATENT DOCUMENTS

765,978 A	*	7/1904	Jigouzo	310/112
4,467,229 A	*	8/1984	Ogita	310/60 A
4,900,965 A		2/1990	Fisher	310/216
4,931,683 A		6/1990	Gleixner et al.	310/89
5,625,276 A		4/1997	Scott et al.	322/24
5,705,917 A		1/1998	Scott et al.	322/46
5,886,504 A		3/1999	Scott et al.	322/45
5,929,611 A		7/1999	Scott et al.	322/46
6,018,200 A		1/2000	Anderson et al.	240/40
6,034,511 A		3/2000	Scott et al.	322/46
6,384,494 B1	*	5/2002	Avidano et al.	310/58
6,441,522 B1		8/2002	Scott	349/156.23
6,744,157 B1	*	6/2004	Choi et al.	310/62
2002/0053838 A1		5/2002	Okuda	310/59

FOREIGN PATENT DOCUMENTS

DE	33 29 720	2/1984
DE	3329720 A1	2/1984
DE	195 13 134	10/1996
DE	19513134 A1	10/1996
FR	2 536 222	5/1984
FR	2536222 A	5/1984
JP	60-118036	6/1985
JP	60118036 A	6/1985
JP	08-322199	12/1996
JP	08322199 A	12/1996

* cited by examiner

Primary Examiner—Dang Le
(74) Attorney, Agent, or Firm—Michael A. Lechter; David E. Rogers; Squire, Sanders & Dempsey, L.L.P.

(57) **ABSTRACT**

An apparatus for converting between mechanical and electrical energy, particularly suited for use as a compact high power alternator for automotive use and "remove and replace" retrofitting of existing vehicles. The apparatus comprises a rotor with permanent magnets, a stator with a winding, and a cooling system. Mechanisms to prevent the rotor magnets from clashing with the stator by minimizing rotor displacement, and absorbing unacceptable rotor displacement are disclosed. The cooling system directs coolant flow into thermal contact with at least one of the winding and magnets, and includes at least one passageway through the stator core. Various open and closed cooling systems are described. Cooling is facilitated by, for example, loosely wrapping the winding end turns, use of an asynchronous airflow source, and/or directing coolant through conduits extending through the stator into thermal contact with the windings.

129 Claims, 43 Drawing Sheets

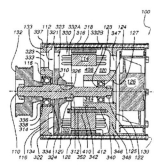

EXHIBIT 5.7
First page of a U.S. patent.

least in the United States, the application must describe the best mode contemplated by the inventor of carrying out the invention.

5.6.3 Claims

The *claims* define the scope of protection provided by the patent. An accused device or process infringes a claim if it includes elements corresponding to *each and every* element of the patent claim. The broader and less specific the terms of the claim, the broader the protection that is afforded by the patent, but also the more vulnerable it is to invalidation based on prior art. The language of the patent claims must, therefore, be drafted with the utmost precision. It is permissible to have a number of different claims in the patent application. In common practice, claims of varying scope, ranging from the most general to the most specific, are submitted. In this way, if it appears after the fact that some relevant piece of prior art exists that invalidates the broad claims, the other, more specific claims are not necessarily invalidated. In this manner, the inventor not only can obtain protection on the broad aspects of his invention, but also on the specifics of the particular product that is put on the market. Claims can also be strategically directed to catch specific potential infringers or to facilitate licenses into different fields of use.

5.6.4 Exclusive Right

As noted above, a patent provides an *exclusive right* to the inventor. A patentee has the right to exclude others from practicing the invention.[20] However, the grant of a patent does not necessarily give the patentee the right to practice the patented invention. It is this distinction that makes the patent system so effective in advancing industry. Through this mechanism, patent protection can be provided for improvements without degrading the protection provided for basic inventions.

An unauthorized item infringes a patent if it includes elements corresponding to *each and every* element in any claim in the patent. It is irrelevant that the item includes additional elements, even if the additional elements or a combination of those elements are patentable in their own right.

To illustrate the "exclusive" nature of patent protection, assume that when the "stool" and "chair" were first invented, there was a patent system in place. Let's say that I obtain a patent on the "stool," claiming: "Apparatus comprising a platform and at least one support member disposed to maintain the platform at a predetermined level from the ground." (See Exhibit 5.8.)

You purchase a stool and determine that it can be improved by incorporating a back support. Ultimately, you invent and obtain a patent on the "chair," claiming: "Apparatus comprising a platform, at least one support member for maintaining the seat at a predetermined level from the ground, and a back extending above the platform." (See Exhibit 5.9.)

Both you and I have patents. However, notwithstanding the addition of the back, shown in Exhibit 5.10, the chair still includes elements corresponding to

5.6 Patents

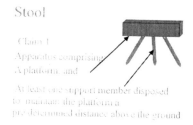

EXHIBIT 5.8
Example of a patent claim on a stool.

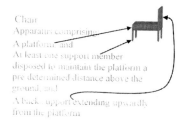

EXHIBIT 5.9
Example of a patent claim on a chair.

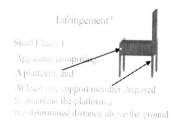

EXHIBIT 5.10
Infringement analysis: application of school claim to chair.

each and every element of the claim of, and thus infringes on, my basic patent on the stool. So you need a license from me to make your chair. Similarly, while I am free to make, use, and sell the stool, I cannot put a back on the stool without infringing your patent on the chair.

In practice, the result is that you and I each obtain a license from the other under the respective patents, and, where before there was only one stool manufacturer, there are now two chair manufacturers.

5.6.5 Patentability

Threshold criteria must be met before an inventor is awarded a patent on an invention: (1) the invention must consist of patentable subject matter, that is, be within certain broad categories of subject matter (*patent-eligible*); (2) the invention must be capable of industrial application (or, in certain countries such as the United States, be useful); (3) the invention must be new (novel); (4) the invention must be nonobvious (it must involve an inventive step); and (5) the disclosure of the invention in the patent application must meet certain formal and substantive standards.

5.6.5.1 *Patent-Eligibility*

The categories of subject matter that are eligible for utility patent protection vary from jurisdiction to jurisdiction (e.g., from country to country), although various treaties tend to set a minimum threshold.[21] The scope of patent-eligibility tends to differ primarily with respect to the extent protection is provided to such things as pharmaceuticals, medical devices and procedures, food and agricultural inventions, business methods, software, game rules, and human genes. In the United States, *patent-eligibility* encompasses essentially "anything under the sun made by man" and excludes only those things that logically could not be subject to exclusivity, or with respect to which exclusivity could not realistically be enforced.[22] For example, laws of nature, natural phenomena, things that are unchanged from the way that they occur in nature, and abstract ideas are not eligible for patent protection.[23] Likewise, mental processes, processes of human thinking, and systems that depend for their operation on human intelligence alone are not patent-eligible subject matter.[24] Claims directed to or encompassing a human organism are expressly not patent-eligible.[25]

There has been considerable controversy on whether computer software or methods of doing business are eligible for patent protection. It is now settled in the United States that they are not per se ineligible for patentability. However, if a patent claim is drafted so broadly that it (1) preempts substantially all uses of a fundamental principle or mathematical algorithm or (2) covers purely mental processes it is not patent-eligible.[26] With these principles in mind, a claimed process is typically patent-eligible if:

- It is tied to a particular machine or apparatus
- It transforms a particular article into a different state or thing

The law in Europe, although stated differently, is substantially very similar. The European patent convention expressly excludes computer programs *per se* and methods of doing business *per se* from patentable subject matter. That being said, however, if there is a technological aspect to a patent claim, such as controlling an industrial process or processing data representing physical things, or an aspect that involves the internal functions of a computer, even though software, it is likely patent-eligible in Europe. In China patentability of software is very similar to that in Europe. In Japan, patent-eligibility tends to depend upon whether a claim relates to data representing the physical or technical properties of an object, in which case it is patent-eligible. If the data represents economic law, a commercial method, or mathematical formula, it is not patent-eligible.

It is important to note that claims intended to cover an invention can be drafted in many different ways, and in some cases, form is exalted over substance. The specific manner in which the claim is drafted can determine whether or not the claim is found to meet the above criteria.

The patentability of genetically engineered organisms, and particularly, human genomes, is another area of controversy. Recent cases in the United States have established that, specifically, a human gene as it occurs in nature (and not created or modified by man), even though isolated, is not patent-eligible, but claims to (1) modified gene sequences that do not occur in nature and (2) *processes* to identify or isolate a particular gene or to modify a gene sequence, or which use the isolated gene to some effect, are patent-eligible.[27]

5.6.5.2 Novelty and Nonobviousness

Novelty and nonobviousness are measured against what is referred to as *prior art*. The applicable patent law defines specific circumstances, referred to as *statutory bars*, which define those things that are considered prior art.[28]

In most countries, subject to specific exceptions that tend to vary from country to country, any publication, public disclosure, or commercialization prior to the effective filing date of the application is prior art against the application.

This is the case in the United States with respect to patents subject to the AIA (post–March 15, 2013); all prior art is established with reference to the effective filing date of the application. Any public disclosure or commercialization of the claimed invention more than one year prior to the effective filing date is prior art and bars a patent. However, patent filings, public disclosures, and commercializations within the year prior to the effective filing date are not considered prior art if they are attributable to the inventor or to someone who derived the subject matter from the inventor, or the third-party disclosure followed an "inventor" disclosure.[29]

For U.S. patents subject to the old U.S. law, statutory bars arise by virtue of relative timing with two dates: the filing date of the patent application and the date of invention.

1. Prior art by virtue of relative timing with the application filing date includes things that were:
    - Commercialized by the applicant more than one year before the application for patent was filed
    - Ascertainable from publicly available information more than one year before the application for patent was filed

- Claimed in a foreign patent granted prior to the U.S. filing of an application filed more than one year before the application for patent was filed
2. Prior art by virtue of relative timing with the date of invention includes things that:
    - The inventor knew from external sources when the invention was conceived
    - Were ascertainable from publicly available information before the invention was conceived
    - Were invented by someone else (and not maintained as a trade secret) before the invention was made

Under the old law, a bar can occur if filing a U.S. application is (1) unduly delayed after making the invention (abandoned)[30] or (2) unduly delayed relative to filing a foreign patent application[31] or (3) the application intentionally misnames the inventors.[32]

5.6.6 The Patent Examination Process

After the patent application is filed with the patent office, it is assigned to a *Patent Examiner* having expertise in the particular technological area of the invention. The examiner then reviews the application to ensure it conforms to formal requirements, conducts an investigation to determine if there is any relevant prior art in addition to that supplied by the applicant, and negotiates the appropriate scope of the claims with the applicant. The examiner's investigation is reflected in an *Office Action* sent to the applicant, which, in brief, lists the prior art (references) considered by the examiner and indicates whether he/she considers the claims to be (1) of proper form, and (2) anticipated or rendered obvious by the prior art.

A response to the Office Action must typically be filed within a specified time period (e.g., three months). The response must answer each and every issue raised by the examiner, arguing against the examiner's positions, amending the claims, or canceling the claims. In effect, the applicant negotiates with the Patent Examiner to determine the exact scope of the claims to which the inventor is entitled. If the examiner agrees with respect to the exact scope of the claims, the application issues a patent.

5.6.7 Patent Pending

Filing a patent application does not, in itself, provide any enforceable rights. That is, no enforceable rights are provided until the patent is actually granted. (Although in certain instances, rights can be retroactive to the date of publication of the patent application.) However, during the period the patent application is pending before the USPTO, a *patent pending* notice can be placed on products including the invention described in the application. A patent pending notice can sometimes scare off certain types of potential infringers. Care must be taken, however, not to mismark; marking a product with a patent pending notice when it is not actually described in pending application can be a violation of the false marking statute.

5.6.8 Patent Ownership

The rules with respect to ownership of patents are far from intuitive and vary from country to country. In the United States, unless there is an express or implied contractual agreement to the contrary, the actual inventor owns the invention and any patent on the invention. Where there are joint inventors, each owns an equal undivided interest in the whole of the invention and any patent on the invention. Absent an agreement to the contrary, all joint inventors are entitled to make, use, and sell the invention, without accounting to the other coinventors.[33] Of course, these are default rules and may be varied by agreement. In practice, the default rules are typically superseded by written agreements.

Under certain circumstances, for example, where an employee is hired to invent, there is an implied agreement that the employer will own all rights to that invention. Similarly, where an invention is made on company time and/or using the businesses' facilities, the business may acquire *shop rights*—a royalty-free license—to the invention. However, in order to avoid any disputes, it is a best practice for a venture to require each of its employees to execute an employee's invention assignment agreement as a condition of employment—explicitly obligating the employee to assign all rights in relevant inventions to the business. Certain countries, however, notably Germany, have laws in place that require, irrespective of any agreements that might be in place, employers to compensate employees for even inventions for which the employee was specifically hired to invent.[34]

5.6.9 International Patents

Patent rights are territorial. For example, the rights granted under a U.S. patent extend only throughout the territory of the United States and have no effect in other countries. Almost every country has its own patent law, and a technology entrepreneur desiring a patent in a particular country must make an application for patent in that country in accordance with that country's requirements. Fortunately, there are several international treaties that aid in applying for patent protection in jurisdictions around the world. Perhaps the most helpful of those treaties is the *Patent Cooperation Treaty*, otherwise known as the PCT.

The PCT facilitates the filing of applications for patents on the same invention in countries that have ratified the treaty. It provides for centralized filing procedures where a single application filed in a member country governmental receiving office constitutes an application for patent in one or more designated member countries. Once the application is filed, one of the eligible PCT governmental searching offices will perform a patent search. When the search is completed, the applicant may then elect to have a PCT governmental office evaluate the patentability of the application pursuant to standards set forth in the treaty. Eventually, the applicant will be required to have the PCT application entered into the national patent office of each of the countries in which patent protection is desired.

Patent maintenance fees (fees that must be paid periodically to keep the patent in force) are required in most countries, but vary in timing and amount from jurisdiction to jurisdiction. In addition, most countries (other than the United States) require that a patented invention be manufactured in that country after a certain period, usually three years. This requirement is generally referred to as the "working requirement." If there is no manufacture within this period, the patent may be void in some countries, although in most countries the patent may be subject to the grant of compulsory licenses to any person who may apply for a license.

5.7 COPYRIGHTS

Copyrights protect artistic expression. They arise automatically as soon as an *original work of authorship* is put into a form that can be read or visually perceived (either directly or with the aid of a machine or device). Neither publication nor registration is necessary to secure copyright protection.

Exemplary categories of original works of authorship include: literary works; musical works; dramatic works; pantomimes and choreographed works; pictorial, graphic, and sculptural works; motion pictures and other audiovisual works; and architectural works. Certain aspects of computer programs are also considered literary works.[35] Copyright protection is also available for original compilations (creative selection or arrangement of preexisting materials or data).[36]

To be "original," a work must be independently created by the author and possess a minimal degree of creativity.[37] Those aspects of a work that are not created by the author are not copyrightable subject matter. For example, facts are said to be discoveries, not creations of the author. The same is true with respect to those elements of a work that are in the public domain and those elements of the work dictated by function. In addition, some types of works are simply not sufficiently creative to warrant copyright protection. These include, for example, numbering schemes for parts, command codes, and fragmentary words and phrases.[38]

A work often includes expressions or descriptions of ideas and concepts, facts, and utilitarian elements. In this context, the term "utilitarian element" means a "useful article" or anything that (1) has an intrinsic function other than just portraying appearance or conveying information or (2) is required to be in a particular form because of some external factor. Examples of utilitarian elements include logical sequences, procedures, processes, systems, and methods of operation. The copyright gives the author exclusive rights with respect to the way in which the ideas, facts, and utilitarian elements are expressed or described. It does *not* give any sort of exclusivity with respect to the use of the ideas, facts, and utilitarian elements themselves.

MINI-CASE
A Classic Example of Copyright Law

A classic example is found in the case of Baker v. Selden.[39] Baker, an accountant, wrote a book describing a double entry accounting system that he had developed. Selden, also an accountant, bought the book and began using the double entry accounting system developed by Baker. Baker contended that the use of his accounting system by Selden constituted copyright infringement. The Supreme Court, however, disagreed. The use of the ideas described in a copyrighted work does not constitute infringement of the copyright.[40]

Expression, which is inseparable from an idea, is likewise not protectable under the copyright law. Expression becomes inseparable from an idea, when, for example (1) protecting the expression would, in effect, give exclusive rights to the underlying idea or because (2) the expression serves a functional purpose or is dictated by external factors (which is the case with respect to the logic reflected in software code). However, an original compilation comprising an arrangement or selection of those unprotectable elements that is itself nonutilitarian and sufficiently original is protected by copyright.

5.7.1 Considerations with Respect to Software

Historically, there was much debate on whether software even qualifies as a "work of authorship" subject to copyright protection. It is now well settled that copyright protection is applicable to human-readable software, machine-readable source code representations of a program, and audiovisual displays in game programs.[41] However, a copyright registration of the underlying program code that generates screen displays may not extend to the displays; arguably, the screens must be subject to a separate copyright. It may be advantageous to register copyrights on the respective screens as audiovisual works separate and apart from the program code.[42]

In any event, however, it must be kept in mind that utilitarian elements of the software are not protected by copyright.[43] A copyright protects against actual copying of substantial portions of copyrighted literal program code. However, categorizing aspects of a computer program other than literal code as protectable expression as opposed to idea or utilitarian is often extremely difficult.[44]

Furthermore, a copyright does not protect the owner against independent creation of a similar program by another, even if the other is generally aware of the copyrighted program. A competitor can, in general, study a copyrighted program, determine the central concept and basic methodology of the program, and then write its own program to accomplish the same results. In practice this is often done using the so-called clean rooms.

5.7.2 Copyrights and the Internet

Just because something is in digital form does not take it out of the realm of copyright law. Unauthorized copying of an MP3 version of a song, or a DVD version of a movie, is a copyright infringement. Using the Internet as a vehicle for transmitting those copies doesn't relieve the copier from liability. The liability of entities that facilitate making illicit copies of content, however, is a bit more complicated.

Contributing to (knowingly encouraging) unauthorized copying of a copyrighted work can create liability for (contributory) infringement.[45] However provisions of the copyright act, The Digital Millennium Copyright Act ("DMCA"),[46] creates certain safe harbors, limiting copyright infringement liability on the part of Online Service Providers ("OSP")[47] arising from certain separate and distinct categories of conduct (transitory communications; system caching; storing information at direction of users; and information location tools). Where the OSP is storing files for its users, or facilitating peer-to-peer file transfers, primary issue with respect to an OSP qualifying for limited liability is whether OSP knows or should know about the infringement and whether the offending files are expeditiously removed once the OSP is put on notice.

MINI-CASE
Viacom Versus YouTube

Viacom sued YouTube for copyright infringement with respect to various video "clips" (~79,000 files) of TV shows that YouTube patrons uploaded to the YouTube website. YouTube took down the offending files once Viacom complained, but Viacom contended that YouTube had actual knowledge of the infringement before the "takedown requests" and accordingly did not fall into the safe harbor of the DMCA. In 2010 a federal district court found that YouTube did not have actual knowledge of the infringement and was immune from liability. However, an appellate court disagreed and sent the case back to the District Court to determine whether, among other things, YouTube willfully blinded itself to the infringement. YouTube ultimately prevailed convincing the court that it did not have the right and ability to control the infringing activity.[48] Google, the parent company of YouTube, settled the suit (on undisclosed terms) prior to the case going back up on appeal. Whether there will be further activity in the courts to settle the issue of what constitutes OSP knowledge of infringement remains to be seen.

5.7.3 Notice
It is advantageous that all published copies of a work bear a *copyright notice*. Basically, the copyright notice includes three elements: the copyright symbol©, the word "copyright," or the abbreviation "Copr"; the named owner of the copyright; and the year of first publication of the work. There are various requirements for where the copyright notice must be placed on the work.[49] However, in general, the placement of the notice should be sufficient if it is placed in a prominent position on the work, in a manner and location as to "give reasonable notice of the claim of copyright."

5.7.4 Copyright Registration
Copyright registration is not a prerequisite for copyright protection. However, registration is significant in several respects. A registration is normally necessary before the copyright on works originating in the United States can be enforced.[50] If the registration is made before publication, or within five years after publication, it establishes the validity of the copyright and of the facts stated in the copyright certificate in court.[51] Unless made within three months after the publication of the work, a registration not made until after the infringement only entitles the copyright owner to be awarded the damages and profits that can actually be proven[52] without an option to set statutory damages and attorneys' fees.[53] A registration is obtained by filing the appropriate completed application form, a specified fee for each application, and two complete copies of the work. In general, two complete copies of the "best" edition must be filed. However, special provisions are made in the Copyright Office regulations for deposit of "identifying portions" of machine readable works in lieu of complete copies.[54]

5.7.5 Copyright Ownership
Copyright ownership under U.S. law is counterintuitive and tends to turn the unwary into casualties. One would think that the person who paid for a work to be created would own it. This is not necessarily so under the copyright law. Unless there is a written assignment or the work qualifies as a *work for hire*, the creator/originator of a work owns the copyright.[55] If the work qualifies as a "work for hire," then the employer of the creator, or the entity that commissioned the work, is considered to be the author and holder of the copyright.[56]

When an independent contractor is commissioned to create a work it is desirable that the work qualifies as the work for hire; with respect to any work other than a work made for hire, the author has an inalienable *right of reversion* that can be exercised after 35 years. Exercising the right of reversion effectively takes back any rights granted (including an assignment) in the work.[57]

A work prepared by two or more authors with the intention that the respective contributions be merged into inseparable or interdependent parts of a unitary work is referred to as a *joint work* by *coauthors*. Absent agreement to the contrary, coauthors are co-owners of the copyright. Each coauthor owns a proportionate share of the copyright and, in the absence of an agreement, is entitled to a share of any royalties received from licensing. A joint owner may generally use or license the use of the work without the consent of co-owners, but must account to the co-owners for their shares of profits derived from any license to a third party.

5.8 MASK WORKS

Under the Semiconductor Chip Protection Act, *mask works* are defined as a "series of related images, however fixed or encoded, that represent three-dimensional patterns in the layers of a semiconductor chip."[58] In essence, the Chip Protection Act protects against the use of reproductions of registered mask works in the manufacture of competing chips. However, there must be copying; a *clean room* procedure can defeat a mask work infringement charge.[59] And, competitors are not precluded from reverse engineering the chip for purposes of analysis or from using any unpatented technology embodied in the mask work.[60]

Registering a mask work involves filing a Copyright Office form together with particular identifying material and a fee. Typically, the owner of a mask work is the person(s) who created the mask work. Presumably, in analogy to the copyright statute, co-creators of the mask work would be co-owners of the mask work protection. However, where the mask work is made within the scope of the creator's employment, the employer is considered the owner of the mask work.[61]

5.9 TRADEMARKS

A *trademark* or *service mark* is used to identify the source or origin of a product or service. It distinguishes goods or services of one company from those of another. Customers connect the goodwill and reputation of the company to its products through its trademarks. Under the law, a competitor is prevented from capitalizing on another venture's reputation and goodwill by passing off possibly inferior goods as those of the venture. In this way, proper use of a trademark can protect the sales value of the venture's reputation, and that of the product, as well as its investments in advertising and other promotional activities used to develop goodwill. However, trademark protection does not prevent the competition from copying or reverse engineering the technical aspects of a product.

5.9.1 Acquiring Trademark Rights

There are three different basic types of trademark law regimes employed around the world: use-based common-law, pure registration, and combination use-based registration.

5.9.1.1 Use-Based Common-Law and Combination Use-Based Registration Systems

In *common law* jurisdictions (e.g., United States, Canada and the United Kingdom) trademark rights are acquired through use of the mark. The first to use a given mark in connection with particular goods or services in a given geographical area obtains the exclusive rights in the mark for use of the mark with those particular goods or services in that particular geographical area. However, if someone else adopts the mark somewhere outside of that geographical area without knowledge of the prior use of the mark, that person would acquire valid common law rights to the mark in the remote area. This is where combination use-based registration comes into play: Obtaining a trademark registration from the appropriate governmental agency provides constructive knowledge of the mark and can prevent subsequent remote users from obtaining rights.

Use of a trademark requires physical association of the mark on or in connection with the product or service. With a trademark, it is sufficient to apply the mark to labels or tags affixed to the product or to the containers for the product, or to displays associated with the product or the like. Trademark usage cannot be established just through use of the mark in advertising or product brochures. However, if it is not practicable to place the mark on the product, labels, or tags, then the mark may be placed on documents associated with the goods or their sale.[62] On the other hand, if services are involved rather than a physical product, use of a mark in advertising is a proper usage for a service mark. A service mark is a mark used in sales, advertising, or services to identify the source of the services.

5.9.1.2 Pure Registration Systems

In those jurisdictions that have adopted a pure registration system, all rights to trademarks derive from registration of the mark with the appropriate government agency. Actual use of the mark within the jurisdiction does not confer any rights (with the possible exception in some jurisdictions of "famous" marks, and certain special circumstances, such as the existence of a particular relationship between the respective parties claiming rights to the mark.) Most jurisdictions in Asia, Continental Europe, and Central and South America have adopted pure registration systems, although the particulars of the law tend to vary from country to country.

> **KEY POINT**
>
> *Trademark Law in China*
>
> Historically, China has had a pure registration trademark system. Effective May 1, 2014, amendments will go into effect that are intended to make "trademark squatting" and counterfeiting more difficult. The new law places time limits within which government actions in the registration process must take place, and, significantly, accords prior users certain rights against subsequent registrants, particularly with respect to well-known marks.

With the Internet now providing essentially instantaneous worldwide access to information regarding the marks used on products, it is particularly important to think in terms of global trademark protection when introducing a product. It is also essential as an applicant to ensure that you have obtained appropriate registrations and have complied with any local requirements to record agreements with the government before introducing a product into, or entering into distribution agreements and so on in, pure registration countries.

5.9.2 Registering a Trademark

Registering a mark generally involves filing an application with the appropriate government agency. The specific criteria for registration and the process for registration tend to vary from jurisdiction to jurisdiction. However, with the exception of a few countries, the government agency generally examines the application to ensure that it meets formal requirements and does not conflict with existing marks prior to issuing a registration.

For example, in the United States, rights in a mark can be created by actual use or by filing an application for registration based upon intent to use the mark.[63] An application may be filed at any time after a *bona fide*, good faith intention to use a mark in interstate commerce can be alleged. However, a registration will not be granted until the mark has actually been used in interstate commerce. Registration provides a number of procedural and substantive advantages.

After the application is filed, a Trademark Examiner reviews the application to determine whether the mark is registrable, that is, whether it is capable of distinguishing the applicant's goods from the goods of others. The examiner also conducts an investigation to determine if the mark is confusingly similar to any mark already being used. Thus, registering a mark and making it available to the Trademark Examiner tends to prevent registration of confusingly similar marks. An example of a U.S. trademark registration (principal register) is provided in Exhibit 5.11.

Once an application is filed, a Trademark Examiner reviews the application to determine whether the mark is registrable, as described above. If all the formalities are met and the examiner finds no conflicting marks, the examiner

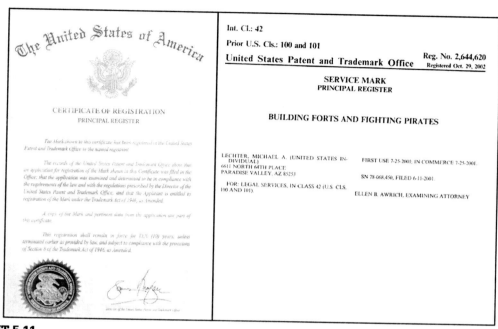

EXHIBIT 5.11
Sample of a U.S. trademark registration.

will publish the mark in the Official Gazette to permit third parties to oppose grant of a registration. Any opposition must be filed within 30 days of the publication date. Assuming that there is no opposition filed, or that the applicant is successful in defending the opposition, the examiner will issue a notice of allowance indicating that, subject to a proper showing of the use in commerce, the mark is entitled to registration.[64] If the application is based upon intent to use the mark, rather than actual use, however, actual use of the mark must commence within a predetermined period of the notice of allowance. Once a registration based upon intent to use is obtained, the registrant is accorded a constructive use priority, meaning, it is as if the mark were actually used with the goods or services on the date of the application for registration.[65]

Even though trademark protection is of potentially infinite duration in common law jurisdictions, registrations must be periodically renewed. There is no limit on the number of times that the registration can be renewed. For example, in the United States, trademark registrations must be renewed every 10 years by filing the appropriate papers and fees within six months before the expiration of the registration. When the registration is renewed, it is necessary to indicate that the mark is being used in interstate commerce. In addition, the renewal application must recite only the goods on which the mark is in actual use at that time.

5.9.3 Trademark Notice Symbols

Once a federal registration has been obtained, the registrant is entitled to use a *registration notice* such as ®, which is typically placed as a superscript to the registered mark. The use of a registration notice is not mandatory. However, the notice does provide constructive notice of the registration and sets off the mark. Setting off the mark signifies that the term or symbol is intended to be a mark that indicates the source or origin of the goods rather than a descriptive term for the goods.

A registration notice is appropriate only when used with the specific term or symbol shown in the registration and only when the mark is used in connection with goods that are within, or are natural extensions of, the definition of the goods set forth in the registration. Accordingly, when a registered trademark is used in conjunction with a new product, it should be determined whether or not the new product falls within the scope of the definition of goods of the registration. If not, a registration notice should not be used and consideration should be given to filing an application for a new registration. In circumstances where there is no applicable registration, a "™" or "SM" symbol may be used. The use of the ™ symbol has no legal significance other than to signify that the term or symbol is intended to be a source-identifying trademark.

5.9.4 International Protection of Trademarks

Historically, trademark protection had to be obtained on a country-by-country basis. However, in recent times, regional protection has become available in certain parts of the world, and, more significantly, international protection can be obtained through the filing of a single application and maintaining a single registration.[66] After a registration is obtained with a trademark in a country that is a signatory to the treaty, an international application can be filed (through the originating country's trademark office), which, assuming no hitches, results in an international registration that has the effect of a national-registration issuing in each of the signatory countries designated in the application.

5.9.5 The Strength of a Mark

The protection afforded by a particular trademark is a direct function of the distinctiveness of the mark and how closely the public associates the mark with the source of the product rather than the product itself. A symbol or word that cannot effectively identify the source of the goods cannot be a trademark. The more descriptive the symbol or word is of the goods or services, the less effective the symbol or word is as a source-indicating trademark. If a symbol or word merely indicates the generic type of goods, or merely describes some characteristic of the goods, it cannot function as a trademark. On the other hand, a symbol or word that has no meaning, or has nothing to do with the goods or services with which it is used, can only function as a source-indicating

Exhibit 5.12 Categories of Marks

Category	Definition	Examples
Fanciful	Term or symbol specifically created to act as a trade mark	"Exxon"; "Kodak"
Arbitrary	Term or symbol that has no relationship to the goods or services	"Apple" for computers
Suggestive	Term or symbol hinting at, but not directly conveying, a characteristic of the goods or services	"Diehard" for batteries; "Building Forts and Fighting Pirates" for legal services
Descriptive	Term or symbol conveying immediate idea of a characteristic of the goods or services	"Windows" for graphical user interface
Generic	Term or symbol referring to the genus of the goods or services	Aspirin; Band-Aid; Cellophane; Thermos; Murphy bed; Dry Ice; Escalator; Lanolin; Laundromat; Scotch Tape and Zipper

trademark. The different categories of marks, in decreasing order of strength, are reviewed in Exhibit 5.12.

Companies with famous marks have to be diligent in policing the way that the public uses their marks. If the primary meaning of the mark to the public becomes generic rather than indicating source, the company will lose its exclusive right to use the symbol or word with its goods or services. For example, the word "Aspirin" was originally a trademark of Bayer AG. A copy of the 1899 label for Bayer Aspirin brand acetylsalicylic acid is shown in Exhibit 5.13.

However, the public began using the word as the generic name for the product, and, at least in the United States, the word ceased to be a trademark.[67] Other examples of former trademarks that are now generic words include: escalator,[68] thermos,[69] cellophane,[70] and pilates.[71] Similarly, the word Kleenex (Exhibit 5.14) is in danger of becoming generic. "Kleenex" is widely used to describe tissue paper, even though it is a registered mark for a Kimberly-Clark product. Other examples of words that are in danger of becoming generic include Styrofoam, Super Glue, Taser, Xerox, Band-Aid, and Google. In the past, some marks that have found themselves in a generic status have, in effect, made a comeback. For example, the marks "Goodyear" and "Singer" at one time were considered to be generic, but eventually reacquired the status of protectable trademarks.[72]

It should be apparent that a given word may be generic in one market and arbitrary in another market; for example, the mark "oscilloscope" for an oscilloscope apparatus is generic. However, the mark "oscilloscope" for chewing gum is arbitrary.

EXHIBIT 5.13
1899 Label for Bayer Aspirin.

EXHIBIT 5.14
Kleenex Logo.

5.9.6 Choosing a Mark

There are certain basic guidelines with respect to choosing a mark. The strongest word mark is a relatively euphonious, easily pronounced, coined word or a word with no relation to the goods or services. However, from a marketing perspective, a mark suggesting the nature or characteristics of the product is desirable. The mark should be simple—a simple mark is not only more easily protected (avoiding the possibility that another could adopt some, but not all, of the elements of the mark), but more easily remembered by consumers.

Before adopting a mark, it is prudent to make sure that no one else is using it. Basically, this involves examining trademark registration files (in the United States, maintained at the USPTO, and accessible at www.USPTO.gov) to see whether any similar mark is already registered to another or whether an application has been made for registration of a similar mark for similar goods or services.

Potential trademark problems arise when the proposed trademark is *confusingly similar* to another mark given the cumulative effect of the differences and similarities in the marks and in the goods or services. Rule-of-thumb criteria for determining whether or not a mark resembles another are

- Do the marks look alike?
- Do the marks sound alike?
- Do the marks have the same meaning or suggest the same thing?

The similarities and dissimilarities of the goods themselves must also be considered. Do the respective goods move in the same channels of trade? Are they sold in the same type of store? Are they bought by the same people? What degree of care is likely to be exercised by the purchasers?

Trademarks used on products purchased by relatively sophisticated purchasers are less likely to cause confusion regarding the source of goods, sponsorship, or affiliation than when used on goods typically sold to unsophisticated purchasers.

5.10 CHAPTER SUMMARY

IP is the great equalizer in the world of business. It is often the primary factor that enables an emerging business to compete successfully against larger, established competitors with vastly more marketing power. IP assets not only can be leveraged to sustain competitive advantage, but also can create credibility in the industry and with investors. Credibility is based not only on a venture's IP assets, but also on the venture's systems and processes to develop and protect them. IP-based strategic alliances and/or licensing can also be an alternative to raising capital, providing a venture the benefit of the co-venturer's resources without having to spend the time and money to develop them on its own.

Technology ventures should continuously and systematically analyze their products and operations to identify potential IP and seek opportunities to develop protectable IP assets. It should develop a comprehensive strategic plan to acquire, maintain, and reap maximum benefit from those assets. All of the various protection mechanisms should be employed, singly and in combination, as part of that plan. At the same time, the venture must be mindful of competitors' IP rights. A few relatively simple procedures and precautions can mean the difference between the success and failure of a technology venture.

In this chapter, we examined the various types of IP: know-how and trade secrets, inventions and patents, works of authorship and copyrights, mask works, and trademarks. The co-venturers in a business must learn to recognize and identify potential IP assets and establish procedures, including keeping contemporaneous records of development activities, to provide a foundation for protecting those IP assets and defending against third-party infringement claims.

KEYTERMS

Intellectual property The intangible asset that comprises or results from human intellect, creativity, innovation, know-how, and reputation and encompasses all rights to technology and rights to benefit from one's repute.

Invention New technological developments or discoveries produced or created through the exercise of independent creative thought, investigation, or experimentation.

Know-how Accumulated practical skill, expertise, data, and information relating to a business and/or its operations, or performing any form of industrial procedure or process.

Open source licenses Standardized licenses that permit use of the licensed software only if the licensee agrees to grant third parties licenses to any modifications or derived works (and in some cases background intellectual property necessary to use the licensee's software) under the terms and conditions of the open source license. The open source license is typically royalty-free, unrestricted with regard to products, platforms, and fields of endeavor, and provides for access to source code, unrestricted redistribution as part of an aggregate of software components from multiple sources. The specific terms of the open source licenses vary from version to version.

License agreement An agreement under which the owner/proprietor of intellectual property rights (the licensor) permits another entity (the licensee) to utilize the intellectual property rights in return for some consideration, typically the payment of royalties.

Interference proceeding A PTO administrative proceeding pertaining to patents filed before March 16, 2013 to determine which of rival claimants was the first to invent a commonly claimed invention.

Derivation proceeding A PTO administrative proceeding pertaining to patents filed before March 16, 2013 to determine whether an inventor named in an earlier filed application derived a commonly claimed invention from an inventor named in a later application.

Trade secret Information, know-how, data, and/or processes that are not readily ascertainable from publicly available information, are subject to reasonable measures to maintain secrecy, and derive value from being kept secret.

Nonproprietary know-how Know-how that is valuable in the operation of a venture, but is generally known in the industry and does not qualify as a trade secret, such as expertise or skill in operating a commercially available machine.

Confidentiality agreement An agreement placing obligation of confidentiality on an entity provided access to confidential information. The agreement also typically limits the use to which the information can be put by the entity.

Non-competition agreement An agreement intended to protect a business from unfair competition arising from proprietary or invested relationships by precluding an entity from engaging in certain competitive activities which would be aided by knowledge of the proprietary information or invested relationships.

Patent A grant by a sovereign government of some privilege or authority. In the context of intellectual property, the grant is typically the exclusive right to make, use, sell, offer to sell, or import embodiments of an invention, design, or asexually reproduced plant.

Patent application A proposed patent submitted to the appropriate patent office for its approval.

Provisional patent application A disclosure document, not subject to the formalities of a regular patent application, submitted to the appropriate patent office, to establish a date with respect to inventions adequately described in the disclosure document, after which subsequently occurring activities do not constitute prior art against which patentability is measured. To get the benefit of the provisional patent application, a regular patent application must be filed within one year of the provisional.

Claiming priority An explicit statement in a patent (or trademark) application claiming the benefit of an earlier filed application containing common subject matter, with the effect that the filing date of the first filed application is deemed the effective filing date of the subsequent application in which priority is claimed.

Continuation-in-part application A patent application that is, in part, substantively identical and contains a specific reference to an earlier filed "parent application," filed while the parent application is still pending, but also including a description of subject matter not included in the parent application.

Continuation application A patent application that is substantively identical (except perhaps in the case of claims) and contains a specific reference to an earlier filed "parent application," filed while the parent application is still pending.

Written description A required portion of a patent application (and patent), typically including a drawing to which the description refers, which provides a detailed description of a preferred embodiment of an invention with a requisite level of detail. In the United States the written description must provide sufficient detail to enable the average practitioner in the relevant technology to make and use the invention.

Person of ordinary skill in the art A hypothetical average practitioner in the field of the invention, having the average level of education and experience and an ordinary level of creativity. In some instances the person is presumed to have known all of the relevant art at the time of the invention.

Patent claim A single sentence defining each of the essential elements of the invention in technical terms, establishing the scope of protection provided by the patent. The patent typically includes multiple claims. Some of the claims (*dependent claims*) may incorporate another claim (*parent claim*) by reference, in effect adding detail to more narrowly define the invention. Claims that do not refer to other claims are known as *independent claims*.

Exclusive right In connection with patents, a right to prevent unauthorized use of a claimed invention by others. An exclusive right, however, does not necessarily give the patentee the right to practice the invention and should be distinguished from a "monopolistic" or "sole" right where the right holder is the only entity that has the right to provide an article or to practice a process.

Patent-eligibility Within certain broad categories of subject matter that are eligible for patent protection.

Prior art Things and activities against which the patentability of an invention is measured.

Statutory bars Specific circumstances set forth in statutes that define prior art.

Patent examiner A patent office employee that considers a patent application and, determines in the first instance whether patent claims will be granted.

Office action A communication notifying the applicant of the examiner's positions with respect to a pending application.

Shop rights An implied license under which an employer may use the invention of an employee made at the place of work during hours of employment.

Patent cooperation treaty (PCT) A multilateral international patent law treaty that permits an inventor in a member country or region to file a single application in a designated language (Arabic, Chinese, English, French, German, Japanese, Korean, Portuguese, Russian, or Spanish) and have that application acknowledged as a regular national or regional filing in any country or region that is a PCT member.

Original work of authorship The product of creative expression (such as literature, music, art, and graphic designs) that has not been copied from another.

Copyright notice A notice placed on copies of published works to place the public on notice of the underlying copyright protection.

Work for hire A work created by an employee within the scope of his or her employment, or certain types of works (e.g., a contribution to a collective work, part of an audiovisual work, a translation, an instructional text, a supplementary work) specifically commissioned by a *written* agreement that designates the work as a work for hire.

Right of reversion An inalienable right given to individual authors under U.S. copyright law to terminate any grant, license, or assignment of a work 35 years from the date of such transfer and reclaim the copyright for the work, as long as the work was not originally created as a work for hire.

Joint work A work prepared by two or more authors (*coauthors*) with the intention that the respective contributions be merged into inseparable or interdependent parts of a unitary work.

Mask works A series of representations of three-dimensional patterns in the layers of a semiconductor chip.

Clean room In intellectual property infringement analysis, a process wherein a first individual or group reverse engineers an item and a second group that does not have direct access to the item develops an analogous item from the report of the first group.

Trademark A word, phrase, symbol, design, or tangible nonutilitarian aspect of a product that signifies the source or origin of the product to the consumer.

Common law A system of laws based on precedential court decisions (originating in England).

Trademark notice symbols Symbols (®, ™, ℠) used to provide notice of a claim of rights in a trademark or service mark, typically placed as a superscript to the mark. The registration notice, ®, may be used only after federal registration has been obtained and provides constructive notice of the legal ownership status of the mark.

ADDITIONAL READING

Lechter, M., 2014. Protecting Your #1 Asset: Leveraging Intellectual Property, second ed. TechPress, Inc., Phoenix Arizona.

Poltorak, A., Lerner, P., 2011. Essentials of Intellectual Property: Law, Economics, and Strategy, second ed. John Wiley & Sons Inc., Hoboken New Jersey.

Revette, K., Kline, D., 2000. Rembrandts in the Attic: Unlocking the Hidden Value of Patents. Harvard Business School Press, Boston Massachusetts.

WEB RESOURCES

The websites below are intended to be destinations for your further exploration of the concepts and topics discussed in this chapter:

http://www.copyright.gov/: This is a site that contains more information regarding copyrights.

http://www.epo.org/patents/patent-information/free/espacenet.html: This website offers the general public free access to worldwide patent information. It has the following main aims: (1) to offer basic patent information to individuals, small and medium-sized enterprises, students, etc.; (2) to increase awareness and use of patent information at the national and European levels; (3) to support innovation and reduce wastage in the innovation cycle; and (4) to supplement existing channels for the dissemination of patent information.

https://www.google.com/?tbm=pts: This is another site that enables people to conduct their own patent searches. The patents are downloadable in PDF format.

http://www.USPTO.gov: This is the website for the U.S. Patent and Trademark Office. It provides search capabilities and has other useful information about obtaining patents and trademarks in the United States.

http://www.wipo.int/portal/index.html.en: This is the website for the World Intellectual Property Organization (WIPO). For technology entrepreneurs interested in building a global business, this site has useful information about developing IP in international markets.

ENDNOTES

1. Economics and Statistics Administration in the U.S. Department of Commerce and the United States Patent and Trademark Office, "Intellectual Property and the U.S. Economy: Industries in Focus," March 2012.

2. European Patent Office and the European Commission's Office for Harmonization in the Internal Market, "Intellectual Property Rights Intensive Industries: Contribution to Economic Performance and Employment in the European Union," September 2013.

3. http://technology-acquisitions.findthedata.org/.

4. http://www.sandisk.com/about-sandisk/press-room/press-releases/2011/2011-05-16-sandisk-announces-agreement-to-acquire-pliant-technology/.

5. http://www.equitrac.com/pr061611.html.

6. http://www.bloomberg.com/news/2011-07-01/nortel-sells-patent-portfolio-for-4-5-billion-to-group.html.

7. http://www.businessinsider.com/most-lucrative-patent-sales-of-2012-2012-11?op=1#ixzz2hH3wqpXL.

8. Ocean Tomo, Historical Impact of IP on VC, 2007.

9. For example, Silicon Valley Bank in San Jose, has repeatedly made loans collateralized by patents, reflected by a security agreement recorded at the USPTO.

10. Profit interests are typically utilized in circumstances where an income stream/profit is directly attributable to the intellectual property, such as when the intellectual property is embodied in the specific product or enables a specific service. A third party pays the company a lump sum in return for, e.g., a percentage of the income stream or profit from the sale or use of a product or service. In many cases the profit interest is capped at a specific amount or terminates after a specific time period.

11. IP securitization is a financing technique whereby a company transfers rights in receivables (e.g., royalties) from IP to an entity, which in turn issues securities to capital market investors backed by the receivables and passes the proceeds back to the owner of the IP. The revenue

from receivables pays the investor/bondholder back with an interest rate over a fixed period. As a practical matter securitization is employed only with intellectual property that has an existing royalty stream and generally involves pooling a number of different intellectual property assets.

12 Examples of such consortia include SEMATECH (semiconductor devices), Semiconductor Test Consortium (STC), RFID Industry Patent Consortium (RF identification) and the Symbian Foundation (platform for converged mobile devices).

13 Prevalent regimes of open source licenses include: Apache License 2.0, Berkeley Software Distribution (BSD) 3-Clause "New" or "Revised" license, BSD 2-Clause "Simplified" or "FreeBSD" license, GNU General Public License (GPL), GNU Library or "Lesser" General Public License (LGPL), MIT license, Mozilla Public License 2.0, Common Development and Distribution License, Eclipse Public License, and so-called "copyleft" licenses. See http://opensource.org, http://www.gnu.org & http://www.fsf.org.

14 In those models, intellectual property is not dedicated to public. Trade secret protection of the technology involved is typically abandoned, it is either provided subject to a license to a select group of entities, e.g., a pool or consortium, requiring a cross license of their intellectual property, or subject to some form of general public license (again relying on underlying intellectual property rights, typically copyright and occasionally patent) making certain requirements of the users. See Jacobsen v Katzer, 87 USPQ2d 1836, 535 F3d 1373 (Fed. Cir. 2008), Wallace v International Business Machines Corp., 80 USPQ2d 1956, 467 F3d 1104 (7th Cir. 2006).

15 California go so far as to prohibit non-competition agreements with employees. Bus. & Prof. Code, § 16,600.

16 See, e.g., Cambridge Engineering v. Mercury Partners, 27 IER Cases 68 (Ill. App. Ct. 2007); H & R Block Eastern Enters. v. Swenson, 26 IER Cases 1848 (Wis. Ct. App. 2007); Whirlpool Corp. v. Burns, 457 F. Supp. 2d 806 (W.D. Mich. 2006); Mohanty v. St. John Heart Clinic S.C., 866 N.E. 2d 85, 225 Ill.2d 52 (Ill. 2006); Coventry First LLC v. Ingrassia, No. 05-2802, 23 IER Cases 249 (E.D.Pa. July 11, 2005); Scott v.Snelling & Snelling, Inc., 732 F. Supp. 1034, 1043 (N.D. Cal. 1990); The Estee Lauder Co. v. Batra, 430 F. Supp.2d 158 (S.D.N.Y. 2006); MacGinnitie v. Hobbs Group, LLC, 420 F. 3d 1234 (11th Cir. 2005).

17 35 U.S.C. §115.

18 35 U.S.C. §120.

19 35 U.S.C. §1.54 (a) (2).

20 Vaupel Textilmaschinen RG v. Meccanica Euro Italia S.P.A., 944 F.2d 870 (Fed. Cir. 1991).

21 For example, the General Agreement on Tariffs and Trade (GATT) was amended in 1994 to include an Agreement on Trade Related Aspects of Intellectual Property Rights (TRIPS) establishing minimum standards for protection of various types of intellectual property, including patents, in members of the World Trade Organization (WTO)

22 Diamond v. Chakrabarty, 447 U.S. 303, 309 (1980).

23 Diamond v. Diehr, 450 U.S. 175, 101 S. Ct. 1048, 209 U.S.P.Q. 1 (1981); In re Bilski, 88 USPQ2d 1385, 1389 (Fed. Cir. 2008).

24 In re Bilski, 88 USPQ2d 1385, 1389 (Fed. Cir. 2008), In re Comiskey, 84 USPQ2d 1670, 499 F.3d 1365, 1371 (Fed. Cir. 2007).

25 35 USC §101 (a).

26 Bilski v. Kappos, 130 S. Ct. 3218, 177 L.Ed.2d 792 (U.S., 2009); CLS Bank Int'l v. Alice Corp. Pty. (Fed. Cir., 2013) citing Gottschalk v. Benson, 409 U.S. 63, 70 [175 USPQ 673] (1972)) and Diamond v. Diehr, 450 U.S. 175, 192 [209 USPQ 1] (1981); UltraMercial, Inc. v. Hulu, LLC (Fed. Cir., 2013).

27 Ass'n for Molecular Pathology v. Myriad Genetics, Inc., 133 S. Ct. 2107, 186 L. Ed. 2d 124, 106 U.S.P.Q.2d 1972 (2013); *Mayo Collaborative Servs. v. Prometheus Labs., Inc.*, 132 S. Ct. 1289, 1293 (2012).

28 See, e.g., 35 U.S.C. §102.

29 35 U.S.C. §102 as amended by the AIA.

30 35 U.S.C. §102 (c).

31 35 U.S.C. §102 (d).

32 35 U.S.C. §102 (f)(old Law).

33 See 35 U.S.C. §262; Willingham v. Star Cutter Co., 555 F.2d 1340, 1344 (6th Cir. 1977); Lemelson v. Synergistics Res. Co., 669 F.Supp. 642, 645 (S.D.N.Y. 1987); Intel Corp. v. ULSI System Technology Inc. ^{27}USPQ2d 1136 (Fed. Cir. 1993); Schering Corp. v. Roussel-UCLAF SA 41 USPQ2d 1359 (Fed. Cir.1997); Ethicon Inc. v. United States Surgical Corp. 45 USPQ2d 1545 (Fed. Cir. 1998).

34 Germany's Act on Employees' Inventions (ArbEG).

35 17 U.S.C. § 101 (a "computer program" is a set of statements or instructions to be used directly or indirectly computed in order to bring about a certain result; "Literary works" are works ... expressed in words, numbers, or other verbal or numerical symbols or indicia, regardless of the nature of the material objects, such as ... film, tapes, disks, or cards, in which they are embodied.); 17 U.S.C. § 117; *Gates Rubber Co. v. Bando Chemical Indus., Ltd*, 9 F.3d 823, 28 USPQ2d 1503, 1513 (10th Cir. 1993).

36 17 U.S.C. § 101 (a "compilation" is a work formed by the collection and assembling of pre-existing materials or of data that are selected, coordinated, or arranged in such a way that the resulting work as a whole constitutes an original work of authorship); 17 U.S.C. §103.

37 *Feist* Publications, *Inc. v. Rural Telephone Services Co.*, 111 S. Ct. 1282, 1296, 18 USPQ2d 1275 (1991).

38 Hutchins v. Zoll Medical Corp., 492 F3d 1377, 83 USPQ2d 1264 (Fed. Cir. 2007), *CMM Cable Rep, Inc. v. Ocean Coast Properties, Inc* 97 F.3d 1504, USPQ2d 1065, 1077–78 (1st Cir. 1996) (copyright law denies protection to "fragmentary words and phrases" ... on the grounds that these materials do not exhibit the minimal level of creativity necessary to warrant copyright protection). See also *Arica Inst., Inc. v. Palmer*, 970 F.2d 1067, 1072–73, 23 USPQ2d 1593 (2d Cir. 1992) (noting that single words and short phrases in copyrighted text are not copyrightable); *Alberto-Culver Co. v. Andrea Dumon, Inc.*, 466 F.2d 705, 711, 175 USPQ 194 (7th Cir. 1972) (holding that "most personal sort of deodorant" is short phrase or expression, not an "appreciable amount of text," and thus not protectable); National Nonwovens, Inc. v. Consumer Products Enterprises, Inc., 78 USPQ2d 1526, 397 F.Supp.2d 245, 256 (D. Mass. 2005) (There are no stylistic flourishes or any other forms of creative expression that somehow transcend the functional core of the directions), *Perma Greetings, Inc. v. Russ Berrie & Co., Inc.*, 598 F.Supp. 445, 448 [223 USPQ 670] (E.D.Mo. 1984) ("Clichéd language, phrases and expressions conveying an idea that is typically expressed in a limited number of stereotypic fashions, [sic] are not subject to copyright protection.").

39 *Baker v Selden*, 101 U.S. 99 (1879).

40 *CMM Cable Rep, Inc. v. Ocean Coast Properties, Inc* 97 F.3d 1504, 1516 41 USPQ2d 1065 (1st Cir. 1996).

41 *See, e.g.*, Asset Marketing Systems Inc. v. Gagnon, 542 F3d 748, 88 USPQ2d 1343 (9th Cir. 2008), *Sega* Enterprises, *Ltd. v. Accolade, Inc.*, 977 F.2d 1510 (9th Cir. 1993); *Atari Games Corp. v. Nintendo of America, Inc.*, 975 F.2d 832 (Fed. Cir. 1992); *Computer Associates, Int'l, Inc. v. Altai, Inc.*, 23 U.S.P.Q. 2d 1241 (2d Cir. 1992); *Johnson Controls v. Phoenix Control Sys., Inc.*, 886 F.2d 1173 (9th Cir. 1989). *Stenograph L.L.C. v. Bossard Assocs., Inc.*, 144 F.3d 96, 100, 46 USPQ2d 1936 (D.C. Cir. 1998). See also *Triad Systems Corp. v. Southeastern Express Co.*, 64 F.3d 1330,

1335, 36 USPQ2d 1028 (9th Cir. 1995) (where defendant's conduct "involved copying entire programs, there is no doubt that protected elements of the software were copied") See, e.g., Lotus Development Corp. v. Paperback Software Int'l, 740 F. Supp. 37 (D. Mass. 1990), MiTek Holdings, Inc. v. Arce Eng'g Co, 89 F.3d 1548, 39 USPQ2d 1609, 1617 (11th Cir. 1996); Digital Communications Assocs., Inc. v. Softklone Distrib. Corp., 659 F.Supp. 449, 463, 2 USPQ2d 1385 (N.D.Ga. 1987).

42. *Digital* Communications *Assoc. v. Softklone Distrib. Corp.*, 659 F. Supp. 449 (N.D. Ga. 1987).
43. *Digital* Communications *Assoc. v. Softklone Distrib. Corp.*, 659 F. Supp. 449 (N.D. Ga. 1987).
44. *Gates Rubber Co. v. Bando Chemical Indus., Ltd.*, 9 F.3d 823, 28 USPQ2d 1503, 1512–13 (10th Cir. 1993).
45. *Intellectual Reserve, Inc. V. Utah Lighthouse Ministry, Inc.* 75 F. Supp. 2d 1290 (D. Utah 1999). (providing links to other sites that were known to contain infringing copies of a work and encouraging the use of the links to obtain unauthorized copies constituted contributory copyright infringement.).
46. 17 U.S.C. §§ 1201 et. seq
47. 17 U.S.C. § 512
48. Viacom Int'l Inc. v. YouTube, Inc., 2013 WL 1689071 (S.D.N.Y. Apr. 18, 2013).
49. 37 C.F.R. §201.20.
50. 17 U.S.C. §411.
51. 17 U.S.C. §410.
52. 17 U.S.C. § 412.
53. 17 U.S.C. §504, 17 U.S.C. §505.
54. 37 C.F.R. §§202.20, 202.21.
55. See 17 U.S.C. §201.
56. 17 U.S.C. §§101, 201(b).
57. 17 U.S.C. § 203 (a) (2009).
58. The Semiconductor Chip Protection Act of 1984, 17 U.S.C. §901(a)(2).
59. Brooktree Corp. v. Advanced Micro Devices, Inc., 977 F.2d 1555, 24 USPQ2d 1401 (C.A.Fed. (Cal.), 1992).
60. 17 U.S.C. §§902(c), 906(a)(2).
61. 17 U.S.C. §901(a)(6).
62. 15 U.S.C. § 1127 (definition of "use in commerce").
63. 15 U.S.C. §1057(c).
64. 15 U.S.C. §1051(d) (1988).
65. 15 U.S.C. § 1057(c).
66. The Madrid Protocol. As of the publication date of this book there are 92 jurisdictions that have signed on to the Madrid Protocol. The United States became a signatory on November 2, 2003.
67. Bayer Co., Inc. v. United Drug Co., 272 F. 505 (S.D.N.Y. 1921).
68. Haughton Elevator Co. v. Seeberger (Otis Elevator Co.), 85 U.S.P.Q. 80 (Comm. Pat. 1950).
69. King-Seely Thermos Co. v. Aladdin Indus., Inc., 321 F.2d 577 (2d Cir. 1963).
70. DuPont Cellophane Co. v. Waxed Prods. Co., 85 F.2d 75 (2d. Cir. 1936).
71. Pilates, Inc. v. Current Concepts, Inc., 120 F. Supp. 2d 286 (S.D.N.Y. 2000).
72. Singer Mfg. Co. v. Briley, 207 F.2d 519 (5th Cir. 1953).

CHAPTER 6

Legal Structure and Equity Distribution

CHOICE OF WRONG LEGAL STRUCTURE ENDANGERS FINANCIAL HEALTH

The distinction between limited and unlimited liability companies is illustrated by the story of the Peroxydent Group. In 1983, Jerome Rudy and Melvin and Jeffrey Denholtz applied for a patent on a peroxide-baking-soda stabilizer toothpaste formulation. The formulation combined peroxide and baking soda in a way that avoids decomposition due to the interaction of peroxide with sodium bicarbonate.

Three years later, the inventors were in need of financing, so they brought in a group of investors—Peter Bohm, Steele Elmi, and Myron Barchas. A partnership was formed called the Peroxydent Group to which the inventors assigned the patent rights. Several years passed, and additional contributions (either in the form of money or effort) were required to take the project to the next stage. Two of the investors, Steele Elmi and Myron Barchas, opted not to risk the additional investment, so control passed to a subgroup of the partnership (two of the inventors, Melvin and Jeffrey Denholtz and one of the investors, Bohm).

For whatever reason, the retiring partners in the Peroxydent Group were not bought out of their ownership interests. Instead, Elmi and Barchas retained their interest in the partnership, and the subgroup of the partnership formed a new entity, Evident Corporation, to which the partnership granted a license under the patents.

In 1997, Evident Corporation brought a patent infringement suit against Church & Dwight (Arm & Hammer) and Colgate-Palmolive. The suit did not go well. The accused infringers counterclaimed, asserting that the patent was unenforceable due to inequitable conduct on the part of the inventors. While the Peroxydent Group partnership was not involved in the initial infringement action, the counterclaim was brought against not only Evident Corporation, but also the Peroxydent Group partnership (as owner of the patent). The court held that "the inventors had been aware of three material references and withheld them from the patent and trademark office with intent to 'deceive,'" rendering the patent unenforceable. This led to the court finding that the case was exceptional, and Church & Dwight were awarded $1.3 million in attorney fees and costs.

To the consternation of the Peroxydent Group partners, the partnership was held to be liable along with Evident Corporation. Equity ownership in the Corporation did not make the shareholders' personal assets available to satisfy the judgment. However, as partners of the Peroxydent group, each of the individual partners, including investors, Steele Elmi and

Continued

Myron Barchas, who had previously opted not to take the risk of proceeding with the Evident Corporation, became personally liable for the entire $1.3 million to the extent not paid by Evident Corporation or the other partners. Elmi and Barchas may not have been personally involved in any of the activities that resulted in the liability, but as general partners of the Peroxydent Group, they were nonetheless each on the hook for the entire liability costs.

Source: *Evident Corporation, and Peroxydent Group, v. Church & Dwight Co., Inc., and Colgate-Palmolive Co., Inc., (Federal Circuit, 399 F.3d 1310, 2005)*

6.1 INTRODUCTION

Building a technology venture is a team sport. At some time during the life of the typical venture, it will need to reach outside of itself for necessary resources. For many ventures, this occurs at inception. The genesis of a venture is often either an idea or a recognition of an opportunity on the part of an individual (commonly referred to as the "founder," "promoter" or, in technology ventures, "inventor"). However, it is a rare individual who alone possesses the skills, expertise, and credibility necessary to develop a viable business around an idea or opportunity. At this stage, he or she must recruit other individuals or entities to supply the prerequisites needed to start a successful venture.

So, how are skills, expertise, and credibility transported into a venture? They can sometimes be brought in by forming a strategic alliance with another entity or by hiring individuals with the needed skills, expertise, or necessary credibility. This, however, presupposes that the founder can actually *afford* to hire those employees. More typically, in start-ups, the individuals who can supply the needed skills, expertise, or credibility are brought into the venture by giving them the opportunity to share in the success of the venture. These individuals are provided an ownership interest, or **equity**, in the venture in lieu of all or part of the compensation that they might otherwise command in the open market and thus become **co-venturers**. The relative ownership interests of the co-venturers, of course, depend upon their relative contributions to the venture.

The foundational choices made by an entrepreneur when setting up a venture can be critical to its ability to attract the interest of co-venturers or sources of capital. In addition, choosing the appropriate type of **investor** and strategically timing capital infusions are critical to obtaining the capital and resources necessary for the success of the venture. Finally, from the perspective of the founders, it is also critical that the cost of obtaining that capital is minimal in terms of interest paid or lost percentage of ownership.

In this chapter, we will explore some of those foundational issues: the various legal structures available for the business entity; tools for attracting the right talent and investors; and equity distribution in the embryonic venture. We begin by exploring some fundamental legal issues that technology venture founders must carefully consider when establishing a new venture.

6.2 OWNERSHIP AND LIABILITY ISSUES

A start-up venture is typically owned by one or more individuals or entities, each making a contribution to or an investment in the business in return for equity interest (percentage of ownership). The individuals or entities that initially form the venture are generally referred to as the **founders**. In general, the individuals or entities that own the venture are referred to as the "principals," "owners," or "equity participants." Individuals or entities that make contributions to the business after it has been formed are referred to as "investors" or "lenders." In the United States, a business entity can take the form of any one of a number of different organic legal structures, including:

- Sole proprietorship
- General partnership
- Limited partnership (LP)
- Limited liability company (LLC)
- Corporation (C-corporation or S-corporation)

Some people would also list various types of organizations that are given special tax treatment, such as a "Cooperative,"[1] or a "Nonprofit"[2] as a business structure. However, these organizations generally are held in the form of one of the structures listed above.

The choice of legal structure can affect the business in many ways, including the potential risk to participants, potential for business growth, availability of benefits, taxation of the business entity along with its individual principals, and the types of **exit strategies** available to principals. One of the primary considerations to make when selecting a particular form of business legal structure is the liability of the principals. Some legal structures limit the liability of the principals and some do not. We turn to that discussion next.

6.2.1 Limited Versus Unlimited Liability

Certain forms of business entities, such as sole proprietorships and general partnerships, are considered to be the "alter ego" of the owners. These are further characterized as **unlimited liability** companies because each owner of the entity (sole proprietor or partner) can be held personally liable for the entire amount of the venture's debts, obligations, and liabilities. Here, liability is not limited to the *pro rata* share of ownership (as Messrs. Elmi and Barchas found to their chagrin with respect to the Peroxydent Group partnership); rather each owner of an unlimited liability company is liable for all of the venture's obligations. All personal assets are at risk to satisfy the debt, obligation, or liability. In legal parlance, each partner is said to be **jointly and severally liable** for all obligations of the business.

Furthermore, the personal assets of each owner of an unlimited liability entity are at risk with respect to not only the debts, obligations, or liabilities that they create through their individual actions, but also those created by an employee or partner, even without any personal involvement or knowledge on the owner's part. In other words, not only is each owner jointly and severally liable for all obligations of the business, but also they are **vicariously liable** for the acts of employees and partners in connection with the business.

This problem with unlimited liability entities has been duly recognized, and most jurisdictions in the developed world (including state governments in all 50 states in the United States) have created other types of entities by statute. These include entities that provide, from a liability perspective and, in some cases, from a tax perspective, a legal existence for ventures that is separate and apart from the individuals participating in the venture, or they limit the liability of the participants to their investment in the venture. These entities, referred to as **limited liability entities**, include corporations, LLCs or Giesellschaft mit beschrankter Haftung (GmbH), and LPs. Limited liability entities are also known as statutory entities. These legal forms limit the principals' liability exposure to the particular assets each contributed to the venture and shield their other assets.

Why would the government want to shield the personal assets of an entity's owners from the debts and obligations of the entity? The reason for the existence of limited liability entities is to promote commerce. Certain types of ventures tend to require substantial infusions of capital well beyond the means of the typical entrepreneur. In order for those types of businesses to be created and sustained, the resources of investors must be brought to bear. The infusion of investor capital can take a number of different forms, including purchase of an equity position in the company. However, unless the venture takes the form of a limited liability entity, merely purchasing or otherwise acquiring even a small equity interest in the venture will make all of the investor's personal assets available to satisfy the debts and obligations of the venture.

Under unlimited liability conditions, it would be difficult to find investors, and if found they would be extremely expensive; the percentage of ownership per dollar invested would have to reflect the amount of risk assumed by virtue of the investment. This is particularly true with respect to wealthier investors because potentially exposing their entire wealth to satisfy the debts of the company would create a disproportionate risk compared to less-wealthy shareholders.

6.2.2 The Extent of Limited Liability

In general, the most important reason for establishing a venture as a limited liability entity is to protect the assets of principals and managers from the consequences of any financial or legal misfortune of the business. As long as these entities are properly formed and are compliant with all necessary formalities, and individuals observe a general fiduciary duty of loyalty and care to the venture in conducting its affairs, they are shielded from personal liability to third parties arising from those business affairs.

Suppose an employee orders goods on behalf of a venture on credit, but the business is unable to make the payments. Under normal circumstances, the employee would have no personal liability for that obligation—the obligation to pay is that of the business not the employee. If the business was an unlimited liability entity, however, the owners of the business *would* be personally liable for that debt. On the other hand, if the business was set up as a limited liability entity, the business entity would be liable, but the owners would not be personally liable for the debt.

It is important to note, however, that individuals do not escape liability for their own improper conduct merely because a business entity was also involved. This personal liability arises not from the individual's equity interest in the business, but rather from the activities of the specific individual. For example, if there had been wrongdoing on the part of the individual placing the order for the goods (such as if the employee knew that payment would not be made), there may well be personal liability based on the individual's culpability. In one case, personal liability was imposed on the president of a limited liability entity when she personally made misrepresentations in order to obtain payment for equipment.[3]

MINI-CASE
Corporate Directors Not Shielded from Consequences of Their Actions

Messrs. Zeh, Gillispie, Krupski, Goodale, Topping, Celic, White, Reeve, Kujawski, Sieminski, and Truskolaski were all non-salaried directors of a duly incorporated Cooperative. The Cooperative entered into a purchase contract for equipment from Aeroglide Corporation provided that Aeroglide would retain title to the equipment until full payment was received, and specified "Buyer agrees to do anything necessary to see that title so remains in the Seller."

A document reflecting Aeroglide's interest should have been filed with appropriate state and/or county agencies—just like recording a mortgage on real property—but for some reason was not filed in this case.

Within a month or so after the purchase contract, an existing loan to the Cooperative from the Springfield Bank came due. The Cooperative had to extend that loan. As a condition to granting the extension, the bank, which did not know that Aeroglide still held the title to the equipment, required that it be given a security interest in the Aeroglide equipment. Without obtaining Aeroglide consent (or advising the bank of Aeroglide's interest) the Cooperative board authorized

Continued

> ### MINI-CASE—Cont'd
>
> a "chattel mortgage" giving the bank its security interest. This effectively was an improper transfer (conversion) of Aeroglide's property to the bank and was in violation of the purchase contract with Aeroglide. Messrs. Zeh, Goodale, Topping, White, Reeve, Kujawski, Sieminski, and Truskolaski, voted to give the bank the security interest, while Messrs. Gillispie, Krupski and Celic, were absent and not involved in the decision. Zeh and Krupski signed the mortgage document.
>
> Ultimately, the Cooperative defaulted on its loan from the bank and the purchase contract with Aeroglide. Because the Aeroglide interest had not been recorded (and the bank was not actually aware of it), the bank's security interest took precedence, the equipment was sold and the proceeds given to the bank.
>
> The Cooperative did not have the money to pay Aeroglide for its equipment and Aeroglide sued the individual directors for the value of the equipment. The trial court held all the directors personally liable. On appeal, the finding of liability was upheld with respect to all of the directors with the exception of Gillespie and Celic.
>
> Ordinarily, equity holders, officers, and directors are not personally liable for the liabilities of a corporation. The individual directors did not become personally liable for the monies owed Aeroglide merely because they approved the purchase of the equipment. And, if the default had occurred through no inappropriate action of the individual directors (even if the default could be traced to unrelated legitimate transactions approved in good faith by the board), they would still be shielded by the corporate structure. That, however, was not the case here. The mortgage to secure the bank loan on the equipment improperly converted rights held by Aeroglide to the bank. Those individual directors and officers who were personally involved in the conversion (voted for the transaction or signed the documents) were personally responsible for the liability. They were liable not because of their positions as officers or directors, but because of their involvement in the "bad acts." Gillispie, and Celic, on the other hand, escaped liability because they neither voted for, nor signed, the mortgage. As it happens, being absent from the meeting when the board voted on the bank mortgage was a lucky break for Gillispie, and Celic.
>
> **Source:** *Aeroglide Corporation v. Zeh, Gillispie, Krupski, Goodale, Topping, Celic, White, Reeve, Kujawski, Sieminski, and Truskolaski,* 301 F.2d 420 (2nd Cir. 1962).

Limited liability status, often referred to as the **corporate veil** (because the corporate structure establishes a protective "veil" between the business and the personal assets of the principal), is almost always disregarded by courts for criminal acts of the officers, shareholders, or directors of a company. Further, federal and state tax laws generally impose personal liability on those individuals responsible for filing sales and income tax returns for the company. Owners, directors, and/or officers of limited liability entities have been held to be personally responsible for unpaid payroll taxes even if they had no personal involvement in the day-to-day function of the company and had no check-signing authority. And, a single-member LLC structure provides no liability protection to the owner for any taxes that have been unpaid by the company, regardless of the involvement or lack thereof.[4]

There are also circumstances under which limited liability entity status will be ignored and liability will be imposed on the business owners, even in the absence of personal culpability, merely by virtue of their equity interests. The imposition of liability on the owners of a limited liability entity is often referred to as **piercing the corporate veil**. The issue typically comes up in the context of a lawsuit against the business, where it does not have sufficient assets to satisfy the damage claims, or the bankruptcy of a LLC when creditors seek relief from its equity holders. When this occurs, the plaintiff will attempt to convince the court that the business structure should not be respected and that the principals of the company, who may have personal assets sufficient to satisfy a judgment, should be personally liable. Piercing of limited liability entities typically occurs only under specific circumstances, such as:

Defective creation of limited liability entity: If the entity was not properly established, the entity will typically be considered a sole proprietorship or general partnership, and personal liability will attach to the owners. Some jurisdictions, however, will forgive technical deficiencies if there was a good faith attempt to create a limited liability entity. The formalities entailed in creating and maintaining the various forms of limited liability entities vary from state to state.

Failure to observe procedural formalities: As will be discussed, certain legal structures require that specific procedural formalities be observed. If those required formalities were not observed, the structure of the entity may be disregarded. For example, if the owners did not treat the business as an entity separate and apart from themselves and the business is nothing more than an alter ego for the owners, limited liability status is likely to be pierced.

Improper intent or fraud: There are boundaries placed upon the extent of the limited liability entity protection on the basis of the intent and purpose of the business. For example, if a corporation or LLC was set up only to shield its owners from liability arising from a fraudulent technology development start-up deal, and the owners siphon out the entity assets to such an extent that the entity is unable to compensate the victims of the fraud, a court is likely to set aside the limited liability entity and allow the victims to recover from the personal assets of the owners.

Statutory basis: Some statutes, most notably the federal securities laws, specifically impose personal liability on persons in control of a limited liability entity.[5]

Examples of activities that may lead to disregarding limited liability entity status include:

- Knowingly incurring company debt or obligations when the company is already insolvent
- Removing unreasonable amounts of funds from the company, endangering its financial stability

- Personally using company funds or assets by the principal owners
- Commingling company assets with the personal assets of the principal owners
- Hiring nonfunctioning employees who do nothing, yet draw a salary
- The assuming of the principal owner's personal debt by the company
- In the case of corporations, a pattern of consistent nonpayment of dividends, or a payment of excessive dividends
- Engaging in activities with the intent to defraud creditors
- Fraudulently making misrepresentations by owners or management
- Making unreasonable loans to company officials and extending unwarranted benefits to them
- Improperly giving corporate guarantees of loans or contracts benefiting an owner manager
- Little or no separation of corporate business from the activities of the dominant owner

There are also instances where personal liability for corporate debt is imposed by statute, such as in cases involving employers' tax withholding obligations, wage and retirement benefits, environmental liability, and violations of the federal securities laws. Personal liability may also arise because of contractual obligations, such as when a shareholder signs a personal guaranty or fails to adequately identify that he or she is signing on behalf of the company as opposed to signing on a personal level.

With these considerations in mind, the technology entrepreneur is ready to make an informed choice about the legal form for the new venture. We will now review the available choices for legal structure.

6.3 CHOICE OF LEGAL STRUCTURE

Each form of business entity has distinct characteristics that come with advantages and disadvantages, depending upon specific circumstances. Here are some factors to take into consideration in order to determine the optimum form of entity for a particular business venture:

- The potential risks and liabilities entailed in the venture: Limited liability entities limit the participant's exposure to the particular assets contributed to the venture.
- Participants: Certain forms of organization have limits on the number of participants and the nature of those participants.
- Capital growth needs and strategy: Certain forms of organization are more amenable than others to raising high levels of capital from multiple sources.
- Management structure: Certain forms of organization are more flexible than others.

- Tax implications: Tax laws in many jurisdictions vary dramatically between the various types of entity structures.
- Regulatory burden: Securities laws have a greater impact on some forms of entity than others.
- Administrative burden, formalities, and expenses involved in establishing and maintaining the various business structures: Some forms of entity require that documents be prepared and filed with a designated state agency, and/or via written agreements between the participants.
- Survivability: Some entities terminate upon certain events.
- Privacy: Certain business and financial information regarding certain entities are required to be made public.
- Exit strategy and liquidity needs: Certain entity structures provide more flexibility for the participants regarding liquidity and/or withdrawal.

As mentioned earlier, legal form choices for a new venture run from sole proprietorships to publicly held corporations. Let us explore each form in detail and weigh in on the various advantages and disadvantages of each.

6.3.1 Sole Proprietorship

The most basic business legal form is the sole proprietorship. A **sole proprietorship** is created by default anytime an individual owns a business without going through the specific formalities required to create a statutory entity. A sole proprietorship can operate under the individual owner's name or under a fictitious name. Most jurisdictions, however, require that fictitious names be registered. A sole proprietorship does not have separate legal status from the owner.

For liability, as well as tax purposes, a business operated as a sole proprietorship is the alter ego of the proprietor in most jurisdictions. A proprietor is personally responsible for the debts and obligations of the business as well as for the actions of employees. If the sole proprietor sells someone an interest of less than 100% of the business, the business by default becomes a general partnership (unless specific steps are taken to create a statutory entity).

A sole proprietorship is owned and managed by one person (or for U.S. tax purposes, a husband and wife). From the perspective of the U.S. Internal Revenue Service (IRS), a sole proprietor and his or her business are one tax entity. Business profits are reported and taxed on the owner's personal tax return and taxed at the personal income tax rate.

Setting up a sole proprietorship is easy and inexpensive because no legal formation documents need be filed with any governmental agency. Once a fictitious name statement is filed in jurisdictions that require one and once other requisite basic tax permits and business licenses are obtained, the enterprise is ready for business.

When a business is operated as a sole proprietorship in the United States, essentially all income generated through the business is self-employment income and subject to the self-employment tax.[6] The primary and significant downside of a sole proprietorship is that it is an unlimited liability entity—its owner is personally responsible for all the business liabilities. Given that fact, why would anyone operate a business as a sole proprietorship? The unfortunate answer is that most of the time they are simply ignorant of the unlimited liability. However, someone might choose to operate as a sole proprietorship because of the relative ease of, and lack of expenses entailed in, establishing the business. Exhibit 6.1 highlights the advantages and disadvantages of the sole proprietorship legal form.

Exhibit 6.1 Advantages/Disadvantages of a Sole Proprietorship

Sole Proprietorship	
Equity holder liability	Unlimited (Equity holder's personal assets are at risk to satisfy all debts, obligations or liabilities of the business.) Vicariously liable for actions by employees.
Equity interest types and terminology	100% owner/sole owner/sole proprietor
Management	Sole proprietor
Organic documents	None (If a name other than that of the sole proprietor is used for the business, some states require a registration of the "fictitious name" under which the entity does business.)
Formation procedure	None
Maintenance formalities	None
Limits on eligible owners	None (Multiple owners creates a partnership.)
Limits on allocations of profit and loss	None
Fiscal year limitations	Must correspond to the tax year of the sole proprietor.
U.S. Taxation	
Income	There is no tax to the sole proprietorship as a business enterprise for enterprise income. All items of income, gain, or loss, pass through and are taxed to the sole proprietor.
Losses	All items of income, gain, or loss, pass through and are taxed to the sole proprietor.
Appreciated assets	The sole proprietor is taxed upon the sale of appreciated assets. Generally, there is no tax upon the distribution of appreciated assets from business enterprise to sole proprietor.
Entity upon liquidation	There is no tax to the sole proprietorship as a business enterprise *per se* upon the sale or distribution of assets.
Owners upon liquidation	Gain realized upon the liquidating sale of appreciated assets by the sole proprietorship passes to the sole proprietor. No gain is recognized upon distribution except to the extent that the money distributed exceeds the sole proprietor's basis in the business.
Self-employment taxes	Sole proprietor is subject to self-employment taxes.

6.3.2 General Partnership

When the profits and losses of a business are shared between more than one individual (or other entities), unless the formalities for creating a statutory entity have been followed, the business is by default a **general partnership**. For liability purposes, a general partnership is considered the alter ego of each of the partners. Any general partner can bind the business to an agreement, and each partner is liable for the acts of the others. A general partnership is an unlimited liability entity. A general partner's liability is not limited to that partner's percentage of ownership. In more legal terms, each general partner is **jointly and severally** liable for the debts and obligations of the partnership. This means that each partner could be held personally liable for the entire amount of the partnership debt and obligations, irrespective of their proportionate ownership in the venture, and all of their personal assets are at risk to satisfy the obligations of the partnership.

Although a written partnership agreement is not required by law, partners typically enter into one to define the internal rules under which the partnership will operate and how profits and losses will be allocated. In the absence of a partnership agreement, each partner has an equal say in management, and profits and losses are typically divided equally among the partners. Equity interests in a partnership are commonly referred to in terms of percentages of ownership or "units" corresponding to a percentage of ownership. And, there is no certificate or instrument (other than perhaps the partnership agreement) representing ownership interests.

From a U.S. tax perspective, a partner cannot also be an employee of the partnership. This does not mean that the partner cannot receive the equivalent of a salary. However, as in the case of a sole proprietorship, all monies received are considered self-employment income for tax purposes and subject to the self-employment tax. Ever since the creation of the LLC legal form, most new businesses that choose a general partnership form of entity do so primarily out of ignorance of the LLC form, which is discussed later in this chapter. Unless the partnership operates without a formal partnership agreement, in most states there is little cost savings in setting up a partnership over setting up an LLC.[7] Exhibit 6.2 highlights the advantages and disadvantages of the general partnership legal form.

Exhibit 6.2 Advantages/Disadvantages of a General Partnership

General Partnership

Equity holder liability	Unlimited (equity holders' personal assets are at risk to satisfy all debts, obligations, or liabilities of the business.)
	Vicariously liable for actions by employees.
Equity interest terminology	Partner
	Varies according to agreement: Percentage, Partnership Units, etc.
	Different classes of partnership interests possible.
Management	Per partnership agreement (In the absence of a partnership agreement, each general partner has an equal say in the management of the partnership.)
Organic documents	None required.
	Written partnership agreement preferred.
	If a name other than that of the partners is used for the business, some jurisdictions require a registration of the "fictitious name" under which the entity does business.
Formation procedure	Formed by agreement (oral, or, preferably, written) between two or more prospective partners or by representing the business as a partnership to the public or in connection with doing business.
	The receipt of a share of profits (other than payment on a debt, interest, wages, and rent, etc.) from a business is evidence of being a partner of that business.
Maintenance formalities	None
Limits on eligible owners	Must be at least two partners; otherwise, no limits or requirements on the number or nature (e.g., individual, partnership, LLC, corporation, or trust) of partners. A "partnership" with a single equity owner is a sole proprietorship.
Limits on allocations of profit and loss	None. Allocations of profit and loss are determined by the partnership agreement. In the absence of an agreement, profit and loss are allocated equally among the partners.
Fiscal year limitations	Must use the tax year of partners having a majority interest in the partnership, or the tax year of all principal partners if there is no majority interest.

U.S. Taxation

Income	There is no tax to the partnership as a business enterprise for enterprise income. All items of income, gain or loss, pass through (per partnership agreement) and are taxed to the partners.
Losses	All items of income, gain or loss, pass through (per partnership agreement) and are taxed to the partners.
Appreciated assets	Generally, there is no tax to the partnership as a business enterprise upon the distribution of appreciated assets. The partners (as individuals or individual entities) are taxed on gains upon the sale of appreciated assets.
Entity upon liquidation	There is no tax to the partnership as a business enterprise *per se* upon the sale or distribution of assets.
Owners upon liquidation	Gain realized upon the liquidating sale of appreciated assets by the partnership passes to the partners (as in the case where there is no liquidation). No gain is recognized upon distribution except to the extent that the money distributed exceeds the partner's basis in his partnership units. Distributed assets will distribute at basis. Liabilities incurred by the partnership increase a partner's basis in his/her partnership interest.
Self-employment taxes	Individuals serving as general partners are typically subject to self-employment taxes.

6.3.3 Limited Partnership

A Limited Partnership (LP) is a legal entity that is created under the laws of a particular jurisdiction, with one or more general partners and one or more limited partners. It is defined by certain documents—specifically, a certificate of LP that is filed with a designated state agency to register the new business and a LP agreement that defines the internal rules under which the LP operates. The LP agreement must make clear which individuals/entities are general partners and which are limited partners.

The general partners are responsible for managing the partnership. The limited partners are, essentially, passive investors. Limited partners can have certain very limited veto and/or approval rights over certain actions, but any substantial involvement with the management or operation of the partnership will convert their status to general partners.

The general partners of the LP are jointly and severally liable with respect to debts of the LP. The limited partners are liable only to the extent of their ownership interest in the LP. In other words, if the LP has a liability, all of the general partners' assets are at risk. In contrast, other than the limited partners' investment in the business, the limited partners' other assets are isolated from liability.

The general partners of an LP often organize as a separate limited liability entity to relieve the burden of unlimited liability. For example, many venture funds are managed by general partners who raise capital from limited partners. The general partners organize as a separate LLC, which manages the LP. By employing a limited liability entity as general partner, only the assets of the LP are at risk.

For tax purposes, an LP is treated as an extension of the partners (general and limited), and profits and losses pass through to the shareholders according to the partnership agreement. These profits or losses are then taxed at the individual income tax rate. It is very important to note that partners are taxed according to their ownership interests *whether or not* profits actually are distributed. The same thing applies to losses. That is, a partner can record those losses on his or her personal income tax that have been allocated by the partnership venture.

Equity interests in a LP are typically referred to in terms of percentages of ownership or units corresponding to a percentage of ownership. However, there typically is not any sort of certificate or instrument representing ownership interests. Exhibit 6.3 highlights the advantages and disadvantages of the LP legal form.

Exhibit 6.3 Advantages/Disadvantages of a Limited Partnership

Limited Partnership

Equity Holder Liability	General partner—unlimited (see general partnership)
	Limited partner—potential liability limited to extent of investment
Equity interest terminology	General partner, limited partner
	Varies according to agreement: General partnership units, limited partnership units
	There may be different classes of partnership interests.
Management	Per partnership agreement
	Limited partners are precluded from taking an active role in management; if a partner takes an active role in management, the partner becomes a general partner with unlimited liability.
Organic documents	Certificate of limited partnership filed with designated state agency
	Limited partnership agreement. If a name other than that of the partners is used for the business, some states require a registration of the "fictitious name" under which the entity does business.
Formation procedure	File a certificate of limited partnership with a designated state agency. Limited partnership agreement defines the internal rules under which the partnership operates.
Maintenance formalities	None. Limited partner may not have any substantial right to be actively involved with the management or operation of the partnership.
Limits on eligible owners	Must be at least one general partner and one limited partner; otherwise, no limits or requirements on the number or nature (e.g., individual, partnership, LLC, corporation, or trust) of partners.
Limits on allocations of profit and loss	None. Allocations of profit and loss are determined by the partnership agreement. In the absence of an agreement, profit and loss are allocated equally among the partners.
Fiscal year limitations	Must use the tax year of partners having a majority interest in the partnership, or, if there is no majority interest, the tax year of all principal partners (defined as owning 5% or more), if no common year-end then that producing the least amount of change in income for partners.

U.S. Taxation

Income	There is no federal tax to the partnership as a business enterprise for enterprise income. All items of income, gain or loss, pass through and are taxed to the partners. May be special state taxes e.g., margin tax or B&O tax.
Losses	All items of income, gain or loss, pass through and are taxed to the partners.
Appreciated assets	Generally, there is no tax to the partnership as a business enterprise upon the distribution of appreciated assets. The partners (as individuals or individual entities) are taxed on gains upon the sale of appreciated assets.
Entity upon liquidation	There is no tax to the partnership as business enterprise *per se* upon the sale or distribution of assets.
Owners upon liquidation	Gain realized upon the liquidating sale of appreciated assets by the partnership passes to the partners (as in the case where there is no liquidation). No gain is recognized upon distribution except to the extent that the money distributed exceeds the partners basis in his partnership units. Distributed assets will distribute at basis. Liabilities incurred by the partnership increase a partner's basis in his or her partnership interest.
Self-employment taxes	Limited partners are not subject to self-employment taxes except for guaranteed payments for services to the partnership. General partners may be subject to self-employment taxes.

6.3.4 Corporation

A **corporation** is a legal entity, created under the laws of a particular jurisdiction (typically under state law in the United States), separate and apart from its owners. The word "corporation" is derived from the Latin *corparæ*, which means to make corporal, or to physically embody.

The corporate entity is defined by certain documents, including articles of incorporation, and these documents are also referred to as a charter in some jurisdictions. The documents are filed with the designated government agency. In the United States, corporations are formed under the laws of the individual states, usually through the state's secretary of state or corporation commission office. A corporation does not have to incorporate in the jurisdiction in which it is conducting business. In fact, many international businesses elect to incorporate in Delaware or Nevada, two states that have governing laws that are favorable to the corporate legal form. In some jurisdictions, however, **foreign corporations** are required to register to do business before they are recognized as business entities in that jurisdiction.

6.3.4.1 *Structure of a Corporation*

A corporation is managed by a board of directors and various officers who are selected and serve at the behest of the board. Officers are responsible for the day-to-day operation of the venture, while the board of directors has fiduciary responsibility to the owners (shareholders) of the venture.

Equity interests in a corporation are managed and tracked via the distribution of stock or shares. There are typically written stock certificates representing the interests of the various owners. Not surprisingly, the owners of a corporation are referred to as shareholders, or stockholders. A corporation is a limited liability entity—shareholders are liable only to the extent that they can lose the value of their shares.

A corporation is considered to be a separate entity from its owners, which is why a shareholder can be an employee of the corporation and can receive a reasonable salary for his or her services. In most countries, that salary is subject to payroll taxes. Other monies paid out by the corporation to the shareholder, such as dividends or distributions, are not subject to payroll taxes or self-employment tax.

Payroll Tax: An employer withholds or pays on behalf of their employees payroll taxes based on the employees' wage or salary. In the United States, the payroll tax includes, in addition to an income tax, the Social Security and Medicare Tax (FICA). The employer withholds one half of the FICA tax from the employee's paycheck and pays the other half of the tax. Some jurisdictions also withhold taxes to cover unemployment compensation and workmen's compensation.

Self-Employment Tax: Self-employment taxes are collected on net earnings. A business owner must pay them to the federal government to fund Medicare and Social Security (FICA), generally analogous to the aggregate of the employee and employer portions of the FICA and Medicare payroll taxes. The business owner is responsible for the entirety of the tax.

6.3.4.2 U.S. Tax Laws and Corporation Types

In the United States, the tax laws have created two particular types of corporations: the subchapter C-corporation and the subchapter S-corporation. Both types of corporation are limited liability entities, but are treated differently from a tax perspective. A C-corporation is taxed as an entity separate and apart from its shareholders.[8] A C-corporation's taxable income is subject to a graduated tax rate that increases with the amount of taxable income.[9] On the other hand, an S-corporation is, for tax purposes, treated as an extension of the shareholders, and profits and losses pass through to the shareholders according to their ownership percentages. There are, however, certain limitations on the classes of stock and number and types of shareholders permitted in an S-corporation (no more than 100 shareholders, no nonresident aliens, and only natural persons can be shareholders).

There are a number of noteworthy consequences that flow from the separate tax entity status of a C-corporation, and the choice of this form for a business has certain tax consequences:

Double taxation: Although salaries paid to corporate employees are deductible by the C-corporation, dividends paid to shareholders are not.[10] The dividends are thus taxed twice, first as income to the company (at the corporate tax rate) and then again as dividend income to the shareholder at the individual's tax rate.[11]

Dividends: Monies are paid to corporate shareholders out of the corporation's profits or reserves, to be distinguished from salaries paid to employees.

Retained earnings: Retained earnings are corporate income (on which the corporate income tax has been paid) that is retained by the company and not paid out to the shareholders. However, there must be good business reasons to retain earnings, such as research and development, expanding facilities, or acquiring another business entity. The IRS tends to view retained earnings as a potential attempt to avoid income tax on the shareholders. If the IRS deems the retained earnings excessive, they are subject to an accumulated earnings penalty tax. It is prudent for the board of directors and officers to document the reasons for retained earnings.[12]

In addition, if a C-corporation form is chosen for certain types of ventures, such as holding companies that receive royalties from technology licenses or consulting companies providing engineering services, the companies may be categorized as **personal holding companies,** or "personal service corporations," and subjected to additional taxes and higher tax rates.[13]

On the other hand, U.S. tax laws also provide specific incentives for particular types of C-corporations that qualify as **small business corporations**. A C-corporation is classified as a small business corporation if the aggregate amount of money and other property received by the corporation for stock does not exceed $1 million. Under certain circumstances, individuals are given favorable treatment with respect to losses from investments in a small business corporation. They are permitted to treat losses as ordinary losses that net out ordinary income as opposed to capital losses, which for the most part can be applied only against capital gains.[14]

C-corporations are typically the form of choice when the entity contemplates a large number of generally passive shareholders and/or a public offering of the shares of the business. Characteristics of a C-corporation and an S-corporation are summarized in Exhibits 6.4 and 6.5.

Exhibit 6.4 Characteristics of an S-Corporation

S-Corporation	
Equity holder liability	Potential liability limited to extent of investment
Equity interest terminology	Shareholder, Stockholder
	Stock
Management	Board of Directors and Officers
Organic documents	Articles of Incorporation (also referred to as a "charter" in some jurisdictions.)
	"By-laws"—define the internal rules under which the corporation operates.
Formation procedure	Articles of incorporation are filed with the designated state agency to "give birth" to the new corporate entity.
Maintenance formalities	Having and observing the bylaws, including meetings
	Minutes of shareholder and board of directors meetings and/or actions
	Separate un-commingled accounts and records
	Board resolutions for significant action/activities
Limits on eligible owners	May not have more than 100 shareholders.
	Subject to certain exemptions, no non-individual shareholders (e.g., corporation, trust). No non-resident aliens
	Some states require more than one shareholder.
Limits on allocations of profit and loss	Special allocations are not permitted. Dividends must be paid on stock ownership.
Fiscal year limitations	Some state require S-corporations to use a calendar year, except under certain circumstances.
U.S. Taxation	
Income	There is no tax to the S-corporation except in two limited circumstances: (1) Recognized built-in gains, and (2) Excess passive net income.
	There may be state taxes.
Losses	Losses may be deducted by shareholders to the extent of their tax basis in their shares (not including any portion of the corporation's debt), subject to certain restrictions, including the basis, at-risk and passive loss limitations.
Appreciated assets	Nontaxable at corporate level until gain is realized, distribution of assets to shareholders is done at current fair market value and can trigger a taxable event for the shareholder to the extent that fair market value exceed basis.

Exhibit 6.4 Characteristics of an S-Corporation—cont'd

Entity upon liquidation	Generally nontaxable at corporate level but taxable at shareholder level through pass-through of corporate tax items.
Owners upon liquidation	Shareholders taxed on gain. Gain is recognized to the extent that fair market value of property distributed exceeds the shareholder's basis in his or her stock.
Self-employment taxes	Self-employment tax does not apply to distributions paid to shareholders. Shareholders pay self employment tax on salary payments provided that compensation for their services is reasonable.

Exhibit 6.5 Characteristics of a C-corporation

C-Corporation

Equity holder liability	Potential liability limited to extent of investment
Equity interest terminology	Shareholder, stockholder. Stock. There may be different classes of stock.
Management	Board of Directors and Officers
Organic documents	Articles of Incorporation (also referred to as a "charter" in some jurisdictions). "By-laws"—define the internal rules under which the corporation operates.
Formation procedure	Articles of incorporation are filed with the designated state agency.
Maintenance formalities	Having and observing the bylaws, including meetings. Minutes of shareholder and board of directors meetings and/or actions. Separate un-commingled accounts and records. Board resolutions for significant action/activities
Limits on eligible owners	There are no restrictions on eligible owners.
Limits on allocations of profit and loss	Special allocations are not permitted. Dividends must be paid on stock ownership.
Fiscal year limitations	May use any fiscal year. Personal service corporations are required to use a calendar year, under certain circumstances.

U.S. Taxation

Income	Taxable income is, in general, subject to a graduated tax rate that increases with the amount of taxable income. Also potentially subject to retained earnings penalty tax (accumulated earnings tax), personal holding company tax, and/or personal service corporation tax.
Losses	Losses may not be passed through to or be deducted by shareholders.
Appreciated assets	There is potential double taxation. There is a tax imposed at the corporate level upon the sale or distribution of appreciated assets. Additionally, there is a potential dividend or capital gains tax upon the distribution of sale proceeds to shareholders.
Entity upon liquidation	Taxed on appreciation in assets upon the sale or distribution of assets. This may result in double taxation as these proceeds are distributed to shareholders in the form of liquidating dividends.
Owners upon liquidation	Shareholders taxed on gain. Gain is recognized to the extent that fair market value of property distributed exceeds the shareholder's basis in his or her stock.
Self-employment taxes	Self-employment tax does not apply to dividends or distributions paid to shareholders. Employees of the corporation are not made subject to self-employment tax by virtue of also being shareholders. The corporation pays the employer's portion of the social security and medicare taxes and withholds the employee portion (which equals ½ the self-employment tax) and remits both to the government.

6.3.4.3 Maintaining Corporate Status

In order to maintain corporate status, certain practices and procedures must be observed. Because corporations are creatures of statute, unless the specific formalities required by the applicable statute are met (e.g., filing appropriate organizational papers with the designated state agency), the enterprise will be considered either a sole proprietorship or, if more than one person has an equity interest, a general partnership. Failure to do so would result in the tendency of the courts to ignore the corporate structure and impose liability on the individual owners and officers of the corporation. Filing the organizational papers and obtaining a charter, however, are merely the beginning.

If various corporate formalities are not consistently observed, the corporation can be easily disregarded and the individuals may be held personally liable. Within the court system, for instance, the corporate status is usually determined by whether or not the principals treat the corporation as a separate and distinct entity and hold it out to third parties as such. Similarly, a corporate structure will almost always be ignored and the corporate veil pierced when the management treats the corporation as its alter egos rather than as a separate entity or when the corporation is found to be a "sham" and has been established only to facilitate fraud against third parties. In general, the following tend to ensure that corporate status is respected:

- Issuing of stocks
- Instituting and observing bylaws
- Filing annual reports with the state
- Holding annual shareholder and board of directors meetings
- Maintaining minutes of shareholder and board of directors meetings
- Promulgating formal written resolutions for significant actions
- Keeping up-to-date corporate records
- Separating and maintaining un-commingled funds and assets of the corporation among major shareholders
- Being adequately capitalized (capital sufficient to operate as an actual business)
- Avoiding dependency on the property or assets of a shareholder not technically owned or controlled by the corporation
- Avoiding personal use of corporate funds and assets by major shareholders except with full documentation

It is particularly important to maintain minutes of board and shareholder actions. Minutes can be the written record of meetings or can reflect the unanimous written actions of the directors or shareholders taken without a meeting. the secretary of the corporation typically prepares and signs the minutes of a meeting. To minimize the possibility of inaccuracies, those minutes are then approved by the board or the shareholders at the next meeting or in the next action. It is a good practice for the minutes to reflect that any requirements established by the bylaws with respect to meetings or actions were observed (e.g., proper notice was given or waived, a quorum was present, and so on). The minutes also typically identify who was present and who was absent, and note any abstentions or dissents on a vote. Any formal resolutions adopted by the board should also be included in the minutes. Many corporations have their minutes reviewed by legal counsel before completing them.

6.3.5 Limited Liability Company

A Limited Liability Company (LLC), like a corporation and LP, is a form of business entity created under the laws of a particular jurisdiction that combines corporate and noncorporate features. The LLC is owned by "members" and is run either by the members or by one or more "managers" (who may or may not be a member).

The LLC as a form of entity is a relatively recent innovation in the United States, although a similar entity, the GmbH, has been available in Germany and various other European nations for some time. The first LLC act in the United States was enacted by the state of Wyoming in 1977.[15] However, it was not until the 1990s (after the federal IRS broadcast a revenue ruling to the effect that an LLC formed under the Wyoming LLC Act would be classified as a partnership for federal income tax purposes even though all of its members had limited liability) that most states embraced the concept and enacted LLC statutes.[16] In 1997, the LLC became even more popular when the IRS established the **check-the-box regulations**, permitting the LLC to choose whether it will be treated for tax purposes as a sole proprietorship, a partnership, a C-corporation, or an S-corporation.[17]

6.3.5.1 LLC Characteristics

An LLC is created by filing a document referred to as **articles of organization** with an appropriate state agency and entering into an **operating agreement** between the members. Equity interests in an LLC are typically referred to in terms of percentages of ownership or units corresponding to a percentage of ownership. There generally is no certificate or instrument (such as a share of stock) representing ownership interests.

The LLC is a limited liability entity and provides liability protection to its members in the same way that a corporation provides liability protection to its shareholders; members are liable only to the extent that their ownership interest in the LLC may lose all of its value. LLCs are not subject to many of the formalities associated with corporations. LLCs are not required to hold annual meetings or prepare annual reports, although it may nonetheless be beneficial to hold regular meetings of the members or to provide in the LLC operating agreement that meetings be held before certain actions are taken.

As noted above, an LLC can elect to be treated for tax purposes as a sole proprietorship, a partnership, a C-corporation, or an S-corporation. If no election is made, the default taxation for a single-member LLC is a sole proprietorship and is a partnership for a multimember LLC. If either C-corporation or S-corporation treatment is elected, a member can be an employee of the LLC and can receive a reasonable salary for his or her services. That salary is subject to payroll tax (e.g., income tax withholding, the Social Security [FICA], and Medicare Tax) deductions. Other monies paid out by the LLC to the shareholder, for example, distributions, are not subject to withholding or self-employment tax (FICA). On the other hand, if either a sole proprietorship or partnership treatment is elected, then, from a tax perspective, a member of the LLC cannot also be an employee of the LLC. Essentially, all monies received from the LLC would then be subject to the self-employment tax on the individual's personal tax return.

In most respects, the LLC compares favorably with the other forms of limited liability entity. However, certain states, notably California, have begun charging relatively high fees to do business as an LLC.[18] Exhibit 6.6 highlights the various advantages and disadvantages of the LLC legal form.

Exhibit 6.6 Advantages/Disadvantages of the LLC

Limited Liability Company (LLC)

Equity holder liability	Potential liability limited to extent of investment
Equity interest terminology	Member
	Membership interests, percentage interest, units. There may be different classes of membership interests.
Management	Managed either by all members, or by specifically designated managers. Members who participate in management are not personally liable.
Organic documents	The entity is created by filing a document referred to as "articles of organization" with an appropriate agency, and entering into an "operating agreement" between the members.
Formation procedure	Filing "articles of organization" with the appropriate agency, and entering into an "operating agreement" between the members.
Maintenance formalities	None; although annual meetings are advisable.
Limits on eligible owners	There are no restrictions on eligible owners.
	Some states require more than one member.
Limits on allocations of profit and loss	None. Allocations of profit and loss are determined by the operating agreement. In the absence of an agreement, profit and loss are allocated equally among the members.
Fiscal year limitations	Some state require LLCs to use a calendar year, except under certain circumstances.

Exhibit 6.6 Advantages/Disadvantages of the LLC—cont'd

U.S. Taxation

Income	The LLC can choose whether it will be treated for tax purposes as a sole proprietorship, a partnership, a C-corporation, or an S-corporation.
Losses	The LLC can choose whether it will be treated for tax purposes as a sole proprietorship, a partnership, a C-corporation, or an S-corporation.
Appreciated assets	Generally, there is no tax to the LLC as a business enterprise upon the distribution of appreciated assets. The members (as individuals or individual entities) are taxed on gains upon the sale of appreciated assets.
Entity upon liquidation	There is no tax to the LLC as a business enterprise *per se* upon the sale or distribution of assets.
Owners upon liquidation	Gain realized upon the liquidating sale of appreciated assets by the LLC passes to the members. No gain is recognized upon distribution except to the extent that the money distributed exceeds the members basis in his or her membership units. Liabilities incurred by the LLC increase a member's basis in his or her membership interest.
Self-employment taxes	Depends upon choice of tax treatment.
	Managers may be subject to self employment tax on their distributed portion of income whether or not actually distributed. A manager-managed LLC electing partnership taxation may be treated similar to a limited partnership where the manager would be subject to self-employment tax and the members would not.

6.3.6 Limited Liability Entities—A Comparison

Establishing the appropriate legal form for an entity is never a black-and-white decision for the entrepreneur. Instead, many shades of gray intercede into the decision, and there are often good arguments to be made for establishing any one form over another. Fortunately, the entrepreneur is not locked into one particular form. Even though changing from a C-corporation to another form of entity under certain circumstances can be problematical, it is otherwise easy enough to change a venture's legal form when its fortunes change or when the ambitions of the founders change.

Still, making a good decision at the beginning is important to overall venture success. In the sections that follow, we look at some of the factors that may enter into decision making about the appropriate legal form of a technology venture.

6.3.6.1 Expense

Sole proprietorships and general partnerships are potentially easier and less expensive to set up than the alternatives, but they are unlimited liability entities. For that reason, they are rarely an appropriate choice for any technology venture unless it is substantially risk-free. One such example may be a single-person technology consultancy that operates principally out of the home. However, even that simple type of business may be better organized as an

LLC or other limited liability entity to protect against potential lawsuits and other types of claims.

6.3.6.2 Shareholder Options

Of the limited liability types, C-corporations typically are the legal form of choice when the entity contemplates a large number of passive shareholders in the near term or anticipates an eventual public offering of stock. A C-corporation can sell shares of itself through private or public stock offerings. This can make it easier to attract investment capital and to hire and retain key employees by issuing employee stock options (EOSs). An S-corporation can also sell shares of stock in private offerings, but it is limited to common stock only. In public offerings, a C-corporation legal form is the form most often used.

6.3.6.3 Taxation

As noted above, the respective types of entities may receive different tax treatment. One form or another may be advantageous depending upon the particular business plan of the entity and the specific circumstances of the principals. For example, even though C-corporations pay taxes at the corporate tax rate on net income, the owners of the corporation are not subject to personal income taxes on **phantom income** (profits that are not distributed). Depending upon the corporate tax rate when compared to the personal tax rates of the principals, this can sometimes be advantageous; also, depending upon the relative corporate and personal tax rates applicable, it is conceivable that a corporation and its owners may have a lower combined tax bill than the owners of a business with a pass-through structure that earns the same amount of profit.

On the other hand, given the personal tax situation of one or more of the principals, losses incurred by the business entity, if passed through to the principal, could be used to offset income from other sources. In that case, as an example, an LLC or S-corporation may be advantageous. An S-corporation does offer a unique advantage with respect to control of employment taxes (such as Social Security and the like). A shareholder of an S-corporation can be an employee of the business and receive a reasonable salary for his or her services. Self-employment tax is paid only on salary payments, and provided the salary reflects reasonable compensation for services, monies can be paid by the corporation to that shareholder. Self-employment tax does not apply to dividends paid to shareholders. At the same time, the S-corporation is not plagued with the double taxation, retained earnings, personal holding company, and personal service corporation issues like a C-corporation is.

6.3.6.4 Distribution of Profits and Losses

The LLC tends to compare favorably with the other forms of limited liability entities; pass through profits and/or losses can be distributed disproportionately from ownership percentages. S-corporations and LPs also provide pass-through of profits and losses for tax purposes. However, the apportionment of profits and losses in an S-corporation must be in accordance with ownership percentages, and a LP imposes limitations on the extent to which the participants can exert management control without losing the benefit of limited liability. To some extent, particular shareholders in a corporation can be accorded disproportionate control of management through an appropriate **shareholders agreement**, which requires the shareholders to elect and designate individuals to the board of directors.

In comparison, in an LLC with the appropriate provisions in the operating agreement, profits losses are not tied to ownership percentages, and limited liability is available irrespective of management control. For example, to entice an investor, the other LLC members may agree to allocate a disproportionate share of any losses to the investor to allow the investor to offset gains on personal income in other investments he or she may have made. The LLC also has the advantage of not being subject to the limitations on numbers and types of owners that are imposed on S-corporations.

6.3.6.5 Formalities

Formation of LPs, corporations, and LLCs all require comparable effort and cost. Exhibit 6.7 provides a comparison of the documents related to the formation of the various limited liability entities.

Exhibit 6.7 Comparison of Organic Documents for Different Types of Limited Liability Entities

	Limited Partnership	Corporation	LLC
Filing required	Certificate of limited partnership	Articles of incorporation/charter	Articles of organization
Required but not filed	Limited partnership agreement	Bylaws	
Not required but desirable		Shareholders agreement	Operating agreement

LLCs are not subject to the same formalities of meetings, minutes, and so on, that are required to maintain a corporation. LPs are likewise not subject to those formalities. However, because LPs require a general partner and the general partner is very often a separate limited liability entity, the LLC structure provides the same advantage without the complexity.

6.4 EQUITY AND EQUITY TYPES

Equity is simply another term for an ownership interest in a business. The particular terminology used to denominate equity varies depending upon the particular legal structure of the entity. As noted earlier, equity interests in partnerships and LLCs are typically referred to in terms of ownership percentages or units corresponding to a percentage of ownership. Equity in a corporation is referred to as "stock" for shares (Exhibit 6.8).

Exhibit 6.8 Terminology for Ownership Interests

Partnership	Corporation	LLC
Units or percentage	Stock or shares	Units or percentage

6.4.1 Corporate Stocks

The articles of incorporation for a corporation specify a maximum total number of shares of stock that it is authorized to issue. The original number of shares of **authorized stock** is largely arbitrary, but it should be selected carefully on the basis of the venture's financial plan and future fund-raising objectives. The number of authorized shares can be changed, but only through a vote by the venture's shareholders. From the authorized pool of shares, the venture then can issue shares to founders and investors. It is the **issued shares** that constitute the ownership structure of the venture. That is, a shareholder's percentage of ownership of the venture is calculated as the proportion of shares held compared to shares that have been issued. Most ventures will reserve a portion of the authorized shares in the company's treasury for later use. The shares held in reserve are called **unissued shares**, or **treasury stock**.

It is also important to note that not all stock is created equal. A venture's articles of incorporation can be crafted to permit it to issue different classes of shares, each having a specific designation and unique preferences, limitations, and/or rights. These share classes are usually associated with financing rounds and are often designated as Class A, Class B, and so on; or as Series A, Series B, and so on. The articles of incorporation either specify the particulars of each class of shares or delegate the authority to the board of directors. In general, the preferences and relative rights of classes of shares relate to voting, payment of dividends, and distribution of corporate assets on dissolution. Limitations of a class of shares typically relate to restrictions on transferability. Shareholders of private ventures normally do not want to leave it up to shareholders to decide to whom they may transfer their shares. Transferability is usually governed by buy-sell clauses in the shareholders' agreement. Below, we discuss two types of stock, common and preferred, that are commonly used to represent ownership interests and rights in a corporation.

6.4.2 Common Stock

Common stock refers to the baseline of ownership in a corporation and a right to a portion of profits from that company. Many ventures do not specifically define common stock. Instead, they refer to classes of stock having different characteristics without ever specifically categorizing any particular class or classes as common stock. The designation for common stock tends to vary from company to company often depending upon whether or not a special class of stock is issued to the founders. If only one class of stock is issued, it will be common stock. As used in this text, the term *common stock* means a class of stock that does not have any preference over another class of stock. Common stock is normally the last in line when dividends are paid and in the event of liquidation. In the event of liquidation, creditors, bondholders, and preferred stockholders typically are paid first and common stockholders are entitled only to what is left.

Owners of common stock usually have voting rights to elect the members of the board of directors who oversee the management and operations of the company. Different classes of stock may have different voting rights, as will be discussed. Profits of the company may be paid out in the form of dividends to the common stockholders. However, dividends on common stock are not guaranteed and usually only occur when the company is profitable. The amount and timing of dividends are determined by the board of directors.

The owners of common stock get the benefits of increases in value as the company grows. On the other hand, the value of their ownership interest decreases if the company is unsuccessful. So, ownership in common stock may provide the greatest upside potential to investors, but it may also provide the greatest risk of loss if the business does not succeed.

6.4.3 Preferred Stock

Technically, any stock that has any sort of advantageous characteristic over other classes of stock is referred to as **preferred stock**. Preferred stock typically has provisions over and above common stock that:

- Provide a specific dividend on a periodic basis (e.g., monthly, quarterly, semiannually, or annually)
- Classify the stock as senior to other classes of stock in the order of distribution (i.e., payments are made to the preferred stockholder before dividends are paid to the other classes or when there is a distribution upon liquidation)
- Indicate that preferred shareholders have limited or no voting rights

Preferred stock tends to be a more conservative investment than common stock; it is not as volatile. Investors purchase preferred stock for the dividends. Unlike a dividend on common stock that is paid out to shareholders essentially at the discretion of the board of directors, dividends on preferred shares often are a fixed amount regardless of the earnings of the company. In some cases, the right to those dividends can accrue on a periodic basis. Although payment of dividends on preferred classes of stock is not guaranteed *per se*, dividends are guaranteed to be paid before payment is made to any junior classes of stock.

Preferred stock can include a wide range of advantages, also referred to as "sweeteners." A few of these advantages are discussed below.

6.4.3.1 Preferred Stock Distributions

A class of preferred stock can be specified to entitle the holders to distribution—**dividends**—and specify the manner in which the distributions are to be calculated. For example, dividends for a class of stock may be cumulative or noncumulative.

Cumulative preferred stock has a provisional right to a dividend payment that accrues periodically, and all accrued dividends must be paid to the holders of that class of stock before any dividends can be paid to any junior class of stock. For example, assume a company issues a preferred class of stock that has a fixed dividend yield to be paid quarterly. However, because of financial problems the company is forced to suspend its dividend payments in order to pay expenses. If the company gets through the trouble and starts paying out dividends again, it will have to pay all of the dividends that have accrued to the holders of the preferred class of stocks before it can pay any dividends to the junior classes.

On the other hand, if the class of shares is noncumulative, dividends do not accumulate, and the shareholder is not entitled to receive dividend payments for the periods in which no dividend is paid. They are, however, typically still entitled to payment of dividends before payment is made to any junior class of stock when the venture is once again able to pay dividends.

6.4.3.2 Convertibility

A class of preferred stocks can also be redeemable or convertible into cash, debt, common stock, or other property. As specified by provisions of the stock, the conversion can be triggered at the option of the corporation, the shareholder, or another person or on the occurrence of a designated event (e.g., a venture capital round of investment). The conversion can be a designated amount, or can be determined in accordance with a designated formula or by reference to some prespecified extrinsic data or events.

6.4.3.3 Participating Preferred

Preferred stock that pays a fixed dividend regardless of the profitability of the company is referred to as **nonparticipating preferred stock**. However, preferred stock can also be made to participate in the profits of the corporation. For example, a **participating preferred stock** may entitle the holder to the right to participate in any surplus profits in addition to a fixed dividend and after payment of agreed levels of dividends to holders of common stock has been made. In some instances, the participation may relate to dividends. For example, the preferred stockholder may receive the greater of the fixed dividend or the dividends paid to a share of common stock (or, if convertible, to the number of shares of common stock into which the preferred stock can be converted).

The participation can also relate to the distribution upon liquidation of the company. In the event that the company is liquidated, holders of **nonparticipating preferred stock** typically receive predetermined preferential return (e.g., an amount equal to the initial investment) plus any accrued and unpaid dividends prior to any distributions to junior classes of stock. However, if the company is being sold at a sufficiently high valuation, holders of common stock may well receive more per share than holders of the preferred stock. If the preferred stock was convertible, then holders of preferred stock would have the option to convert their shares into common stock, giving up their preference in exchange for the right to share pro rata in the total liquidation proceeds. On the other hand, if the preferred stock was participating, the preferred stockholders could in essence have their cake and eat it too. Participating preferred stock might not only receive the predetermined preferential return, but also participate on an "as converted to common stock" basis with the common stock in the distribution of the remaining assets.

6.4.3.4 Voting Rights

Although at least one of the classes of shares must have voting rights (and, at least in combination with the rights of other classes of shares, voting rights that cover all matters defined in the articles of incorporation that require votes), a preferred stock class may be given special, conditional, or limited voting rights or, for that matter, no right to vote at all.

One vote per share, typically, tends to be the norm for common stock, but a corporation can issue different "classes" of common stock, distinguished by the specific voting rights associated with each share. The voting rights associated with common stock can range from no voting rights at all to multiple votes per share.

6.4.3.5 Founder's Stock

Founder's stock is used to refer to stock issued to the founders of the corporation. There is, however, no universal definition of founder's stock in terms of its characteristics. If the corporation issues a single class of stock, founder's stock is simply common stock issued to the founders. However, founder's stock can sometimes be a separate class of stock with its own specific characteristics. For example, the company founders may issue two classes of stock: Class A stock, which has one vote per share, and Class B stock, which has 10 votes per share. The founders would offer the Class A stock to investors and issue the Class B stock to themselves, with the intent that the super-voting rights would ensure that they retain decision-making control of the company. In addition to the super-voting privilege, there have been a number of characteristics proposed for founder's stock, including:

- Having the right to convert shares into any other class of stock when that class of stock is offered to investors
- Assigning liquidation preferences at a designated amount (such as actual monies invested by the founders, a specified percentage of the founders holdings, or the valuation of the company prior to a successive round of financing)
- Making the founder's stock preferred participating

Often, when part of the founder's contribution to the venture is sweat equity, founder's stock is subject to a **vesting** requirement in accordance with a predetermined, usually time-based, schedule or formula. Actual ownership of stock does not transfer until the vesting condition is met.

Of course, there tends to be a tension between having a special class of founder's stock with protective characteristics and the ability to attract future investors for subsequent financing rounds. In many instances, when a special class of founder's stock is in place, investors will attempt to negotiate away the founder's preferences as a precondition to investment.

It is important that founders understand the tax consequences of receiving stock in return for services. As a general proposition, such stock is considered income to the founder. Under the U.S. tax law, the founder has the option to recognize income (the difference between fair market value of and the price paid for the stock) either when the stock is issued or when the stock vests.[19] However, if the founder/employee elects to recognize the value of the stock as income when the stock is issued, an 83(b) election must be filed with the IRS no later than 30 days after the stock is issued. Because the value of the stock is likely to increase over time, it is generally beneficial for a founder/employee to make an 83(b) election. The 83(b) election starts the one-year capital-gain holding period and freezes ordinary income (or alternative minimum tax) recognition to the issue date.

6.5 RAISING CAPITAL

As previously noted, at some time during the life of the typical venture, it will need to reach outside of itself for necessary resources. One of those necessary resources is often capital.

For a transactional analysis of the process of forming a venture and of subsequently bringing investors into the venture, it is convenient to use a **co-venturing paradigm**: the business venture is considered a separate and distinct entity from each individual participant or investor (co-venturer), and each co-venturer makes a contribution to the venture in return for some form of consideration.

The consideration provided by the respective co-venturers to the venture can take a number of forms, such as, for example, money, intangible assets such as intellectual property, tangible assets, or services. Likewise, the consideration provided by the venture to the co-venturers can take one or more of a number of forms, including equity in the venture, money (e.g., interest on a loan), or a **profit interest** in all or part of the venture. Exhibits 6.9 and 6.10 illustrate such an analysis for the formation of a business venture and the introduction of investors to an existing business, respectively.

The consideration given to an investor in return for the investment is, in effect, the cost of the investment to the preinvestment venture principals. Financing tends to be categorized in terms of the particular consideration given to the investor. For example, debt financing involves taking out a loan or selling **bonds** (debt instruments) to raise capital, in effect paying interest to the lender for the use of the lender's money; the lender contributes money to the venture in consideration of a promise to repay the money to the lender plus a fee (interest). Equity financing involves, in effect, sale of equity (part of the ownership) to the investor; the investor contributes money to the venture in consideration of an equity interest.

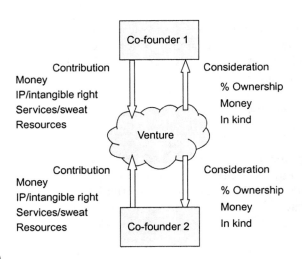

EXHIBIT 6.9

Transactional analysis of business formation. *Source: Lechter, M., 2010. OPM, Other People's Money: The Ultimate Leverage, second ed. TechPress.*

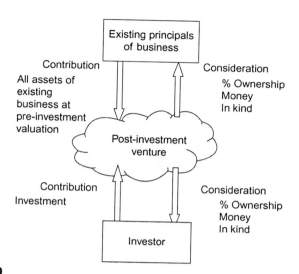

EXHIBIT 6.10

Transactional analysis of investment in existing business. *Source: Lechter, M., 2010. OPM, Other People's Money: The Ultimate Leverage, second ed. TechPress.*

Equity financing should be distinguished from debt financing. The cost of debt financing is typically finite, and the lenders will not share in any appreciation in the value of the business. However, in the short term, repayment of the loan can place constraints on the start-up venture's cash flow; both principal and interest must be repaid, and repayment is required whether the company is successful or not. The requirement to make periodic payments to service debt can make it difficult for the start-up venture to manage its cash flow.

Equity financing entails selling part of the ownership of the company. Equity financing is more flexible than debt financing with respect to cash flow concerns. There are typically no short-term repayment requirements, and equity investors share the risk. However, equity investors also share in any appreciation in the value of the business. And, although profit interests can be separated from governance and control, anytime equity investors are added, there is some loss of independence on the part of the preinvestment ownership team.

6.6 EQUITY DISTRIBUTION IN THE START-UP VENTURE

We have discussed the various legal structures that a business can take and different forms of equity. We now return to the inter-relational dynamics of the creation and growth of the business. When one or more founders form a business, each makes a contribution to the business in return for some consideration. Initial contributions by principals are typically made in return for, at least in part, an equity interest. The particular form of that equity interest depends upon the particular legal structure selected for the venture. The question we now address is: how should the equity be distributed among those individuals or entities?

When the business is founded by more than one person (or entity), the perceived relative values of the contributions that each makes (or will make) to the venture are typically reflected by the relative percentages of ownership. The issue is simplified if each of the founders makes a strictly monetary contribution to the venture in return for his or her equity interest. In that case, the relative percentage of ownership is easily determined. Each principal would hold an ownership interest equal to the ratio of the amount of money contributed by the individual to the total money contributed.

To the extent that founders make nonmonetary contributions to the venture, the venture can provide separate compensation, other than equity, for such contributions. For example, one or more of the founders will often be involved in the management and operation of the business. In that case, each individual will provide those services as an employee or independent contractor and will be separately compensated. Most often, the venture pays a salary or fee for those services. However, compensation can also be in the form of trade, for example, the services are exchanged for an equivalent fair market value of services or product from the business. Likewise, a founder can also provide resources or IP to the entity for compensation other than equity. In that case, the founder provides the IP or resources in the role of an independent entity (e.g., a lessor or licensor) and is compensated with something other than equity, such as a lease or royalty payment.

Despite these alternative compensation options for nonmonetary contributions, such contributions often are compensated with equity. When nonmonetary contributions must be figured into the equity distribution equation, the determination of relative ownership percentages is more complicated. For example, when nonmonetary contributions are made for equity in the business, the relative value of that contribution must be determined. In many instances, this value can be difficult to determine. Nonmonetary contributions can take any number of other forms:

- Intellectual property
- An intangible right or relationship, such as a contractual right (e.g., the right to purchase property or a purchase order from a potential customer)
- Sweat equity (providing time and effort without drawing a "salary" or at a reduced "salary"; services provided at market rates are not a contribution to the venture)

- Resources (facilities, distribution, R & D, and management)
- Credibility (e.g., credit, reputation) or access to sources of capital (network)

When the business is founded by more than one entity, the perceived relative values of the contributions that each makes or will make to the venture are typically determinative of the relative percentages of ownership. There are standard techniques for valuation of an existing business. However, in many instances, those standard valuation techniques are not applicable to nonmonetary contributions. In fact, quite often with start-up ventures, the issue of relative value of contributions is simply a matter of the negotiation skills of the contributors. Factors tending to influence the valuation of a nonmonetary contribution include:

- The relevance of the contribution to the venture. For example, there is often a premium for contributing the idea or premise around which the venture is built.
- The intrinsic or demonstrable merit of the contribution. For example, consider the negotiating position of an inventor with an unproven idea versus that of an inventor who can provide a prototype.
- Favorable assessments by independent third parties.
- Legal exclusivity or protections. For example, consider the negotiating position of an inventor with an unproven idea versus an inventor with a patent.
- Uniqueness of the contribution.
- Demonstrable expertise in the technology or processes that form the basis of the venture.
- Relevant experience and track record.
- Relevant contacts and network.
- Reputation and goodwill.
- Efforts already taken that increase the likelihood of success of the venture.

In practice, pure force of personality and skills in negotiations often tend to play a significant and sometimes definitive role in determining the relative equity positions of the founders.

6.6.1 Employee Stock Options

ESO plans are a flexible tool used by many companies to reward employees for performance and attract and retain a motivated staff.[20] In effect, options preserve cash for the business while giving employees a stake in the success of the company. The popularity of ESO plans has seen remarkably steady growth since the 1980s. ESO plans became particularly popular during the days of the dot.com boom and indeed, for a period of time, were an expected component of a compensation package for employees of high-technology companies. Today, ESO plans continue to be a relatively popular method of compensating employees and garnering loyalty for the employer. One survey estimates that over 11.2 million employees are participants in such plans.[21]

In essence, an ESO plan is a contract between employer and employee that gives the employee the right to buy a specific number of shares in the company at a specified price, referred to as the strike price, over a specified period of time, referred to as the exercise period.[22] When the employee exercises the option, he or she will get the benefit of the difference between the fair market value of the stock at the time of exercise and the strike price. In a typical ESO plan, there are specific provisions regarding the employee's eligibility to earn options under the plan and various restrictions on how the option is exercised and that transferability of the stock after the option is exercised. Several common provisions and restrictions are explored next.

6.6.2 Vesting

In the context of an ESO, vesting is a process or condition that must take place before an option can be exercised by the employee. For example, an ESO plan may require uninterrupted employment for a specified period of time (or some other length of service) before the employee is "vested." The vesting condition is most frequently the passage of time, but it may also be other things, such as the occurrence of an event or a milestone in the growth of the company. ESO plans may incorporate simple or elaborate vesting schedules. A typical vesting schedule for employees of a technology start-up is incremental over four years, with the employee having no exercisable rights until he or she has remained with the company for at least 12 months, at which point the right to exercise the option on the first 25% of the shares vests and approximately 2% vest each month thereafter. No further vesting occurs after an employee leaves the company.

6.6.3 Restrictive Clauses

ESO plans typically include restrictions or limitations on the transferability of the stock after the option is exercised to protect the company's interest. For example, an employee may be precluded from making any transfer of stock for a period of time after exercising the option. Or, an employee who is leaving the company may be required by the plan to sell his or her stock back to the company. Some plans may also require that the option or the shares be offered back to company on a right of first refusal basis whenever a stock sale is contemplated to an outside or unrelated third party.

6.6.4 Tax Issues

In the United States, the Internal Revenue Code creates two basic categories of ESO plans: incentive stock options (ISOs), also referred to as qualified stock options, and nonqualified stock options. In order to qualify as an ISO, the plan must meet relatively stringent requirements of the Internal Revenue Code. For example, the strike price cannot be discounted from fair market value, options cannot be available to nonemployee directors, and the exercise period cannot exceed 10 years.[23] In view of the restrictions of the qualification rules, most ESO plans are nonqualified.

In general, exercise of a nonqualified ESO has tax consequences for the employee. The spread between the market price and option price at the time the option is exercised is considered ordinary income to the employee. Qualified ESO plans however qualify for special tax treatment. Neither the grant nor exercise of an ISO is a taxable event for the employee. Instead, the tax is deferred until the employee sells the underlying stock. At that point, any gain that the employee has made on the stock sale is taxed. If the sale occurs at least two years after the option is granted and at least one year after the option is exercised, the income is taxed at favorable long-term capital gains rates. If these holding-period requirements are not met, the spread at exercise is taxed as ordinary income and only the subsequent appreciation of the underlying shares is taxed as capital gain.

Nonqualifying stock options can create anomalous tax consequences for the employee in the event that the employee does not immediately sell the stock after exercising the option and the value of the stock goes down. Exercise of the option is a taxable event, that is, the employee is taxed on the spread even though the actual gain (or loss) is less than the fair market value at the time the option was exercised. On the other hand, exercise of a qualified option is considered a preference item for purposes of calculating the alternative minimum tax.

6.7 CHAPTER SUMMARY

This chapter examined the various legal forms available for a new technology venture: unlimited liability (e.g., sole proprietorship or general partnership) and limited liability (e.g., LP, corporation, or LLC). Limited liability entities are creatures of statute; to form a limited liability entity it is necessary to file certain documents with the appropriate governmental agency, and certain formalities must be followed to maintain the limited liability entity. The particular documents that must be filed and formalities vary depending upon the particular entity. Unlimited liability entities are the default form of business entity; unless the necessary documents are filed and formalities observed for a limited liability entity, the entity will be either a sole proprietorship, or, if there is more than one owner, a general partnership. Even when not required, it is desirable to draft appropriate governing documents, such as partnership, operating and shareholders agreements, to avoid misunderstandings and establish good operating procedures.

In an unlimited liability entity the owners, merely by virtue of being an owner, are *jointly* and *severally* responsible for all liabilities of the entity (including those created by its employees in the course of their employment), whether or not they were personally involved in creating the liability. On the other hand, although an unlimited liability entity will not shield an individual from responsibility for their personal actions, principals, officers and directors do not become personally responsible for liabilities of the entity merely by virtue of their position. They are liable only to the extent that they may lose the value of their investment in the entity; their personal assets not invested in the entity are not at risk. It seems unwise for any business owner today to choose a legal form that includes the disadvantage of unlimited liability.

Each particular form of limited liability entities has advantages and disadvantages depending upon the nature, number, and tax circumstances of the principals participating in the venture, the management structure of the venture, and the venture's capital management plan and its tolerance for regulatory and administrative burdens.

Allocating percentage of ownership between founders (or to an investor) is a function of the relative value of the contributions of the individual founders (or investor) to the venture. In practice, however, determining the relative value of contributions is often a difficult process because standard valuation techniques tend to be inapplicable to start-ups. This is especially true when the founders make nonmonetary contributions, such as IP, resources, and services (sweat equity) to the venture. In any event, it is desirable to have written documentation of the nature and extent of each participant's contribution to the venture, and the consideration (e.g., percentage of ownership) to be provided for that contribution.

KEYTERMS

Authorized stock The maximum number of shares of stock that a corporation is legally permitted to issue, as specified in its articles of incorporation.

Bond A document (debt instrument) evidencing the loan of funds (the purchase price of the bond). The bondholder is periodically paid interest at a fixed rate (coupon), and the principal is to be repaid at a specified date (maturity date). Bonds are commonly used by business (and government) entities to finance particular projects and activities.

Common stock A class of stock that does not have any preference over another class of stock.

Corporate veil The protective "veil" established by the corporate structure, separating personal and business assets.

Keyterms

Corporation A legal entity, created under the laws of a particular jurisdiction separate and apart from its owners.

Co-venture A generic term covering not only actual business entities, such as partnerships, corporations, and so on, but also virtual business entities created through contractual relationships such as licenses and profit-sharing arrangements.

Co-venturers The participants in the co-venture. Also used as a generic (entity independent) term for the principals of a venture (e.g., partners, shareholders, and members).

Co-venturing paradigm An analytical model for a business venture where the business venture is considered a separate and distinct entity from each individual participant or investor (co-venturer) and each co-venturer makes a contribution to the venture in return for some form of consideration.

Cumulative preferred stock A type of preferred stock to which dividends are to be paid on a periodic basis, with a right to be paid any of those periodic dividends that have been omitted in the past, before junior classes of stock can receive dividends.

Dividends A distribution of a portion of a company's earnings, decided by the board of directors, to a class of its shareholders, typically described in terms of the dollar amount each share receives (dividends per share).

Employee stock option plan An agreement giving an employee the right, during a specific time period, to buy a specific number of shares of company stock at a specified price.

Equity A generic term for ownership interest in a business entity.

Foreign corporation In a given jurisdiction, a corporation formed and organized in some other jurisdiction. For example, an Arizona Corporation doing business in Florida would be considered a foreign corporation in Florida.

Founder, or promoter The individual or individuals that recognize an opportunity and are the primary forces initiating the creation of a business entity.

Founder's Stock Stock issued to the founders of the corporation. It may be common stock or preferred stock.

General partnership A business created by default anytime two or more individuals own a business without creating a statutory entity.

Inventor In a technology venture, the individual(s) primarily responsible for conceiving the technology.

Investor An individual or entity that makes a contribution to the business after it has been formed. The consideration for the contribution often includes an equity interest in the business entity. Founders that do not actively participate in the management or operations of the venture are also sometimes referred to as investors.

Issued shares or outstanding shares The stock that is actually held by shareholders of the corporation. The number of outstanding shares will increase if the corporation issues additional shares and will decrease if it buys back its shares.

Jointly and severally liable Where each member of a group is responsible for the entirety of a liability and the liability of an individual member is not limited to that member's pro rata share. The liability can be collected entirely from any one of the members of the group, or from any and all of the members in various amounts, until paid in full.

Limited partnership A legal entity with one or more general partners and one or more limited partners.

Nonparticipating preferred stock Preferred stock that pays a periodic fixed dividend regardless of the profitability of the company.

Operating agreement An agreement between the members of an LLC defining the operating rules of the entity. The operating agreement typically governs membership, management, rights, duties, operation, capital contributions, voting rights, transfer of interest, and dissolution and distribution of income and losses of the company.

Participating preferred stock Preferred stock that pays, in addition to any periodic fixed dividend, the profit interest in the corporation.

Personal holding company Under U.S. tax law (26 USC §542), a corporation with more than 50% of its stock owned by five or fewer individuals and more than 60% of its income (after certain adjustments) coming from dividends, rents, royalties, and so on, and/or certain personal service contracts. Personal holding companies are subject to, in addition to regular corporate tax, a significant penalty tax.

Phantom income Revenue that is reported to the Internal Revenue Service as income for tax purposes, but is not actually received by the entity or individual paying the taxes. For example, retained earnings in a partnership or an LLC taxed as a partnership may be considered income to the individual members even though there is no distribution of the income to the members.

Preferred stock Any stock that has advantageous characteristics over other classes of stock, typically a priority (preferred) claim on the assets and earnings of the corporation than on other classes of stock.

Profit interest A right to a percentage of the profit made on a defined interest base. The interest base can be, for example, the entire profits of the business venture, profits made from a particular line of business, profits made from a particular product, profits made from a particular distribution channel, and so on.

Shareholders agreement An agreement between the shareholders of a corporation, generally analogous to the operating agreement of an LLC, typically governing the shareholder's relationship; management and operation of the company, obligations, privileges, and voting rights of the shareholders; and restrictions on the sale of stock. Shareholders agreements are common in closely held corporations.

Small business corporation A C-corporation where the aggregate consideration (money and other property) received by the corporation for stock does not exceed $1 million.

Sole proprietorship A business created by default any time an individual owns a business without creating a statutory entity.

Treasury stock or unissued shares Shares of stock that have been authorized in the articles of incorporation, but not actually issued.

Vesting A process or condition that must take place before an interest is transferred or before the right or option can be exercised.

Vicarious liability and imputed liability This is where, because of a relationship between the respective entities, one entity is strictly liable for harm or damages caused by another. For example, an employer is vicariously liable for harm or damages caused by an employee acting within the scope of employment.

ADDITIONAL READING

Emerson, R., 2009. Business Law, fifth ed. Barron's Educational Series, Inc., Hauppauge, NY.

Glenn, B., 2013. Business Entity Basics. Zinerva Publishing, LLC.

Harris, L., 2012. Mastering Corporations and Other Business Entities. Carolina Academic Press, Durham, NC.

Lechter, M., 2010. OPM: Other People's Money: The Ultimate Leverage, second ed. TechPress. Phoenix, Arizona.

WEB RESOURCES

The websites below are intended to be destinations for your further exploration of the concepts and topics discussed in this chapter:

http://corp.delaware.gov/: The website of the Delaware Division of Corporations.

http://www.sba.gov/category/navigation-structure/starting-managing-business/starting-business/choose-your-business-stru: A discussion of the various types of entities by the Federal Small Business Administration.

http://nvsos.gov/index.aspx?page=425: The Nevada Secretary of State website.

http://www.sec.gov/: The Securities and Exchange Commission (SEC) website.

http://soswy.state.wy.us/Forms/Publications/ChoiceIsYours.pdf: A comparison of various types of Wyoming entities provided by the Wyoming Secretary of State.

ENDNOTES

1. A "Cooperative" is a business or organization owned by and operated for the benefit of (provides services to) its members, also known as user-owners. Profits and earnings generated by the cooperative are distributed among the members. Cooperatives are often organizations of businesses that pool their resources.

2. A "Nonprofit" is an organization that uses any revenues for its own expenses, operations, and programs; any revenues in excess of expenses are used to achieve its goals rather than being distributed to its owners or trustees as profit or dividends.

3. Advantage Leasing Corp. v. NovaTech Solutions, No. 03–216 (Wis. Ct. App. March 24, 2005) (unpublished opinion).

4. Littriello v. United States, 484 F.3d 372 (6th Cir. 2007).

5. Securities Act of 1933 §§§ 11,12,15.

6. A federal "self-employment tax," analogous to the taxes on employee wages related to the Social Security and Medicare system, is imposed on the net "self-employment income" of anyone carrying on a trade or business as a sole proprietor, an independent contractor, a partner in a partnership, a member of a single-member LLC, or is otherwise self-employed. Under the Self-Employment Contributions Act Tax (SE Tax Act). 26 USC Ch 21. This tax is, however, not applicable to dividends and distributions.

7. California is a notable exception because of fees charged with respect to LLCs. See endnote 16.

8. The Internal Revenue Code, Title 26 United States Code (USC).

9. 26 USC (IRC) §11. As of 2008, the graduated rate was: (A) percent of so much of the taxable income as does not exceed $50,000 (B) percent of so much of the taxable income as exceeds $50,000 but does not exceed (C) percent of so much of the taxable income as exceeds $75,000 but does not exceed $75,000 $10,000,000 (D) percent of so much of the taxable income as exceeds $10,000,000.

10. Because of the deductibility of salaries, the IRS scrutinizes the compensation paid to owner-executives. If the compensation is found to be excessive, the IRS can reclassify it as a dividend and increase the taxable income of the corporation.

11. As of 2008, the tax on dividend income was 15%.

12. 26 USC (IRC) §531 et seq, 26 CFR § 1.531–1 et seq.

13. 26 USC §541 et seq, 26 CFR § 1.531–1 et seq, 26 USC §§541 et seq., 26 USC §11.

14 26 USC §1244.

15 Wyoming Statute, Title17, Chapter 15.

16 Rev. Rul. 88-76. 1988-2C. B. 360, September 2, 1988.

17 26C.F.R. §§ 301.7701-1 - 301.7701-3.

18 In California, a Franchise Tax Board has been established to administer Personal Income Tax and the Corporation Tax. For the privilege of doing business in California, an LLC must pay not only an annual franchise tax (as of 2008, a minimum of $800 per year), but also, if it has total annual income above a specified amount (as of $250,000), it must pay a fee to the California Franchise Tax Board (as of 2008, ranging from $500 to $11,790, depending on the amount of the LLC's total income).

19 26 USC § 83.

20 Employee Stock Option Plan should not be confused with a type of retirement plan referred to as an Employee Stock Ownership Plan (ESOP).

21 The National Center for Employee Ownership survey performed by the National Opinion Research Center (NORC) of the University of Chicago relying on survey information and statistics from the Department of Labor, Bureau of Labor Statistics survey. See details at http://www.nceo.org/library/eo_stat.html.

22 An ESO plan can be part of a negotiated comprehensive employment agreement with an individual employee.

23 The requirements include, among other things:

 1. The option may be granted only to an employee (grants to non-employee directors or independent contractors are not permitted).

 2. The option must be exercised (a) within 10 years of grant and (b) while still employed or no later than three months after termination of employment (unless the optionee is disabled, in which case this three-month period is extended to one year).

 3. The option exercise price must be at least the fair market value of the underlying stock at the time of grant.

 4. The ISO cannot be transferred by the employee (other than by will or by the laws of descent) and cannot be exercised by anyone other than the option holder.

 5. No more than $100,000 in ISOs can become exercisable in any year.

CHAPTER 7

Developing and Implementing the Technology Business Plan

TaskRabbit SUCCESS DUE TO EFFECTIVE PLANNING

Who would have thought that a girl raised in a town so small that it had no street lights and who graduated from a small women's college in rural Virginia where dairy cows were a part of the campus would come up with a disruptive technology? But that is the background of Leah Busque founder of TaskRabbit in San Francisco. Leah built the original TaskRabbit product and created its original bunny logo. The company is a type of eBay for errands for anyone who has a need to find someone to take care of things for them from walking a dog, to cleaning dishes, to waiting in line for the new iPhone 5, to assembling IKEA furniture. Driven by the desire to help people (neighbors) connect with other people (neighbors), Busque wanted to develop a platform that allows people with skills and abilities to be able to share them with others who need them. The company started as RunMyErrand.com with advisor Scott Griffith, chief executive officer (CEO) of Zipcar, in an office space in Zipcar's company headquarters in Boston, Massachusetts. But the company did not get any traction, despite the plan, until Busque was selected to attend Facebook's 12-week startup boot camp run by Dave McLure, an angel investor, in Silicon Valley. During this boot camp, along with 24 other would-be entrepreneurs, Busque built a plan to launch a better version of the company that would fit into the very busy San Francisco lifestyle. In 2009, she met Tim Ferriss, author of *The 4-Hour Workweek*, who embodied the need for her services. Becoming an advisor, Ferriss introduced Busque and her plan for TaskRabbit to venture capitalists at Baseline Ventures and Floodgate. These venture capitalists became the seed investors of the company which still did not position itself totally as a disruptive technology but more as a company that could help people save time by employing others. This changed 2 years later after a venture capitalist in the Founders Fund heard Busque speak at Le Web in Paris. The Founders Fund, founded by an eccentric billionaire, Peter Thiel who co-founded PayPal with Luke Nosek, only backs startups with life-changing disruptive technologies that have the basis of a multibillion-dollar company with a large network effect. The Founders Fund was an early investor in Facebook. In June 2012, TaskRabbit received $13 million in additional Series C financing, $10 million of which was from the Founders Fund.

Today TaskRabbit has about 11,000 employees who have cleared background checks in 10 markets and are willing to do tasks desired by people in their market. Although a nice number, the business plan still needs to evolve to meet the goals established while continuing to attract more

Continued

people. It, as most companies, has refocused several times. One of the first was to make sure men were included after the initial focus on women. Now, the fast growing customers of TaskRabbit are small businesses who want to hire people from part-time receptionists to data-entry people. Competition is intense. On one side are the hire a gofer startups such as Fancy Hands, Gigwalk, and Zirtual. And on the other side there are more focused suppliers such as Lyft (offering services as a cabbie) or Exec (offering services as a housecleaner). Will TaskRabbit succeed and be acquired profitably by some giant like Amazon.com? Only time will tell.

Sources: *https://www.taskrabbit.com/, http://www.fastcompany.com/3012593/taskrabbit-leah-busque#1*

7.1 INTRODUCTION

A business plan or its modification is an important part of the new venture creation process and is often called the entrepreneur's roadmap for implementing the vision and the strategy of an entrepreneur such as Leah Busque. You would not think of building a house or a new product or service without a plan, so why do some entrepreneurs believe they can start something as complex as a new business venture without a plan?

A business plan is particularly needed in today's complex, global, hypercompetitive rapidly changing technology-driven environment. Today's environment, perhaps more than ever before, requires an understanding of the complexity of the inputs and outputs so that a new enterprise can be effectively established, launched, modified, and grown particularly for a technology entrepreneur.

This chapter explores aspects of writing a plan, the purpose of the business plan, the elements of a solid technology business plan, the methods for estimating and forecasting costs, and the development and update of the business plan.

7.2 WRITING THE BUSINESS PLAN

Creating and building a successful technology venture requires effective planning. Although indeed the process of thinking this through and developing the necessary strategies is important, the process and the discipline required to putting this in writing makes the thinking process more effective and give the venture a better opportunity for success. Often something conceptualized in the mind of a technology entrepreneur does not make sense once it is committed to writing.

A frequently asked question by a technology entrepreneur is how long a **business plan** should be (How many pages should it have?). Although there is no strict answer to this question, oftentimes it is too long.[1] The main body of the business plan, which is discussed later in this chapter, should be about 30–40 pages.[2] This does not include exhibits or appendices which contain additional market statistics, the actual patents, the résumés of the management team and scientists, and other material that can add another 20–100 pages of material. And, a two-page executive summary is needed to use as a door opener if capital needs to be raised.[3] One factor that impacts the length of the plan is its purpose, which is discussed in the next section.

Even though a technology entrepreneur needs to write the plan itself, he or she should get help with editing and laying out the plan in the most favorable way. No matter how good the content, the way it looks and is presented affects its evaluation. Any spelling errors, mathematical errors, numbers not making sense, or poor graphics impact the evaluation of the plan to some extent. And, a knowledgeable investor will easily discern whether the plan was written, at least initially, by the technology entrepreneur or prepared externally by someone hired to do so.

MINI-CASE
Startup Valley Provides Crowdfunding Platform

StartupValley is a crowdfunding website for technology startups. It is a similar platform to Kickstarter.com, but with a more narrow focus on technology start-ups. Any entrepreneur can present his or her business plan for a technology start-up to share with potential investors. Instead of relying on enough funds to reach its set funding goals, startup companies that successfully gain their needed investment are charged a flat 5% by the service. The site is currently in Beta pending the finalization of SEC regulations. Anyone making $200,000 within the last two years may be an accredited microinvestor in a technology venture of their choice. Once SEC regulations are finalized, investment will be open to unaccredited investors as well.

7.2.1 Purpose of the Plan

Besides providing a roadmap or direction for the new technology venture, a business plan can have several specific reasons for being written. The first and by far the most usual reason is to obtain financing.[4] Most technology entrepreneurs do not have all the resources needed to develop and launch a technology venture. External financing is needed whether provided by financial institutions, private individuals, venture capital firms, or private equity firms. This nonpublic capital called **enterprise capital** is essential to start and grow technology ventures. Each of these sources of capital, which are discussed in Chapter 8 of this book, needs a business plan in order to provide the capital needed. Keep in mind when writing a business plan for this purpose the private investor or venture capitalist who is receiving it may receive 50–100 plans each week. Of 100 business plans received, only 10 will not be discarded through the reading of the two-page executive summary. And, of these 10 remaining, only one or two will receive funding. This means that the business plan of a technology entrepreneur must be crafted to be a good selling document to survive the scrutiny that occurs before funding.

Another reason for writing a business plan is to determine all the resources needed including financial. This is usually an end result of a good technology business plan. A careful review of the current existing resources of the technology entrepreneur and the enterprise and the resources needed at various stages of development such as financial, human, technological, supply, or distribution provides insight into the plan that must be in place to make sure these resources are available in a timely fashion.[5] Potential suppliers of the resources need to be specifically identified along with a solid approach to each supplier.

The third reason for writing a business plan is to provide a direction for the new venture. In the development, launch, and growth of a technology venture there are many different directions and opportunities available.[6] A good business plan allows these to be more easily evaluated and best direction or opportunity selected.

Being able to evaluate the results of the venture is the fourth reason for writing a business plan.[7] Even though this may not be the most popular reason for the technology entrepreneur to write a business plan, particularly when the results may not be favorable, the business plan points out problems that need attention and focus. Pro forma income statements discussed in the next section are forecasted numbers which eventually become actual numbers once the plan is implemented and the time period reached. There is always a positive or negative deviation between the numbers forecasted and the numbers achieved.[8] When this difference is large and negatively affects the technology venture, corrective action is needed. Because a technology entrepreneur usually never has enough time or money, by focusing on these deviations attention can be paid to the most critical problems. This is called **management by deviation**.

The final reason for writing a business plan is to obtain a co-venturer.[9] A business plan lays out the plan for the business for perspective co-venturers and helps them decide whether to be involved. Whether obtaining a member for the board of advisors, a firm in the distribution channel, a supplier firm, or a firm that could be a joint venture partner, the job is easier when a business plan has been prepared. Most seasoned entrepreneurs and business persons will not consent to be a member of any advisory board of a technology venture without having evaluated at least a preliminary business plan.

KEY POINT

Business Planning Software Can Help in Plan Writing

LivePlan is business plan writing software that has been used by over 200,000 entrepreneurs. The software asks questions about aspects of the business and creates financial spreadsheets and illustrated charts from the answers. It includes a feature to create a lean business plan pitch that encourages start-ups to stay adaptive and nimble in their early stages in order to refine the model when core assumptions are validated or dismissed. The feature further helps communicate the highlights of the business in a 60-second pitch and creates a customizable web link to get noticed by potential investors.

7.3 ELEMENTS OF A BUSINESS PLAN

Although there are some variations, most technology business plans have the same elements.[10] as indicate in Exhibit 7.1. These can be grouped into three main general sections:

MINI-CASE
Socialcast Founder Raises Capital with Five Slides

Socialcast is a social network platform that allows you to connect with people and online resources needed to get work done. Its founder, Tim Young, explained how he obtained investments for his company. He used a lean presentation of only five slides shown off of his laptop instead of a projector to offer a more intimate presentation to investors. He detailed guidelines for keeping the summary presentation relevant and efficient. He further researched their backgrounds and understood who he was pitching his business to. By having a well-organized business plan and efficient presentation, and understanding the investors he was pitching to, he was able to raise $10 million through three rounds of funding in just 1 year.

Sources: http://techcrunch.com/2010/11/02/365-days-10-million-3-rounds-2-companies-all-with-5-magic-slides/;

7.3.1 Section 1

Section 1 contains the title (cover) page, table of contents, and **executive summary**. The title (cover) page is an important part of every business plan because it has:

- The company name, address, telephone, fax, e-mail address, and website
- Name and position of each identified member of the management team
- The purpose of the plan, the amount of money needed, and funding increments
- At the bottom of the title page: "This is confidential business plan number ____." Put in a low number and then keep track of when and who received this numbered plan for a 30-day follow-up (Recognize, however, that this legend does NOT in itself make the business plan confidential. See Chapter 5).

The first page after the title (cover) page is the table of contents. This follows the usual format and lists at least the major subsections in each section and the corresponding page numbers, as well as each figure, table, and exhibit. Preferably each major subsection and smaller subsections should be labeled as 1.0, 1.1, 1.2, 2.0, 2.1, 2.3, and so on. The executive summary precedes the numbering and therefore either has no numbers, smaller letters, or Roman numerals. The tables and figures should have a separate list as should the exhibits (appendices).

The last item in Section 1, following the table of contents, is the all-important two-page executive summary. This is by far the most important document in the business plan because it is the screening section. Many readers, especially potential providers of capital, never read beyond the executive summary. One head of a very successful venture fund, who is now managing his eighth fund,

Section 1: Title page
        Table of contents
        Executive summary

Section 2:
1.0 Description of business
- Description of the venture
- Product(s) and/or service(s)
- Mission statement
- Business model

3.0 Technology plan
- Description of technology
- Technology comparison
- Commercialization requirements

5.0 Financial plan
- Sources and applications of funds statement
- Pro forma income statement
- Pro forma cash flow statements
- Pro forma balance sheet

7.0 Organization plan
- Form of ownership
- Identification of partners and/or principal shareholders
- Management team background
- Roles and responsibilities of members of organization
- Organizational structure

9.0 Summary

2.0 Description of industry
- Type of industry
- Future outlook and tends of industry
- Analysis of competitors
- Trends and market forecasts

4.0 Marketing plan
- Market segment
- Pricing
- Distribution
- Promotion
- Product or service
- Sales for first 5 years

6.0 Production (outsourcing) plan
- Manufacturing process (amount subcontracted)
- Physical plant
- Machinery and equipment
- Suppliers of raw materials
- Outsourcing aspects

8.0 Operational plan
- Description of company's operation
- Flow of orders and goods
- Exit strategy

Section 3: Appendices (exhibits)
- Exhibit A – résumés of principals
- Exhibit B – market statistics
- Exhibit C – market research data
- Exhibit D – competitive brochures
- Exhibit E – competitive price lists
- Exhibit F – leases and contracts
- Exhibit G – supplier price lists

EXHIBIT 7.1
Elements of the business plan.

indicated that he receives about 1500 business plans a year, discards 1400 based on the cover page or executive summary, and, of the remaining 100, discards 80 based on the initial one- to two-hour examination. Of the remaining 20, about four to six will receive a term sheet and probably an investment from his fund. So the executive summary needs to be really well written to invite further reading of the business plan. Some capital providers only want to initially have a copy of the executive summary. If this passes the first evaluation, then the entire business plan is requested.

The executive summary should have the name of the company and address at the top just as it appears on the title (cover) page. It should begin with defining the nature and size of the problem existing. In the case of TaskRabbit, the problem is a large one—the need for someone to do a task on a personal basis. The larger and more critical the problem, the more interest there will be on the part of investors and others.

Your proposed solution should follow this problem.[11] Again, for TaskRabbit, this is making available several individuals who have had a background check and are capable and willing to do the task on a fee basis. In this section, all competitive ways to solve the problem should be discussed, showing the uniqueness or the unique selling propositions of your solution. These would include: personal, capable, willing, certified, and having passed a background check for TaskRabbit.

Following the solution is the size of the market, trends for at least 3–5 years, and future growth rate. The market needs to be large enough and accessible to deliver the sales needed for the profits and returns expected by investors. The need for help for busy professionals to provide free time makes for a very exciting perspective for TaskRabbit.

The entrepreneur and team who will deliver these sales and profits then need to be described. The education, accomplishments, and industry experience of each known member of the top management team needs to be described. The individuals involved in TaskRabbit are very noteworthy and include: CEO Leah Busque and advisors Tim Ferriss (author of *The 4-Hour Workweek*) and Peter Thiel (founder of Founders Fund).

The team needs to deliver sales and profits, which should be summarized over a five-year period in the following format:

	Year 1	Year 2	Year 3	Year 4	Year 5
Total revenue					
Cost of goods sold					
Gross margin					
Operating expenses					
Profit (loss) before taxes					

These numbers are taken directly from the pro forma income statement summary in the financial plan in Section 2. Note the exact calendar year is not used, but rather year 1, 2, 3, 4, and 5, with 1 indicating the first year of company operations after the investment is received.

The two-page executive summary closes with a statement of the resources needed, the increments of capital accepted, and contact information. An example two-page executive summary is provided in Appendix 7.1.

7.3.2 Section 2

Following the executive summary, which is the end of Section 1 of the business plan, Section 2 starts on a new page with its first part: **Description of the Business**.

7.3.2.1 *Description of the Business*

In this section, the nature of the venture is described to provide an understanding on how the venture will operate and deliver the products or services to solve the problem identified.[12] Information on the products or services should be in enough detail to be easily understood; this will be expanded on in Section 3 if it is a technological product or service that employs a new technology and in the product section of the marketing plan (Section 7.3.2.4). The mission statement of the company should be described as well as the business model[13]—the entire picture of how the company does business—and if this business model significantly differs from the way business is presently being done in the industry.

7.3.2.2 *Description of Industry*

This section discusses the type and size of the industry, the industry trends for the last 3–5 years, future outlook and growth rate, and a thorough analysis of competition presently fulfilling the same need that the new idea does.[14] This is a large section with significant use of data from authoritative sources. Sometimes there is so much data that only part of it appears in the body of the plan, and the rest appears in an Appendix at the end of Section 3. Graphs, charts, histograms, and other graphics should be used to thoroughly explain the industry, its growth projection, and the competitors. A graph showing the market growing is important based on the trends of this market to date. The market, the market segment, and target market for the first year will be further discussed in the first section of the marketing plan.

7.3.2.3 *The Technology Plan*

Following the description of the industry is the **Technology Plan**. Some business plans where there is not a technological advancement in the product or service being offered would not have a technology plan. For example, one author founded a rainbow decal and sticker company with no significantly new technology so there was no technology plan in the business plan of the company. Whenever the product or service has a patent, there will always be a technology plan because the patent adds value to the venture. A general rule is if you are having a hard time deciding whether to have a technology plan, then put one in because it is better to have one than not. The technology plan describes the state of the technology presently available and how the new technology revolutionizes the way things are done. The patent or patent filing should be discussed and the document itself put in an Appendix (exhibit) in Section 3.

7.3.2.4 The Marketing Plan

The **marketing plan**, the next section, begins with a discussion of the market segment and target market for the product or service.[15,16] It defines, usually through using one or more segmentation techniques, the most appropriate market and its size. Of the many available segmentation techniques[17] (demographic, geographic, psychological, benefit, volume of use, and controllable market elements), the two most widely used ones particularly for technology entrepreneurs are demographic and geographic because this is the way secondary data is published. If the venture is a business to consumer (B2C), the most important market data is the demographics of the selected geographic market. The most widely used demographic variables to determine the size of the market and a typical customer profile are age, income, and gender. For a business to business (B2B) venture, the business market needs to be identified using the classification (country) system of the country for the industrial (business) customer being served. The NAIC code in the United States, the SIC code in Korea, and the SIC code in China each uses a numbering system to classify each industry and specific products and services in that country. A sum of all the output of these numbers is the gross national product of the country. This procedure will provide the trends, size, and growth rate of the particular industry market that can be used to develop the typical customer profile.

Following the delineation of the target market, a marketing plan needs to be developed to successfully reach and sell to that target market. The marking plan has four major areas—product or service, price, distribution, and promotion (see Exhibit 7.2). These are thoroughly discussed in Chapter 10, "Marketing Your Venture." The product or service part describes the characteristics and quality of the offering, the assortment of items to be offered, the guarantee, any servicing provided if needed, and the packaging. The latter can be very important for technology entrepreneurs in the B2C market because it can be a major area of distinctiveness as well as a sales tool in the distribution center(s) used.

The second variable, price, is closely related to the product or service, particularly the quality level. The price, the worst executed of the marketing areas by entrepreneurs, needs to reflect the competitive prices, the costs, and the consumer reaction to the price. If a distribution system is used, then there will be a chain of markups on the cost.

The distribution area has two major aspects: distribution channels and physical distribution. The distribution channels include entities handling the product such as retailers, wholesalers, or representatives. The physical distribution or logistics is becoming an increasingly important area and includes transportation, storage (warehousing), and inventory.[18]

Product/service
- Quality
- Assortment
- Guarantee
- Servicing (if needed)
- Package

Price
- Price/consumer reactions relationships
- Price/cost relationships
- Price/competitive reactions relationships

Distribution mix

Distribution channels
- Retailers
- Wholesalers
- Representatives

Physical distribution
- Storage
- Inventory
- Transportation

Promotion
- Advertising
- Personal selling
- Publicity
- Sales promotion
- Social media

EXHIBIT 7.2
Elements of the marketing plan.

The final area of the marketing plan is the promotion area, which is composed of advertising, personal selling, publicity, sales promotion, and social media. The latter three are particularly important for technology entrepreneurs because they can be used to produce multiple exposures cost effectively. Social media, including the website of the new venture, is a particularly useful part of the promotion area. A marketing budget needs to be prepared for the first year indicating where the money will be specifically allocated to promote the company and achieve the initial sales of the first year. This first-year sales figure concludes the marketing part of the business plan and is a good start for the next section—the financial plan.

7.3.2.5 The Financial Plan

The **financial plan**, the next part of Section 2, focuses on a discussion of the created statements indicated in Exhibit 7.3. The financial information contained in the financial plan consists primarily of these 12 financial statements. All but one of these become actual statements after the business is launched. Although these statements have the same content, they are different from the actual ones filed in that they are forecasted—pro forma—statements which will, upon the end of the time period, become actual statements.[19] How to determine estimates (forecasts of revenues and costs) are discussed in the next section of this chapter. The one new statement is the first one—the sources and

> **Exhibit 7.3** Financial Statements in the Business Plan
>
> **Sources and Uses of Funds Statement**
> Pro forma income statement—5-year summary
> Pro forma income statement—first year by month
> Pro forma income statement—second year by quarter
> Pro forma income statement—third year by quarter
> Pro forma cash flow statement—5-year summary
> Pro forma cash flow statement—first year by month
> Pro forma cash flow statement—second year by quarter
> Pro forma cash flow statement—third year by quarter
> Pro forma balance sheet—year 1
> Pro forma balance sheet—year 2
> Pro forma balance sheet—year 3

uses of funds statement—which describes how much money is needed (uses) and where it will come from (sources). The uses part often includes money for renovations, inventory, working capital, and reserve for contingencies. Each use statement will include working capital—the money needed until the venture positively cash flows, the point in time when the revenues from operations exceed the cost of operations. Sources of money will always include the entrepreneur and usually friends and family. The other sources of finance include banks, private investors, venture capitalists, and grants, which are discussed in Chapter 8, "Raising Capital."

7.3.2.6 The Production Plan

Following the financial plan is the **production or outsourcing plan** that indicates how the offering will be developed and produced. Some service ventures will not have this part in their business plan. Each individual cost needs to be specified so that an understanding is provided of the actual costs involved in the final offering and how much this can be reduced through economies of scale. All suppliers or outsourcing firms should be described in detail.

7.3.2.7 The Organizational Plan

The **organizational plan** discusses primarily two aspects of the venture: the form of ownership and lines of authority and responsibility. The selection of the general ownership form is country specific, but needs to take into account several aspects. These include taxation, number and location of investors, liability issues, and number and type of employees' perceived fringe benefits that will be paid.

Generally, the ownership form selected should have the lowest possible tax consequences and minimum liability. In some countries, there are also liability issues with the hiring and firing of employees and with closing the company.

Also, in some countries a foreign (resident) partner may need to have a specific ownership position, which in some cases can be over 50%. These are the overall organizational structures in the United States:

Individual legal entities:

- Proprietorship
- Partnership

Organization legal entities:

- LLC
- SC
- C-Corporation
- Professional Corporation
- Not-for-Profit Corporation
- Hybrid Corporation

Due to the organization becoming the legal entity in terms of liability, an entrepreneur should not establish his or her company as a proprietorship or partnership. And, due to the tax laws, most entrepreneurial companies in the United States are started as an LLC or SC.

The second aspect—lines of authority—can best be viewed in terms of an organizational chart. Although each functional area needs to be specified with a description of the duties and responsibilities, most of these will not have a specific person mentioned until the company becomes operational and sometimes, in terms of a chief financial officer and human resources manager, not until years after the start-up. The founding individuals of the company should be in the positions they are most capable of. Of course, the new venture needs a CEO/president, someone responsible for the operation of the venture.

7.3.2.8 The Operational Plan

Following the organizational plan is a short section—**the operational plan**. This describes in detail how the company will operate, including the flow of goods and orders. An important aspect discussed here is the exit strategy by which investors will get their equity and a return on equity hopefully in a 5–7 year period of time from the initial investment. There are only three mechanisms for having the capital to provide this exit and return desired: (1) generating retained earnings of the venture; (2) selling to another financial institution or firm; or (3) going public and being a publically traded company. The most likely exit avenue is selling to another company and, if this is mentioned, then three to four likely exit companies need to be identified and discussed. Section 2 concludes with a brief summary that completes this section of the business plan.

7.3.3 Section 3

Section 3 contains all the backup material to support areas in Section 2. This includes secondary support data, any research data, contracts or leases, the patent document, and most notably the résumés of the entrepreneur and any known members of the management team. Nothing new should be introduced in this section.

7.3.3.1 *Estimating (Forecasting) Revenues and Expenses*

Coming up with an estimate of revenues, expenses, and start-up costs is not an easy task for the technology entrepreneur, particularly an estimate based on a disruptive technology that is very unique and new. If the technology entrepreneur has worked in the industry, which is usually the case, he or she already has at least a sense of the possible volume of sales. Remember, this is usually optimistic because a new venture will typically not be able to achieve the level of sales of other companies already in the industry for a period of time. This is similar when it comes to expenses. The knowledge and experience of the technology entrepreneur as well as that of his or her personal network of people who can be trusted is the basis for the estimates of the sales and expenses.

7.3.3.2 *Estimating Sales (Revenue)*

The sales forecast should be calculated first and is an appropriate way to end the marketing plan as a segue into the financial plan, which typically follows in most business plans. With a brand new or breakthrough technology product, the estimate is often based on market data gathered by others, on comparison of the sales of similar products, and on the opinions of "experts." As such, these estimates are difficult to make and often not very reliable.[20]

To improve the sales estimates for the first year, the technology entrepreneur should first think in units (numbers) of sales, not dollars or other currency sales. The unit sales multiplied by the price in currency such as US$ results in the revenue. Scenario and sensitivity analysis are useful tools for predicting consumer product or service (B2C) sales as well as industry product or service (B2B) sales.

To forecast a new consumer product or service, the technology entrepreneur should find any products or services that presently fill the same need and determine to the extent possible their sales. This can be facilitated through using the country code of the product (e.g., NAIC code in the United States and SIC code in China) to get total product category sales and any specific sales figures in government and trade publications. Care must be taken to make sure there is no significant difference occurring between the amount of product sold to the distribution (sell-in) versus the amount of product purchased at the distributor by the consumer (sell-through).

Once the sales in units for the first year for the new technology product or service have been determined, then future years (2, 3, 4, and 5) can be determined using as a basis a revision of the growth rate of the industry, which can be found in government publications, trade journals, and trade associations.

Estimating sales for technology industrial products or services (B2B) is somewhat easier than estimating sales for a new B2C product or service because the market is not as heterogeneous or geographically dispersed. Talking with experts or several potential customers, obtaining sales figures from products or services filling the same need, and talking with distributors can provide a good basis for estimating the unit sales in the first year.

7.3.3.3 Estimating Expenses

Once the unit sales for the first year have been determined as well as the unit sales for years 2, 3, 4, and 5, it is usually easier to determine expenses.[21] The most important and usually the most difficult expense to determine is the initial cost of the technology product. This is particularly the case when it is a radically new breakthrough innovation. Estimates from trusted engineers or outsource manufacturers can provide solid information for determining costs of the initial production that needs to take into account economies of scale of purchasing, manufacturing, and management knowledge that will reduce the cost in subsequent years. This cost of goods sold should be figured on a per-unit basis, which then can be multiplied by the number of units sold to obtain the cost of goods sold for the time period. Even a technology service can have a figure for cost of goods sold, which is preferable for control purposes, by determining the time-equivalent expenses involved in delivering the service.

The other general selling and administration expenses (operating expenses) are not as difficult to obtain and usually include such things as management salaries and fringe benefits, employee salaries and fringe benefits, vehicle lease and costs, packaging costs, distribution costs, lease/maintenance expenses, supplies, depreciation or any owned equipment, promotion expenses, travel expenses, manufacturer representative expenses, telephone expenses, utilities (water and electricity) expenses, office supplies, and insurance. Keep all financial spreadsheets and support figures as clean and as clear as possible to provide a rich understanding of the expenses of the new venture so that these can be effectively managed.

7.3.4 Business Plan Development and Update

The business plan is a very important document both for providing direction to the new venture and as a tool for raising financial resources. It is important that it be well written and edited. The best way for a technology entrepreneur to proceed is to write down all the information in a draft format and then go back and rewrite it. Keep in mind during this process the audience for your plan, and arrange the material in a way, such as the one suggested, that items flow smoothly from start to finish. Clear and concise writing is needed, and all numbers need to be consistent. If possible, have a friend or colleague critique the final business plan. If needed, you can always pay a professional writer to edit the plan to make sure the plan text flows smoothly.

Besides being a blueprint for the launch, direction, and growth of the venture for the first five years and a presentation of a complete, comprehensive picture of the business, the business plan is a living document. It is based on estimates and assumptions on what the technology entrepreneur expects will happen once the business starts. Of course, it is impossible for any technology entrepreneur to exactly predict the future. Parts of the business plan, particularly the sales, costs, and financial projections, will not be right on the mark. A technology entrepreneur needs to review and update the business plan periodically the first year and then annually thereafter. In this way, the business plan reflects what the business is actually doing and where it is going and becomes a living guide for the technology venture.

7.4 CHAPTER SUMMARY

Every new venture needs a business plan to set the direction for the firm and to obtain financial resources. The essential elements of a business plan are contained in three sections, with the main elements being in Section 2. The most important document in the plan is the executive summary because most business plans are not read beyond this by potential investors.

Each business plan needs to be well written and organized and address as many anticipated questions as possible. It needs to flow smoothly and consistently without errors so that the reader has a clear understanding about the details and future success of the new venture.

KEYTERMS

Business plan Written document describing all relevant internal and external elements and strategies for starting a new venture.

Enterprise capital External financing provided by financial institutions, private individuals, venture capital firms, or private equity firms.

Management by deviation Corrective action needed by focusing on these deviations when there is a large difference between the numbers forecasted and the numbers achieved.

Executive summary Highlights the key points in the business plan.

Description of industry The type and size of the industry, industry trends, outlook and growth rate, and analysis of the competition.

Description of the business Provides complete overview of the product(s), service(s), and operation of the new business.

Marketing plan Describes market conditions and strategy related to how the product(s) and/or service(s) will be distributed, priced, and promoted.

Technology plan Describes the state of the technology presently available and how the new technology will revolutionize the way things are done.

Financial plan Projections of key financial data that determine economic feasibility and necessary financial investment commitment.

Production plan Details how the product(s) will be manufactured.

Organizational plan Describes the form of ownership and lines of authority and responsibility of members of the new business.

Operational plan Details how the company will operate.

ADDITIONAL READINGS

Cassar, G., 2009. Financial statement and projection preparation in start-up ventures. Accounting Review 84 (1), 27–51, The article details a study that explores the elements of preparing financial statements and financial forecasting for start-up entrepreneurs. It explains the importance of financial forecasting and sales projections for start-ups in high-tech industries.

Layton Turner, M., 2012. Short but sweet. Entrepreneur 40 (10), 105–9, This article describes the trend toward short business plans over lengthier plans for new business ventures. It provides advice for drafting short business plans as well as the role of interested investors in embracing shorter business plans.

McAdam, M., Marlow, S., 2011. Sense and sensibility: the role of business-incubator client advisors in assisting high-technology entrepreneurs to make sense of investment readiness status. Entrepreneurship & Regional Development 23 (7/8), 449–68, The article describes the problem of professional client advisor's lack of expertise in the field of technology. It examines how entrepreneurs starting ventures in the technology industry and business-incubator client advisors can combine their respective expertise to create a business plan.

Shelters, D., 2013. Start-Up Guide for the Technopreneur: Financial Planning, Decision Making and Negotiating from Incubation to Exit. John Wiley and Sons, Singapore, This book is a guide to fundraising and financial planning for the technology entrepreneur. It gives advice on business planning and strategic decision making relevant to the fast-paced technology industry.

Wilkins, P., 2012. The foundations of a successful strategy. EMDT: European Medical Device Technology 3 (3), 30–1, The article describes areas that medical technology start-ups need to focus on to ensure future success. It addresses the necessity of a timely business plan and how companies must respond to changes in investor strategy.

WEB RESOURCES

www.bplans.com: This site provides a wide sample of business plans in a number of technology and other industries.

http://www.sba.gov/smallbusinessplanner/index.html: This website has been set up by the U.S. SBA. It offers a number of references and other useful services to small and entrepreneurial business ventures.

www.businessplans.org: Similar to the bplans website, this site has a number of sample business plans and useful templates for creating your own business plan.

http://tech.seas.harvard.edu/home: This is the home page for the Technology and Entrepreneurship Center at Harvard University. It is a good starting point for understanding the relationship between high technology and entrepreneurial success in the global economy.

ENDNOTES

1. Layton Turner, M. 2012. "Short but Sweet." *Entrepreneur* 40(10): 105–109.
2. 2010. "Writing a Thorough Business Plan." *Wearables*, 14(3): 29.
3. Kwicien, J. 2012. "Put It in Writing." *Employee Benefit Adviser*, 10(1): 30.
4. Boni, A.A. 2012. "The Pitch and Business Plan for Investors and Partners." *Journal of Commercial Biotechnology*, 18(2): 38–42.
5. Cordeiro, W.P. 2013. "Small Businesses Ignore Strategic Planning at Their Peril." *Academy of Business Research Journal*, 3: 22–30.
6. Enman, C. 2013. "Plan for Growth; It Won't Happen by Accident." *SDM: Security Distributing & Marketing*, 43(5): 42.
7. Fernández-Guerrero, R., L. Revuelto-Taboada and V. Simón-Moya. 2012. "The Business Plan as a Project: An Evaluation of its Predictive Capability for Business Success." *Service Industries Journal*, 32(15): 2399–2420.
8. Benson, B.W., W.N. Davidson III, W. Hongxia and D.L. Worrell. 2011. "Deviations from Expected Stakeholder Management, Firm Value, and Corporate Governance." *Financial Management (Wiley-Blackwell)*, 40(1): 39–81.
9. Pollock, J. and D. Sumner. 2012. "A Close Eye on Business Partners." *Internal Auditor*, 69(6): 43–46.
10. Ross, F. 2012. "Key Elements of a Business Plan." *Green Industry PRO*, 24(4): 28–29.
11. 2010. "Are You Ready to Start Your Own Business?" *T+D*, 64(7): 21.
12. Tifft, V. 2013. "Synchronizing Your Business." *Smart Business Akron/Canton*, 23(7): 7.
13. Sweeney, D. 2013. "What's Your Mission?." *Smart Business Orange County*, 8(9): 8.
14. 2013. "Key Findings from the 6th Annual Industry Trends Survey." *Trusts & Estates* 152(1): 1–8.
15. 2013. "Marketing Plan Template: Exactly What to Include." *Promotional Marketing*, 3.
16. Darlington, H. 2013. "The Marketing Plan: Part 2." *Supply House Times*, 56(3): 42–46.
17. Viselgaitė, D. and M. Vilys. 2011. "Peculiarities in Construction of Segmentation Models: Theory and Practice." *Business, Management & Education/Verslas, Vadyba Ir Studijos*, 9(2): 171–184.
18. Naughton, D.M. 2012. "How to Sell to the World's Biggest Company." *Harvard Business Review*, 90(7/8): 106–107.
19. Cassar, G. 2010. "Are Individuals Entering Self-employment Overly Optimistic? An Empirical Test of Plans and Projections on Nascent Entrepreneur Expectations." *Strategic Management Journal*, 31(8): 822–840.
20. Fernández-Guerrero, R., L. Revuelto-Taboada and V. Simón-Moya. 2012. "The Business Plan as a Project: An Evaluation of Its Predictive Capability for Business Success." *Service Industries Journal*, 32(15): 2399–2420.
21. Thomas, B.G. and S. Bollapragada. 2010. "General Electric Uses an Integrated Framework for Product Costing, Demand Forecasting, and Capacity Planning of New Photovoltaic Technology Products." *Interfaces*, 40(5): 353–367.

PART 3

Into the Breach

CHAPTER 8

Capital and Capital Sources

INVESTOR CAPITAL FLOWING TO HYDRO-POWER

Small investors have discovered the potential for cash generation that comes from power generation. In particular, hydro-electric power driven by the flowing water in a river, stream, or from a waterfall has begun to attract intrepid small investors who sell their power to the local grid. Colorado investor Sam Perry bought a derelict hydro plant on Washington's Burton Creek for $150,000 in December 2012. He expects that he can generate 480 kilowatts per hour, enough to sell up to $400 per day or $12,000 per month for as long as the Creek continues to run. No small feat, however, was the work required to refurbish the plant that had been out of commission for several years. Mr. Perry is one of a new breed of young investors who have been captivated by green energy, sustainability, and the environmental movement. Glenn Phillips bought his Sheep Creek Hydro plant and today it generates 1,600 kilowatts/hour at peak capacity, and generates checks from the local utility that average $20,000 per month. Investors are betting that utility requirements to have a certain percentage of their capacity generated from "green" sources will continue to generate demand for their hydro power. In the meantime, however, investing in hydro power is paying off for some. For now, Mr. Perry says, "Cash flow is entirely dependent on water flow".

Source: Adapted from Joel Millman, "Bargain Hydropower Plants have Small Investors Chasing Waterfalls," The Wall Street Journal, October 4, 2013.

8.1 INTRODUCTION

All new ventures need some cash resources to start. Technology ventures are no exception. In fact, some types of technology ventures require more capital than any other kind of venture. For example, Tesla Motors required a huge outlay of original capital to begin building its brand of electric vehicles. On the other hand, some technology ventures can be launched with very little capital. Successful technology ventures such as Dropbox and Airbnb were launched via the technology accelerator Ycombinator with less than $30 thousand of investor capital.

Technology entrepreneurs have a number of options when it comes to how and where they obtain the capital they need to launch and operate their ventures. Capital needs change over time, and the potential sources of capital also change as the venture grows and matures. The pros and cons, the benefits and costs of capital differ depending upon the particular type and source of capital raised by the venture.

This chapter explores the various sources of capital that are available to the technology entrepreneur. It delves into managing capital needs over time, the factors to be considered in developing a capital management plan for a technology venture, various fundraising tools and techniques for raising capital, as well as alternatives to raising debt or equity capital.

8.2 THE CAPITAL MANAGEMENT PLAN

Consider capital formation in the context of the co-venturing paradigm: Each co-venturer makes a contribution to the venture in return for some form of consideration. When a capital contribution is made (unless a gift), some form of consideration must paid by the venture for that contribution. In other words, raising capital from external sources comes with a cost; the consideration given in return for the contributed capital is a cost to the venture (and, specifically, to the pre-contribution equity owners of the venture). That cost can be a percentage of equity ownership (diluting the percentage of ownership of the pre-contribution principals) or in the form of an inflexible obligation to make payments to service debt (irrespective of the success and cash flow needs of the venture) along with restrictions on the activities and capital management of the venture.

As we will discuss, there are a variety of funding types and sources available to the technology entrepreneur. However, in practice, depending upon the circumstances, some forms of capital may be easier to obtain or less costly than others. At a given stage of development, a venture may not meet the criteria for certain types of funding sources. And, even if available, the terms under which capital is offered may not always be acceptable to the venture. In any event, the nature, amount, and timing of capital infusions to a venture can have profound effects on the ownership, control, and capital management of the venture. It is imperative that the technology entrepreneur, (particularly as the founder of a venture) develop a plan to minimize the impact of raising capital on retained ownership and control, and the ability to manage cash flow in the venture. This capital management plan should address four basic issues:

- Whether to **self-fund** or seek capital from external sources;
- If external capital is sought, what type of capital (e.g., debt or equity) to seek;
- When to seek capital; and
- How much capital to seek.

Self-funding, if feasible, permits you to avoid the cost of external capital, as well as various potential securities laws considerations. The issue with self-funding, however, is whether or not sufficient funds can be amassed, and if so, whether the funds can be amassed in time to take advantage of windows of opportunity. In addition, even if self-funding is a viable option, there are a variety of reasons why the principals of the venture may wish to use external capital: to share the risk entailed in capitalizing the venture; to leverage their funds to permit them to pursue other opportunities; or to take advantage of particularly advantageous valuations of the venture (an investor will, occasionally, adopt a greater than market valuation because of synergies between the venture and others of their holdings). Of course, in many (if not most) situations, self-funding is simply not a realistic option.

Once a venture determines to seek external capital, it must then determine what type of capital to seek. In general, capital is categorized into two basic types (based on the consideration given by the venture in return for the capital): equity or debt. **Equity financing** entails "selling" an ownership interest in the venture. A capital contribution is made by an equity investor in return for a percentage of ownership. **Debt financing** involves taking out a loan or selling "bonds" to raise capital; the venture gets the use of the lender's money in return for a relatively inflexible obligation to repay the money plus an additional fee (interest). There are pros and cons to both equity and debt financing. An equity co-venturer shares proportionately in both the risk and the rewards. (The ownership percentage taken by an equity investor is reflected as the dilution of the ownership percentages of the pre-investment principals.) Equity financing is particularly advantageous in the short term; there is no expectation of payments in the short term, and the venture is given flexibility to manage its cash flow. And, the equity co-venturer shares the risk; if the venture fails, their capital contribution will not be repaid. Typically, an equity co-venturer is not able to get their contribution back from the venture, unless and until there is a "liquidity event," such as the acquisition of the venture by a third party or a public offering. However, "equity is forever." The long-term cost of "equity money" contributed to a successful venture is potentially infinite. Equity investors share in the rewards (and, typically anticipate that the long-term returns may be substantial).

On the other hand, the monetary cost of debt financing is finite–ultimately, the loan is paid off. If the venture is successful, the ultimate long-term cost of debt financing is typically considerably less than the ultimate cost of a comparable amount of equity financing. However, lenders do not assume risk; the debt must be repaid irrespective of the success of the venture. And, debt financing can place significant short-term stresses on the venture; it may subject the venture to difficult payment obligations and restrictive loan covenants. The obligation to service the loan is typically inflexible and repayment is required irrespective of the success or cash flow requirements of the venture. Repayment of the loan principal, with interest, often begins immediately upon receiving the loan.

In practice, the choice of funding source may not always be entirely in the hands of the technology entrepreneur; certain types of capital may not be available to a venture during certain stages of its development. The sources of the different types of capital tend to have different investment criteria. The venture may not meet those criteria during certain stages of development. For example, during the initial startup phases of a venture (e.g., seed stage), where there is no income stream, a venture may not be able to meet a lender's criteria, and, realistically, debt financing would not be available. On the other hand, during the **growth stage** of development, for instance, where positive market reaction to the venture's offerings is requiring it to grow, the venture may well qualify for a loan and debt financing may then become available to the venture.

The issues of when and how much capital to seek are inextricably interrelated. In determining when and how much capital to raise, the entrepreneur must address two competing concerns: the adequacy of the amount raised to meet the venture's needs and the cost of the money raised. There should be a specific reason for raising capital–an immediate goal. For a start-up venture, the goal may be to bring the business to "self- sufficiency," or to a point where it is feasible to raise additional funds at a lower cost. The amount must be adequate to accomplish the goal, and, preferably, to deal with unforeseen contingencies and take advantage of opportunities that might arise. On the other hand, the venture should not pay too steep a price for something that it doesn't immediately need. The cost factor is particularly important if the capital is obtained in return for equity interests in the business. Raise enough capital to avoid a cash shortage, but raise too much "equity money" too early, and you will unnecessarily dilute the ownership of your business.

The cost (in terms of percentage ownership per dollar) is primarily a function of the valuation of the venture at the time you seek "equity money" (**Pre-investment valuation**). For that reason, fundraising should be correlated with critical growth stages and milestones in the development of the venture. Ideally, you identify specific milestones or events that would cause a significant increase (appreciation) in the valuation of the venture, and raise only enough money to get the venture to the next milestone that significantly increases its value.)

In other words, you build up the value of the venture as much as possible before each successive round of financing and do not raise money until the venture actually needs it (taking into account the lead time necessary to raise funds). In this way, the cost (dilution) of bringing in equity investors can be minimized.

Of course, that is not as easy as it sounds. Nonetheless, it is possible to develop a reasonable capital management plan that changes over time with the needs of the growing venture.

Although this may sound counterintuitive, the entrepreneur should avoid raising *too much* capital. There are two primary problems associated with **overcapitalization**. First, with too much capital on hand the entrepreneur may not develop disciplined internal cost controls and operational efficiencies. There is nothing that motivates the entrepreneur to manage costs more than the threat of running out of cash. Second, overcapitalization may mean the venture has sold more equity (ownership shares) in the company than it needed to, or it has taken on more debt than necessary.

Technology entrepreneurs also need a variety of tools to raise capital. For example, an up-to-date business plan is an essential component of the fundraising process. Technology entrepreneurs are advised to develop and routinely update their business plan so that it is ready to present to investors or lenders. There are a number of other tools, and some tried and tested techniques for fundraising that will be discussed later in this chapter.

Raising funds also requires that the entrepreneur develop what is referred to as a "deal". "Deal" is the term used to refer to the conditions that the entrepreneur and the investors or lenders agree to in the transfer and use of funds. These conditions are usually embodied in a promissory note in the case of debt financing, or, with respect to equity financing, in an operating or shareholders agreement, a **Private Placement Memorandum** (PPM), or sometimes in a stock offering itself. Investors and lenders have different expectations regarding the funds they provide to the technology venture. Investors are concerned with the returns they can earn on their invested capital; while lenders are concerned with whether the venture will be able to make the interest and principal payments on the loaned amount. The two different sources of capital require different conditions or terms in the deployment and use of the capital they provide. Technology entrepreneurs must learn how to structure deals that meet and exceed the interests of these various sources of capital.

Technology entrepreneurs also have opportunities to acquire capital from non-traditional sources. For example, the federal government has grant programs in key technology areas that have funded a large number of startup ventures. The U.S. Government's Small Business Innovation Research (SBIR) grant program has awarded more than $2 billion to a variety of technology entrepreneurs.

Let's first consider the equity financing options that are available to the technology entrepreneur.

8.3 REGULATION OF FINANCING ACTIVITIES

Because of past abuses, many jurisdictions stringently regulate the sale of passive investment instruments–such as stocks and bonds. In general, the regulations are imposed through what are referred to as Securities Laws. In the United States, there is regulation both at the federal level (through the U.S. Securities and Exchange Commission (SEC)), and at the state level through various analogous agencies. The federal securities laws govern all sales of securities. The term **security** is very broadly defined, covering, in essence, any situation that involves using someone else's money based on the "promise of profits," prospects of a "return" from other than the investors own efforts, or makes money subject to the entrepreneurial or managerial efforts of others. In other words, it covers not only stocks and bonds, but also other forms of investment in a venture, where the investor is not actively involved in the management of the venture.

> ### KEY POINT
>
> *Not All Equity Interests Are Securities*
>
> One of the reasons for the popularity of the Limited Liability Company (LLC) form of entity for startups is that it provides limited liability, while at the same time, without necessarily implicating the securities laws. Equity interests in an LLC are typically not securities, **so long as all members are actively involved in the management of the venture**. General Partnership interests are likewise outside of the definition of security, but do not provide limited liability. Corporate stock, however, even if sold to management, are expressly defined as Securities under U.S. Federal law.

Any time fundraising activities involve the sale of a security, certain rules relating to the disclosure of information must be observed. You are required to make "full disclosure" to the potential investor of all "material facts." Information is a "material fact" if it is something that investors would typically want to consider before making the decision to purchase the security.

And, unless the activities fall within very specific exemptions, not only must disclosure be complete and accurate, but the security (or transaction involving the security) must be registered with the SEC (and sometimes also with analogous state agencies) and written information must be provided to prospective investors in a specific format (a prospectus in a prescribed form) that has been "deemed effective" by the SEC. Transactions that fall within the specific exemptions are referred to as "private placements", and those that do not are referred to as "public offerings." The first time that a venture makes a public offering is referred to as an "Initial Public Offering" (IPO). The registration process is sometimes referred to as "going public."

The registration process tends to be very expensive, and once "public" a venture is required to maintain certain internal audit controls, and publicly disclose certain information relating to business operations, financial conditions and management. The administrative burden on Public companies imposed by the securities laws and the costs of compliance is significant. The IPO has historically been a favored exit strategy (liquidity event) for certain types of investors. In view of the cost under present-day regulation, the IPO has become much less popular. In any event, "going public" is generally not a viable option for startups; if for no other reason, at least two years of audited financials are required for registration.

It is desirable (if not imperative), as a general proposition, that efforts to raise capital for startup technology ventures fall under one of the exemptions to the registration requirements in all applicable jurisdictions (e.g., in the US both at federal and state levels). Exemptions under the federal law are defined by regulations promulgated by the SEC.[1] In general, the respective exemptions have different criteria, and apply under different circumstances. In general, in order to qualify for the various exemptions the venture and offering must meet (depending upon the particular exemption):

- Limitations on the dollar amount of the offering (or other offerings within a predetermined time period);
- Limitations on the number and nature of the entities to which the security can be offered;
- Restrictions on the types of entities offering the securities (issuers) to which the exemption is available;
- Restrictions on general solicitation or advertising;
- Requirements as to the nature of the disclosure that must be made and/or documents provided to the offerees;
- Requirements as to the nature of filing with the SEC; and
- Requirements for restrictions on stock transfer;

Some federal exemptions provide relief from (preempt) state registration requirements, others expressly require that state exemption criteria be met.

The securities laws draw distinctions based upon the wealth and sophistication of the entities to which the security can be offered. **Accredited investors** are defined as (in addition to various institutions such as banks, investment companies, retirement plans, and the like) individuals whose net worth at the time that the securities were purchased (individually or jointly with spouse) exceeds $1 million (exclusive of primary residence), or whose individual net income was over $200,000 (or $300,000 jointly with spouse) for each of the two years prior to the purchase, and reasonably expects the same income level in the current year.[2] A "sophisticated investor" is someone who has, or is represented by someone (a "purchaser representative[3]") who has, (or you reasonably believe to have) "such knowledge and experience in financial and business matters that he is capable of evaluating the merits and risks of the prospective investment".[4]

A small offering (no more than $5 million) is exempted from Federal registration requirements if it is private (no advertisements or general solicitations) and made strictly to accredited investors.[5] Various other exemptions have limitations on participation of non-accredited investors, and specific disclosure requirements in respect of non-accredited investors.[6]

There is also a general "private placement" exemption[7] for private offerings, although the precise requirements to qualify for the exemption are not specified in the statute. Private placement "safe harbors" are, however, specifically defined by what is known as Reg D,[8] "Rules Governing the Limited Offer and Sale of Securities without Registration under the Securities Act of 1933." Reg D actually defines a number of different exemptions, with different criteria/limitations according to, e.g.,: maximum size of the offering permitted; participation of non-accredited investors; general solicitation/advertising; and/or applicability of state regulations to the offering.

Other exemptions are provided based on (along with specified criteria and limitations): filing what is effectively a mini-registration (Reg A exemption)[9]; strictly intra-state offerings[10]; and offerings strictly outside of the US.[11]

In 2013, the SEC proposed to revise its rules according to the Jumpstart Our Business Startups or JOBS Act that was passed in 2012. This Act was intended to encourage private investment in small startups, and to open the door to funding that otherwise might be closed for many startup ventures. Some of the principle proposed changes that are consequences of this Act include:

- Increasing the number of shareholders a company may have before being required to register with the SEC;
- Exemption for small companies from the rules prohibiting public offering of equity shares. This provision would enable small companies to publicize a stock offering, including making use of publicly available online **crowdfunding** portals that link small offerings with people. Portals that offer non-equity "investing" at the time this book is going to press include Kickstarter, CrowdCube, Seedrs, CrowdCheck, CircleUp and others.
- Relief from some regulatory requirements for companies designated as **emerging growth companies**. This provision primarily exempts such companies from some of the disclosure requirements that were enacted following corporate scandals of the early 2000's.
- Allowing larger fundraising under the SEC's Regulation A of up to $50 million (increased from its previous limit of $5 million).

The JOBS Act went into effect in late 2013 and will most certainly provide more opportunities for growing technology ventures to generate visibility for their fund raising efforts if some of the key provisions that it intends are passed by the U.S. Congress. It will be interesting to see how crowdfunding and the ability to publicly solicit investment opportunities will affect fund raisers and investors. The Mini-Case below provides an example of a technology entrepreneur who was able to launch his venture based on a successful Kickstarter crowdfunding campaign:

MINI-CASE
Pebble E-Paper Smartwatch Raises $10M on Kickstarter

One of Kickstarter's most successful fundraising campaigns was engineered by the founders of the Pebble smartwatch. The Pebble watch is designed to connect to Android or iOS smartphones via Bluetooth technology. For example, the watch can alert users about incoming calls via a silent vibration. The Pebble smartwatch also can download apps of its own turning it into, for example, a bike computer that displays speed, distance, and pace. The use of an "e-paper" display permits you to read the Pebble watch output even in direct sunlight. Users can customize their Pebble smartwatch with colored wristbands and apps unique to their lifestyles. Currently available apps are as diverse as a range finder for golf enthusiasts to apps that let users customize the watch face to their individual taste.

Initially intending to raise just $100,000, Pebble's Kickstarter campaign garnered nearly 70,000 contributors who contributed over $10M! Pebble offered a variety of non-equity items in return for contributions of various amounts, e.g.,: a subscription to an exclusive newsletter "on all things pebble" for contributions of $1 or more; various models of the pebble watch for contributions running up from $99 – $220 or more); or early delivery of a prototype smartwatch, plus later delivery of a production model for contributions of $235 or more.

Source: Adapted from Victor Marks, "Pebble E-Paper Watch for iPhone and Android: Review," ABlogtoWatch.com, September 28, 2013; Chris Morris, "Kickstarter's 10 Biggest Success Stories," CNBC.com, August 20, 2012.

Even if exempted from registration, the full disclosure requirements is applicable to all securities offerings. The full disclosure to investors required by the securities laws for other than public offerings is typically provided through a document referred to as a Private Placement Memorandum (PPM).

The PPM resembles a business plan in that it will describe the business, including its products and services, management team, financial projections, competitors, and the target market. In addition, the PPM goes far beyond the business plan in providing the investor with disclosure of the risks associated with an investment in the venture. Risks discussed in the PPM might include the potential for patent filings to be rejected by the patent office, misjudgment of the market potential of the venture's products and services, inability to raise sufficient capital in the future to operate, and many other contingencies. In general, technology entrepreneurs should involve legal counsel in preparing any PPM.

8.3.1 Costs of Equity Financing

The costs associated with conducting a traditional private equity offering (e.g., an offering *not* using the new crowdfunding approach) are usually greater than applying for and receiving an equivalent amount of debt. The greater costs involved in a traditional equity offering are part of the technology entrepreneur's deliberations when deciding on the type of financing to pursue.[12] Of even greater influence on this decision is the degree to which equity financing will **dilute** the ownership positions of the existing shareholders. Anytime a venture sells shares to raise capital, the ownership positions of the existing shareholders will be affected. The degree to which dilution occurs is a function of the venture's **valuation** at the time of the financing.

A discussion of methods for determining a venture's valuation will follow later. However, here we will simply note the *effects* of different valuation levels on venture ownership. Suppose, two individuals launch a technology venture, and they decide to split ownership equally between them, with each now owning 50 percent of the company. They have invested their own funds to launch the venture, and so they have no other shareholders. Now, suppose they have developed their technology past the prototype stage and have received positive feedback from test markets. These two entrepreneurs are ready to expand the company, but they need capital to pursue that end. They decide that they will need to raise $1 million. What effect will that amount of capital raised have on their respective ownership positions?

The answer is: "It depends." It depends on their ability to persuade investors regarding the value of the venture. Let's assume the two entrepreneurs convince each other that their company—today, before they've raised any money—is worth $2 million. In the language of deal making, this is referred to as the **pre-money valuation** of the venture. That is to say, the pre-money valuation is what the venture, arguably, is worth prior to any new money coming in. There are a number of factors to consider in establishing the value of a private company, but we will put off that discussion until later. For now, let's assume that the $2 million pre-money valuation is reasonable, and that the two owners are able to persuade potential investors that this is a reasonable valuation of the company. At that valuation, each owner's 50 percent stake in the company is worth $1 million (50 percent of $2 million).

If the technology entrepreneurs are successful in raising the $1 million, the venture would have a **post-money valuation** of $3 million. Note the post-money valuation is simply the pre-money valuation plus the amount raised. What is the dilutive effect of this raise to the two original shareholders? It's plain to see that the investors (regardless of how many) who put in the $1 million will own 1/3 of a venture that has a post-money valuation of $3 million. The two entrepreneurs, who previously owned 50 percent each, have been diluted by 1/6 apiece to a stake of 33.3 percent, respectively. Together, they maintain a controlling interest, 67 percent, in the company, but they have been diluted. Note, however, that despite the dilution in the percentage of their ownership, the dollar value of their shares remains $1 million (i.e., 1/3 of $3 million).

8.3.2 Sources of Equity Financing

The technology entrepreneur can pick from a wide range of sources for equity financing that will vary depending on the venture's stage of development. All ventures can be described by virtue of their development stage, which in itself conveys useful information to lenders and investors, enabling them to sort through the deals they encounter. A common breakdown of the various stages of development is presented in Exhibit 8.1.

In the **seed stage**, most ventures are funded by the founders, and anyone else they can convince of the potential for their success. For most technology ventures in seed stage, this broader group of potential investors is limited to what is often referred to as "FF&F"—which stands for "friends, family, and fools". This little memory device is only partly facetious. Scholarly research has demonstrated that most pre-startup technology ventures are funded by the entrepreneur, family members, friends, and business partners.[13] The other "F"—fools—is used primarily as a reminder that startup ventures are very risky investments.

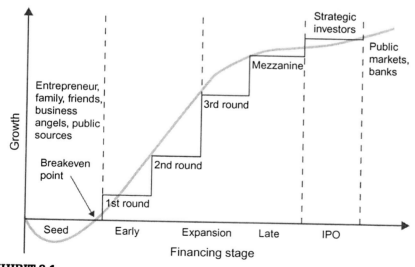

EXHIBIT 8.1
Stages of venture financing.

Seed stage ventures, especially those that have not yet begun generating revenue from sales, are at a very high risk of failure. Because of this, investors and lenders have little incentive to put their capital to work at this stage. Research into the sources of funds for seed stage ventures is very clear. In most cases, funds are derived from the assets of the founders. The good news is that most entrepreneurial ventures are launched with less than $20,000 in seed funds.[14] The founders may tap into their personal savings, leverage a line of credit on a credit card, take out a home equity loan, and adopt other strategies to accumulate the needed startup funds.

8.3.3 Angel Financing

As technology ventures launch and enter what is referred to as the **early stage**, most continue to be funded by founders and their network of family and friends. However, if the venture has matured beyond the prototyping stage and is preparing to, or actually beginning to, sell into the target market, then angel financing may be available.

Angel investors are high net worth individuals who are interested in allocating a portion of their overall portfolio to high-risk and, potentially, high-return private ventures. Angel investing has become a significant source of capital for seed and early stage technology ventures in the United States. Angel investors provided nearly $23 billion in financing in 2012.[15] They often act alone, investing both their own and their family's wealth. However, many angels aggregate in groups to review and discuss deals, and to co-invest with others in order to have a larger impact on the ventures they prefer.

According to the Angel Capital Association, there are more than 350 angel investor groups throughout the United States and Canada.[16] In the United States, most angel groups are concentrated on the East and West coasts. However, each state has at least several angel investing groups, primarily concentrated in the larger cities. For example, there are three angel investor groups in the state of Arizona, one of which is based in Phoenix and focuses exclusively on technology ventures. The Arizona Technology Investor Forum (ATIF) is an angel investor group that meets several times per year and reviews three or four ventures at each meeting.[17] Meeting formats are fairly standard across angel groups. Entrepreneurs generally are allotted time for a brief (20–30 minute) presentation, which is usually followed by 15–20 minutes of question and answer. Some groups also perform extensive **due diligence** on the ventures that present. For groups that provide due diligence for members, the findings of the individual or team that performed the due diligence are also reported at the meeting.

Few angel groups specialize in any single industry, and most have interests and expertise among their members in a variety of areas. Angel groups mainly invest in the range of $150,000 to $500,000 per deal. However, some groups can go higher than that. In 2012, there were more than 250,000 active angel investors in the United States, but only 15,000 of them were members of angel groups.[18] Thus, most angel investors are independent of angel groups. Determined technology entrepreneurs need to find these non-affiliated angels to provide the highest probability of acquiring necessary funding. Of course, given our definition of angels as high-net worth individuals (more than $1M in net worth), there are some obvious places they might be found: golf clubs, charity events, and entrepreneurship networking events are three locations that seem likely. Another good technique for finding these unaffiliated angels is to talk to other entrepreneurs. They may have some leads to angel investors and may be able to assist with an introduction.

8.3.4 Venture Capital

For the technology venture that receives angel capital, the challenge becomes growing the venture. Ideally, the technology entrepreneur wants to grow the venture, increasing the enterprise value, which would result in a **return on investment** for the angel investors. If the growth is substantial and future growth projections even greater, then the technology venture may become attractive to **venture capital**.

Venture capital is managed by venture capital (VC) firms. VC firms are typically structured around a team of **general partners** who often are successful entrepreneurs, or, at least, financially experienced individuals. VC firms raise money from high net worth individuals and financial institutions (such as insurance companies, pension funds, and the like) who are considered to be **limited partners** to the VC firm. These individuals pledge an amount of capital to the VC firm, and are required to put as much as 20 percent of it immediately into a fund that will be managed by the general partners. This capital is then invested in ventures according to the interests and expertise of the firm. As the fund is invested over time, the limited partners will be called upon to invest additional portions of their pledged amount. The typical VC firm structure is illustrated in Exhibit 8.2.

Most venture capital firms have specific industries and venture types that they prefer. For example, one of the top venture capital firms in Silicon Valley, Benchmark Capital, invests primarily in high-growth Internet companies. The firm has invested in a wide range of well-known Internet ventures.

Once the VC firm receives a return from its investment, the VC must return the capital back to its investors based on the economic terms in the **limited partnership agreement** (LPA). Before a VC firm can receive any of the capital, it must return the original principal in its entirety to the limited partner investors plus any guaranteed return that was agreed upon in the LPA. The VC firm and the limited partners split the net profit, which is the capital that remains after

CHAPTER 8 Capital and Capital Sources

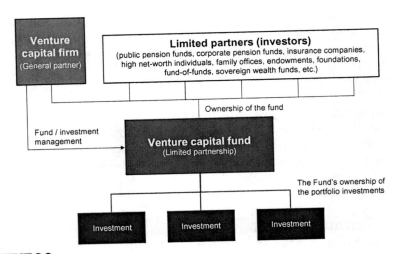

EXHIBIT 8.2
Typical VC firm structure.

the principal and the guaranteed return are returned to the limited partners. Below is an example of how the gains are split between the VC firm and the limited partners with an investment of $100.

- Principle investment in VC firm: $100
- Guaranteed return to Limited Partners: 8%
- Split of carried interest to VC: 20%
- Split of carried interest to Limited Partners: 80%

Assume a $300 return from this $100 investment. The steps to this transaction are:

1. VC firm receives $100 from its limited partners to invest in startups.
2. VC firm takes this capital and invests $98 into various startups (2% is management fee).
3. Startups use this capital to grow their business. They later sell the business and provide a return of $300 to the VC firm.
4. VC firms takes the $300 and returns the original principal of $100 to the Limited Partners plus the required return of $8 (8% on the original $100).
5. Net Profit of $192 is left after returning $108 to limited partners ($300-$108)
6. Net Profit is split 20% to VC Firm ($38=20% of $192) and 80% to Limited Partners ($154=80% of $192).

In summary, the limited partners receive a total of $262 getting their principal back, their required return, and a percentage of the net profit. The VC firm receives $38, a percentage of the net profit that is referred to as

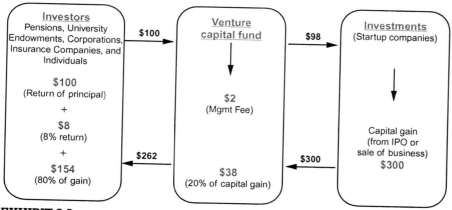

EXHIBIT 8.3
Illustration of venture capital returns.

carried interest. The typical compensation model for a VC firm is referred to as "2 and 20". That is, 2% of the committed capital is used for the management fee and 20% of the fund's net profits is used as carried interest. Exhibit 8.3 illustrates this set of transactions.

8.3.5 What do Venture Capital Firms Look For?

Venture capital firms look for many specifics in technology ventures when deciding whether to make an investment. First and foremost, they seek out experience and talent in the management team. Venture capitalists often state that they invest in the management team, not the technology.[19] This is because no technology, no matter how revolutionary or disruptive, can be commercialized successfully without a solid management team. In addition, experienced VCs know that most technologies they invest in will change radically over time in response to changing market reactions and conditions. It is important to recognize that, when VCs evaluate the management team, they do not require that the team be devoid of past venture failures. In fact, most VCs recognize the contingent nature of venture success and do not automatically consider a past failure in a negative light.[20]

Another primary area of consideration for VCs is the market potential for the technology.[21] The VC wants to know if the venture is "market ready", and they will attempt to forecast the likely marketplace demand. This will include forecasting the willingness of the customer to switch to the venture's products based on the venture's stated value proposition. It will also include an analysis of the size of the market opportunity. Obviously, the VC will be more interested in large market opportunities than small ones.

Another item that VCs commonly consider to be important is the time to an **exit event** or **liquidity event.** An exit event is the manner in which shareholders can convert their stock into cash (see Chapter 13 for a thorough discussion of exit strategies). This is usually achieved via acquisition by a larger company, or by going public on a major stock exchange. Either way, the capital invested by the VC is tied up until this event occurs.[22]

In general, VCs are more risk averse than the founders of a technology venture seeking capital. In their respective evaluations of the investor readiness of the venture, the founders generally will have a more favorable perspective. VCs will evaluate a venture from both a risk and return perspective. Risk and return are evaluated in terms of the risk of venture failure and potential profitability.[23]

The technology entrepreneur tends to have a strong interest in innovation, and will often discount the factors that VCs find important. Most technology entrepreneurs place a higher importance on the technological aspects of the opportunity and consider the management team and marketability to be of lesser importance. In fact, research has identified certain cognitive biases, such as "overconfidence", that are common among technology entrepreneurs.[24] Recognizing these biases is an important part of raising VC funds. The entrepreneur who does not recognize the factors that VCs consider important may not receive an adequate hearing. On the other hand, the technology entrepreneur who tailors the VC presentation to address the issues they are most concerned with enhances his or her chances of receiving funds.[25]

8.4 DEBT FINANCING

Debt financing is normally provided to technology ventures via an **institutional lender,** such as a bank. However, this may not always be the case. Some entrepreneurs structure deals where they take loans from private parties. While this is not a common practice, some estimates indicate that as many as 10 million people in the United States have accepted private loans from people they know.[26] Borrowing from family and friends can provide easier terms than an institutional lender, including a longer repayment period and a lower interest rate. However, borrowing from family and friends has some risks that are dissimilar from institutional lenders.[27] For example, if the entrepreneur does not repay the loan within a reasonable amount of time, the trust of the family and/or friend lenders could be lost. Worse, if the entrepreneur defaults on repaying the loan, relationships may become strained, family ties could be severed, and lawsuits could be filed.

It is important to carefully document the terms of the business loan between friends and relatives who lend money to a venture. This can help ensure that these individuals are less apprehensive about how their money is being used.

It can also prevent messy or awkward situations in the event the venture has difficulties living up to the loan covenants. If details regarding remedies for late payments and even for complete default are discussed and agreed to in advance, the impact of such eventualities can be lessened.

8.4.1 Institutional Lender Requirements

Institutional lenders, such as banks, can be a source of financing for young technology ventures. However, banks will require more documentation than friends and family normally require; they will also require that the venture be able to secure the loan by pledging an equivalent amount of **collateral**. That is, most institutional lenders reduce the risk of their loans by gaining a **security interest** in property owned by the borrower that is adjudged to be equivalent in value to the loan principal amount. Collateral can be any property owned by the borrower, including personal property.

It is also not uncommon for technology entrepreneurs to provide—or to be required to provide—a **personal guarantee** on the amount of the loan. That is done when the venture has few assets to use as collateral, but the entrepreneur has personal assets that will suffice. For example, some entrepreneurs will use their personal savings, home equity, or real estate as collateral for loans from a bank for business purposes. In the case of default, the bank would then be able to exercise its claim on the collateral in an effort to recover whatever portion of the loan remains unpaid.

8.4.2 Loan Rates, Payment Methods, and Lender Types

In addition to pledging collateral in the amount of the loan's principal, institutional lenders normally require that the entrepreneur begin to pay back the loan immediately, with an added interest charge. Payments on a loan are typically made in monthly increments. The rate of interest charged to a small technology startup for a loan will exceed the rate charged to large, well-established, companies. The latter receive preferential rates from banks—usually called the **prime rate**—because they are less risky borrowers than a startup venture. Startup ventures will receive a rate that is termed prime plus. That is, the rate will be some measure above the prime rate depending on the risk profile of the venture and, to some extent, its principals and the collateral they are able to pledge.

Most banks will have a similar prime rate, but they will vary significantly from one to another in the rates they charge to entrepreneur borrowers. For example, some banks specialize in originating real estate or construction loans. They will typically have loan officers who are familiar with the industry and will have experience in judging the risks involved in lending to this type of venture. A bank that specializes in real estate lending may not be able to evaluate the risk involved in lending to a technology startup. It's important for the technology entrepreneur to conduct research prior to approaching a lender to determine whether it is capable of understanding the nature of the venture and its risks. Silicon Valley Bank, for example, specializes in loans to technology ventures.

Loans made by institutions also require payback to occur over a certain period of time referred to as the **term** of the loan. Short-term loans will carry slightly lower interest rates than long-term loans, but they will also require larger monthly payments. Long-term loans are generally considered to be those that have a term of 10 years or greater. Different lenders use different methods to calculate loan repayment schedules depending on their needs, borrowers' needs, the institution's interest rate policy, the length of the loan, and the purpose of the borrowed money. Normally, business loans are paid back on a monthly schedule in equal payments throughout the term of the loan.

Institutional lenders will also usually require that the startup venture adhere to certain **restrictive covenants** in order to remain in good standing on the loan. Restrictive covenants enable the lender to maintain some control of the venture by specifying performance targets. In the event these performance targets are missed, the lender would have the option to **call the loan.** That is, it can demand complete payment of the outstanding principal before the term of the loan is completed. Restrictive covenants are of two types: positive and negative. Positive covenants specify performance targets that must be attained in order for the borrower to remain in good standing. For example, a bank and borrower may agree to establish positive covenants targeting total sales, cash flow, profitability, or others. Negative covenants establish performance floors below which the venture may not fall in order to remain in good standing. For example, the bank and the borrower may agree that the venture may not fall below target figures in sales, cash flow, the ratio of debt to equity capital in the venture, and others.

Restrictive covenants are usually determined through negotiations between the lender and the borrower, but many lenders have lending guidelines that establish in advance the covenants that they must put into any lending agreement. Technology entrepreneurs must be aware of the covenants that banks require, and determine whether they will impede the venture's ability to succeed. For example, many businesses are subject to fluctuations in sales based on business cycles and other factors. A restrictive covenant that did not account for these sales fluctuations may result in an unnecessarily premature decision by the lender to call the loan. In a time of decreased sales, such an action could be fatal to the startup venture.

Another type of loan facility that is exceedingly useful to the startup venture is the line of credit or revolving loan. A line of credit is simply a pool of money that is set aside by a lender for a business to use as needed. The borrower can draw down the line of credit for business expenses without having to fill out a loan application each time funds are required. This saves the lender and the borrower a lot of time. Lenders provide lines of credit using the same risk calculations as any other loan, and they will also usually require collateral in the amount of the credit line. Borrowers benefit from a line of credit in that they only need to pay back the amount withdrawn, and pay interest only on the withdrawn capital. In addition to the interest collected, most banks will charge an annual maintenance fee to provide a line of credit to a company.

8.4.3 Small Business Administration (SBA) Loans

Another type of loan commonly used by technology startups is the **Small Business Administration Loan**. The name of this loan type is confusing to some. The Small Business Administration (SBA) is an agency of the United States federal government. However, the SBA does *not* originate the loan made to the venture. A technology entrepreneur who wants to secure an SBA loan must do so through a commercial bank that provides such loans. The bank *originates* the loan, and the SBA *guarantees* payment on the loan up to a predefined percentage of the principal amount. In this way, the SBA provides a form of collateral and takes a good deal of risk out of the loan for the bank.

While the SBA provides a credible and useful debt facility for many entrepreneurs, an SBA loan also has significant disadvantages. SBA loans generally require more documentation and disclosure on the part of the small business than does a traditional loan. SBA loans also frequently have higher interest rates than would a loan provided directly by a commercial lender. Still, there are many advantages to getting an SBA loan, not the least of which is the fact that many startup ventures have no other options. The SBA has more loan programs than the primary ones listed above, including some that are designed specifically to support minority and women-owned enterprises.

8.5 FUNDRAISING TOOLS AND TECHNIQUES

Fundraising for a technology venture is never a sure thing, but there are some tools and techniques that can improve the odds of acquiring needed capital. One of the primary tools of fundraising is the business plan. It is rarely possible to raise money outside of the friends and family network without a business plan. Angel investors will almost always ask to see the venture's business plan before investing. Venture capitalists and institutional lenders will always require a business plan. Entrepreneurs should develop and maintain a business plan at all times because most startup ventures are in near-constant fundraising mode. A good business plan will describe the venture and its technologies in terms the investor or lender can understand and evaluate.

Another tool that should be part of the technology entrepreneur's fund raising arsenal is the executive summary. The executive summary is an abridged version of the business plan, normally condensed down to a single page.

8.5.1 Private Placement Memorandum

Private equity fundraising requires that the venture prepare what is referred to as a private placement memorandum (PPM). The PPM is a legal document that specifies all the risks associated with investing in the venture, including the potential for complete loss of all invested capital. The PPM also specifies the amount of capital that is to be raised, the type of security being offered, and the rights and privileges associated with investing. Most early stage fundraising will divide the amount to be raised into investment **units**. For example, if a venture is raising $500,000 and selling its stock for $1.00/share, it does not want 500,000 individuals each purchasing one share. That would be an administrative nightmare. Instead, the venture would sell the $1.00/share stock in units of, say, $25,000. That means that anyone interested in investing would need to purchase at least one unit.

In addition to stating the amount that is going to be raised, the PPM will often specify the minimum amount that needs to be raised in order for the venture to be able to use the funds. This is often referred to as the **min/max**. This concept can be understood most clearly from the perspective of the investor. If an investor puts money into a venture, he or she wants to be sure that the venture has a fair chance of success. However, if the venture raises only a small portion of what it needs, it may fail. The min/max specifies the minimum amount of money that is required for the venture to have a fair chance to succeed. The venture does not use any of the funds it raises until it achieves the minimum amount. Funds raised prior to reaching the minimum are held in trust in a bank account. Once the minimum has been raised, the venture is allowed to **break the bank** and begin to use the invested capital.

8.5.2 Subscription Agreement

The final document to include in equity fundraising is the **subscription agreement**. A subscription agreement is a document that a potential investor signs, indicating an intent to invest at a certain amount. Even though a subscription document is not considered to be binding on the potential investor, it creates a psychological commitment on the part of the investor. For example, imagine that a technology entrepreneur completes a lengthy presentation to an angel investor in their office. The investor indicates an interest, but would prefer to look over the business plan and PPM before making a decision. If the entrepreneur walks out of the office with no signed commitment, the potential investor may not be interested enough to continue thinking about the deal. It is far more influential to have a signed agreement when conducting follow up conversations, including the penultimate conversation where the entrepreneur asks for the check, than it is merely to have a promise.

8.5.3 Elevator Pitch

Technology entrepreneurs should also practice their investment "pitch", which is sometimes referred to as an **elevator pitch**. This term is used to conjure what it would be like to meet a potential investor in an elevator and, in the limited time available, describe the business in a manner that captures the investor's attention. An elevator pitch should articulate the venture's offering (product and/or service), its business model (how it will make money) and the size of the opportunity. In a full investor presentation, the entrepreneur will need a slide deck (PowerPoint) that clearly articulates the business, the market, the value proposition, financial outlook, and the deal (including projected returns) for the investor.

8.6 ALTERNATIVES TO DEBT AND EQUITY FINANCING

While the primary sources of capital for the startup venture are the sources of debt and equity discussed above, there are alternatives. Many technology ventures are able to get started and fund operations using government grants as their primary revenue source. Next, we look at two government grant programs available to the technology venture, and also at bootstrap financing, which relies on internal cash only to grow the venture.

8.6.1 Small Business Innovation Research (SBIR)

One particular type of government grant that is commonly used by technology ventures is the Small Business Innovation Research (SBIR) program. The U.S. Small Business Administration's Office of Technology administers the SBIR program. SBIR is a competitive program that encourages small businesses to explore their technological potential and provides the incentive to profit from its commercialization. Since its enactment in 1982, and as part of the Small Business Innovation Development Act, SBIR has helped thousands of small businesses to compete for federal research and development awards. Small businesses must meet certain eligibility criteria to participate in the SBIR program:

- American-owned and independently operated
- For-profit
- Principal researcher employed by business
- Company size limited to 500 employees

Each year, eleven federal departments and agencies are required by SBIR to reserve a portion of their R&D funds for award to small business investments, including:

- Department of Agriculture
- Department of Commerce
- Department of Defense
- Department of Education
- Department of Energy
- Department of Health and Human Services
- Department of Homeland Security
- Department of Transportation
- Environmental Protection Agency
- National Aeronautics and Space Administration
- National Science Foundation

These agencies designate R&D topics and accept proposals. Following submission of proposals, agencies make SBIR awards based on small business qualification, degree of innovation, technical merit, and future market potential. Small businesses that receive awards then begin a three-phase program.

- Phase I is the startup phase. Awards of up to $100,000 for approximately six months support exploration of the technical merit or feasibility of an idea or technology.
- Phase II awards of up to $750,000, for as many as 2 years, expand Phase I results. During this time, the R&D work is performed and the developer evaluates commercialization potential. Only Phase I award winners are considered for Phase II.
- Phase III is the period during which Phase II innovation moves from the laboratory into the marketplace. No SBIR funds support this phase. The small business must find funding in the private sector or other non-SBIR federal agency funding.

8.6.2 Small Business Technology Transfer (STTR)

Another grant program offered by the SBA is the Small Business Technology Transfer Program (STTR). Central to this program is expansion of the public/private sector partnership to include joint venture opportunities for small business and nonprofit research institutions. The STTR program reserves a specific percentage of federal R&D funding to offer awards to small business and nonprofit research institution partners. The STTR program combines the strengths of both entities by introducing entrepreneurial skills to high-tech research efforts. The idea is that the small business partner is able to transfer from the laboratory to the marketplace the technologies and products developed within the non-profit organization. Small businesses must meet certain eligibility criteria, just like the SBIR program to participate in the STTR Program:

- Must be American-owned and independently operated.
- Must be for-profit.
- Principal researcher does not need to be employed by small business.
- Company size is limited to 500 employees.

The nonprofit research institution partner must also meet certain eligibility criteria. It must:

- Be located in the United States.
- Meet one of three definitions.
- Be a nonprofit college or university.
- Be a domestic nonprofit research organization.
- Be a federally funded R&D center (FFRDC).

Each year, five federal departments and agencies are required by STTR to reserve a portion of their R&D funds for awarding to small business or nonprofit research institution partnerships:

- Department of Defense
- Department of Energy
- Department of Health and Human Services
- National Aeronautics and Space Administration
- National Science Foundation

These agencies designate R&D topics and accept proposals. Following submission of proposals, agencies make STTR awards based on small business or nonprofit research institution qualifications, degree of innovation, and future market potential. Small businesses that receive awards then begin a three-phase program.

- Phase I is the start-up phase. Awards of up to $100,000 for approximately one year fund the exploration of the scientific, technical, and commercial feasibility of an idea or technology.
- Phase II awards of up to $750,000 for as many as two years expand Phase I results. During this period, the R&D work is performed and the developer begins to consider commercial potential. Only Phase I award winners are considered for Phase II.
- Phase III is the period during which Phase II innovation moves from the laboratory into the marketplace. No STTR funds support this phase. The small business must find funding in the private sector or find other non-STTR federal agency funding.

8.6.3 Bootstrap Financing

Another way that start-up technology ventures finance their growth is through what is often referred to as bootstrap financing. Here, the company uses its own sales and cash flows to invest in its growth. This type of internal growth is also referred to as organic growth. That is, the company grows only by virtue of its own ability to sell, control costs, and reinvest profits.

Bootstrap financing has the advantage of helping the firm steer clear of the dilutive effects of equity financing and the debt burden effects of debt financing. The primary disadvantage of this type of financing is that it limits the venture's ability to grow rapidly. That could be a major disadvantage for technology ventures in highly competitive industries where acquiring market share is the key to long-term success.

8.6.4 Licensing

One of the tools available to technology-based companies to self-finance their growth is licensing third parties to use their technology in return for some consideration. That consideration typically takes the form of money (although it can be other things such as, e.g., access to needed resources, or a cross license to use the third party's technology).

If done strategically, licenses to third parties can be granted without diminishing the competitive advantage that your company obtains from the licensed technology. For example, the license can be limited to a particular market other than your target markets. Or, the license can be for a limited time period and also limited to a geographic region outside of the region that you can reasonably anticipate being able to service during that time period. Or, the license can be limited to sales channels that you have no reasonable capacity to service, or product lines that, for whatever reason, you have no interest in pursuing.

In some instances, licensing can be a viable alternative to raising capital, and, in fact, can sometimes be the basis for a company's business model.

8.7 CHAPTER SUMMARY

This chapter examined the sources of financing available to technology ventures throughout their lifetimes. During the early stages of the venture's life, friends and family are the primary sources of capital beyond the founders' own finances. As the firm begins to grow and develop sales, other avenues of financing might open. Angel financing becomes available when the firm is able to tell a more compelling story about its growth prospects, especially if it already has significant sales and well-known clients. In addition, firms that have developed collateral such as physical assets or highly regarded contracts may be eligible for debt financing. Commercial lenders, such as banks, are risk averse and lend only when there are assurances that the venture can pay back the principal with interest.

Raising capital is normally never ending for technology venture entrepreneurs. Maturing ventures seeking equity capital from angel investors must prepare legal documentation and sell shares in compliance with Securities and Exchange Commission rules. This chapter examined several rules that enable raising private equity from accredited investors. The most common capital raises for growing ventures, conducted under the SEC's Regulation D. Rules 504, 505, and 506, offer different frameworks for raising equity capital. Each rule shelters the venture from overbearing disclosure requirements that are required of public companies that sell shares to the investing public. The rules do require that ventures develop standard disclosure documents such as a private placement memorandum (PPM). The PPM will disclose all of the risks associated with investing in the venture and detail the terms of the stock offering. In addition to the PPM, technology entrepreneurs typically also provide a business plan and executive summary to interested investors when raising capital.

Raising equity capital requires that the venture establish some reasonable valuation to determine ownership percentages. Valuation is an imprecise science, with several acceptable methods available. Each method is likely to produce a different valuation. Ultimately, the entrepreneur and investors negotiate a valuation they can agree to. The agreed valuation is referred to as the premoney valuation. After the investment is made, the postmoney valuation is used to calculate the relative ownership shares of founders and investors. Founders who hold stock at the time of the investment will have their ownership percentages diluted based on the amount of investment received.

The chapter also discussed options to debt and equity financing, focusing on several opportunities for nonrecourse government grants. The Small Business Innovation Research (SBIR) and Small Business Technology Transfer (STTR) programs are excellent sources of financing for ventures that are in the research and development stage.

KEYTERMS

Self-funding Financing the venture from monies generated by the venture's activities.
Equity financing Venture financing provided by investors in consideration of an equity interest in the venture.
Debt financing Venture financing provided by lenders who expect to be paid back, with interest, on the amount of the loan.

Growth stage The stage where positive market reaction is causing a venture to grow, sometimes exponentially.

Overcapitalization Ventures that raise too much capital; they have a tendency to be less concerned about sound financial management.

Private placement memorandum A document specifying the terms associated with an equity offering. In the United States the PPM provides the equity investor a full disclosure as required by the U.S. Securities and Exchange Commission (SEC).

Security An instrument sold as a passive investment, covering not only stocks and bonds, but also other forms of investment in a venture where the investor is not actively involved in the management of the venture.

Accredited investor An individual whose net worth exceeds $1M at the time of the investment or who has a net income over $200K for each of the two years prior to the investment.

Crowdfunding A new form of fundraising that uses web-based campaigns.

Emerging growth companies The JOBS Act defines this as companies with less than $1M in revenue in the prior fiscal year.

Dilution Selling new equity in a venture that dilutes the ownership percentages of existing shareholders.

Valuation The accepted value of a venture at the time of an investor transaction.

Premoney valuation The value of a venture immediately prior to a new equity investment.

Postmoney valuation The value of a venture immediately following a new equity investment.

Seed stage The stage of a venture's lifecycle where it is not yet generating revenue, but has launched and is trying to grow.

Early stage The stage of a venture's lifecycle where it has begun generating revenue and serving customers.

Angel investors High net worth individuals who allocate some of their wealth toward investing in entrepreneurial ventures.

Due diligence A process where investors thoroughly examine a venture prior to making an investment.

Return on investment The return on invested capital in an entrepreneurial venture.

Venture capital Money managed by venture capital firms and invested during the expansion or growth stage of a venture's life cycle.

General partners The managers of a VC firm, who are the general partners and make all decisions regarding capital allocation.

Limited partners Investors in a VC fund who are limited partners and contribute capital, but are not involved in capital allocation decisions.

Limited partnership agreement The pact between a VC firm and its investors regarding allocation of returns on invested capital.

Carried interest The returns accruing to the managers of a VC fund that are "carried" by capital provided by the limited partners.

Exit event The event where a venture's founders and investors sell the venture or go public, converting their shares to cash.

Institutional lender A bank or other organization that makes loans to entrepreneurs.

Collateral Property deemed to be equivalent to a loan's value and that will become property of the lender in the case of default.

Security interest The interest a lender has in collateral pledged to cover a loan.

Personal guarantee Some loans require the entrepreneur to guarantee repayment, including use of personal funds if necessary.

Prime rate The interest rate lenders charge their top customers.

Term of a loan The length of time specified to pay back the loan with interest.

Restrictive covenants Loan provisions that must be adhered to by the borrower.

Call the loan If restrictive covenants are violated, the right of lenders to have the loan paid back immediately.

Small Business Administration loan A loan originated by a bank and backed by the U.S. Small Business Administration (SBA).

Units The blocks of share that equity offerings often sell rather than single shares. Each block is referred to as a unit.

Min/Max A minimum closing and maximum amount that will be accepted by many equity offerings.

Break the bank The specified minimum an investment reaches when the entrepreneur can begin to use the invested capital.

Subscription agreement An agreement signed by an investor specifying the amount they intend to invest at some future date.

Elevator pitch A short description of a venture that an entrepreneur might make to a potential investor during a short elevator ride.

ADDITIONAL READING

Feld, Brad, Mendelson, Jason, 2012. Venture Deals: Be Smarter than Your Lawyer and Venture Capitalist. John Wiley & Sons, Hoboken, New Jersey.

Cohen, Brian, Kador, John, 2013. What Every Angel Investor Wants You to Know: An Insider Reveals How to Get Smart Funding for Your Billion Dollar Idea. McGraw-Hill, New York.

Lechter, Michael, 2010. OPM: Other People's Money, The Ultimate Leverage, second ed. TechPress, Inc., Phoenix, Arizona.

Wilmerding, Alex, 2006. Term Sheets and Valuations: A Line by Line Look at the Intricacies of Term Sheets and Valuations. Aspatore Books, Eagan, Minnesota.

WEB RESOURCES

www.kickstarter.com: This website provides entrepreneurs, product developers, and designers with an opportunity to raise capital without trading away equity. Kickstarter fundraising campaigns are focused on rewarding capital providers in other ways, such as discounts on products.

www.secondmarket.com: This site enables entrepreneurs to sell equity interest in their venture based on the 2013 JOBS Act. Entrepreneurs provide their business plans and deal terms, and investors decide whether they want to jump in.

www.microventures.com: Another site offering to connect angel investors and entrepreneurs, this company also promises to conduct due diligence on every listed venture, possibly providing more security to investors if the venture should pass the DD process and become listed.

http://www.sec.gov/info/smallbus/qasbsec.htm: The U.S. Security and Exchange Commission's site discussing small business compliance with the securities laws.

ENDNOTES

1. 17 CFR §230.
2. See, e.g., Securities Act §2(a)(15), 17 CFR §§230.215, 230.501(a)17 CFR §§230.501(a)(5)-(6).

3. 17 CFR §230.501(h). In general, purchaser representative is required to be independent of the issuer, sophisticated with respect the subject matter of the investment, and acknowledged in writing to be the purchaser's representative in the transaction.
4. 17 CFR §§230.506(b)(2)(ii).
5. Securities Act §4(a)(5).
6. 17 CFR §§230.504, 230.505, 230.506.
7. Securities Act §4(a)(2).
8. 17 CFR §230.500 et seq., and particularly, §230.506.
9. 17 CFR §230.401 et seq.
10. Securities Act § 3(a)(11); 17 CFR §230.147.
11. 17 CFR §901 et. seq.
12. Kelvin W. Willoughby. How do Entrepreneurial Technology Firms Really Get Financed, and What Difference does it Make? *International Journal of Innovation & Technology Management*, 5(1): March 2008; p. 1–28.
13. Scott Shane. The Illusions of Entrepreneurship: The Costly Myths that Entrepreneurs, Investors, and Policy Makers Live By. (New Haven, CT: Yale University Press, 2008).
14. Ibid, p. 79.
15. Jeffrey Sohl, "The Angel Investor Market in 2012: A Moderating Recovery Continues", Center for Venture Research, April 25, 2013.
16. Robert Wiltbank and Warren Boeker. Returns to Angel Investors in Groups. Ewing Marion Kauffman Foundation, Kansas City, MO. November 2007.
17. For more information on this group, see http://atif.asu.edu.
18. Spectrum Academic Studies, 2012.
19. Katleen Baeyens, Tom Vanacker, and Sophie Manigart. Venture Capitalists' Selection Process: The Case of Biotechnology Proposals. *International Journal of Technology Management*, 34: 2006; p. 28–46.
20. Jason Cope, Frank Cave, and Sue Eccles. Attitudes of Venture Capital Investors towards Entrepreneurs with Previous Business Failure. *Venture Capital*, April-September 2004, p. 147–172.
21. Jonathan T. Eckhardt, Scott Shane, and Frederic Delmar. Multistage Selection and the Financing of New Ventures. *Management Science*, 52(2): February 2006; p. 220–232.
22. Richard Stanley. What a VC Looks for in Start-Ups. *Drug Discovery & Development*, 7(8): 2004; p. 15.
23. J.B. Roure and R.H. Keeley. Predictors of Success in New Technology Based Venture. *Journal of Business Venturing*, 5(4): 1990; p. 201–220.
24. R.A. Baron. Cognitive Mechanisms in Entrepreneurship: Why and When Entrepreneurs Think Differently than Other People. *Journal of Business Venturing*, 13: 1970; p. 275–294.
25. Alex Proimos and Wayne Murray. Entrepreneuring into Venture Capital, *The Journal of Private Equity*, Summer 2006; p. 23–34.
26. Glenn Townes. Financing a Business with Loans from Family and Friends. *The National Federation of Independent Business*, August 18, 2005. http://www.nfib.com/object/IO_24228.html.
27. L. Collins. Finding Funds. *Engineering Management*, 16(5): October 2006; p. 20–23.

CHAPTER 9

Launching the Venture

OAKLEY FOUNDER DOES IT AGAIN WITH RED

In the 1970s when Jim Jannard was in his 20s, no one would have envisioned him as a technology geek, or a future member of Forbes' wealthiest individuals. Jim was a biker who sold motorcycle parts out of his car. He was not a brilliant, hard-driving engineer working in an electronics or photography company. In 1975, Jim Jannard founded Oakley which manufactured motorcycle goggles that were not only good looking, but had toughness and optical clarity. This successful product was followed by even more successful ski goggles, and then sunglasses. Oakley went public in 1995 and was purchased by an Italian company, Luxottica, for $2.1 billion. Before the sale, a skull and crossbow flag flew at the company's militaristic-looking headquarters just outside of Los Angeles, California. The sale allowed Jim Jannard to buy several islands and private planes with Forbes citing his worth at $2.8 billion.

His passion for photography, personally owning over 1,000 cameras, allowed Jannard to shoot most of the photos and videos used by Oakley in its marketing campaigns, while creating the desire to push the limits of imaging chip technology and build a digital video camera that would have an output that looked as good as film.

In order to develop this camera, Jannard hired Ted Schilowitz. It was a daunting task—build a digital video camera that would be smaller and cheaper than a film camera and yet have output just as good. Particularly difficult would be to build a light-sensitive chip that would replace film in capturing an image. With an additional 200 employees working on the project, some of whom were physicists and mathematicians, a non-working prototype with specifications was displayed at the 2006 National Association of Broadcasters (NAB) showing a price of $17,500, one fifteenth the cost of a traditional film camera. Over 500 people left deposits of $1,000 to place an order for the camera and many more followed online.

It was not until early 2007, that two working prototypes emerged which were tested by Peter Jackson, who was the director of *The Lord of the Rings* trilogy. Jackson produced a 12-minute mini movie which the company showed at the 2007 NAB show. The camera, Red One, became available at a price of $30,000 with everything needed to shoot a movie 8 months later. And, with production limitations, most buyers with deposits did not receive their camera until the end of 2008. RED followed the same tactic announcing a newer model, Scarlet, at the 2008 NAB show for a price of $3,000. The camera was finally available for purchase in 2011, at a price of $10,000.

Continued

Peter Jackson purchased 50 RED cameras, and many others did as well. Steven Soderbergh used three RED cameras in shooting two *Che* films in Spanish jungles.

Of course, competition has emerged looking to take advantage of one of the problems of the camera—the images produced are hard-edged and cold, right for only certain types of filming. Panavision introduced Genesis, a much lower resolution camera, which produced better overall film quality, and ARRI in 2010 released a formidable digital competitor—the Alexa. The Alexa was used to shoot *Skyfall* and *Life of Pi*. Also, Canon in 2011 launched the C300, a digital cinematic camera. Many film makers feel that the C300 is a better camera than RED. Today, Canon's C500 and Sony's F55 are as sharp as and perhaps better than RED.

Overpromising on the delivery date and pricing are not the best ways to successfully launch a new technology product; even one so cutting edge like RED and Scarlet. In spite of this, the company has many devoted buyers and appears to have solidified a position as a player in the movie camera industry.

Sources: *http://www.red.com, http://www.forbes.com/profile/james-jannard/*

9.1 INTRODUCTION

One of the most critical strategies that needs to be developed by the technology entrepreneur is the one for launching the venture.[1] Strategies exist at four different levels within the technology venture: enterprise-level, corporate-level, business-level, and fictional-level. Enterprise-level strategy focuses on the relationships between the technology venture and society. This is usually not of paramount concern at the start of a venture, but needs to be reflected in the mission statement and the business model, as well as in the other strategies. The corporate-level strategy focuses on such issues as diversification and managing the portfolio of products and markets of the technology venture. Like the enterprise-level strategy, the corporate-level strategy comes more into play after the venture has been launched. The business-level strategy will be incorporated right away because it is involved in competing within a single industry or a few industries. It involves the acquisition and employment of resources to launch and grow the venture. And finally, the functional-level strategy is the base of all the other strategies and involves marketing, accounting, finance, and human resource policies that directly impact the technology venture throughout its existence.

This chapter will focus on business-level and functional-level strategies in launching the new venture. First, entry positioning will be discussed, followed by a presentation of penetration strategies and a discussion of the first mover advantages and value chain analysis. The chapter closes with a discussion of building a contingency plan, planning for change, and implementing corporate venturing.

MINI-CASE
TaskRabbit

TaskRabbit is an online market for outsourcing tasks and small jobs, founded by IBM software engineer Leah Busque, in 2008. The idea originally came about for small errands that could be outsourced online for a mutually agreed upon price. The firm started as RunMyErrand.com in Boston, with about 100 errand runners.[2] In 2009, the firm acquired $1.8 million in seed funding. The firm grew quickly, officially changing its name to TaskRabbit in 2010, and launching a mobile phone app in 2011. By that year, TaskRabbit had moved into several major cities and received an additional $17.8 million in funding. Today the company has received just under $40 million in venture capital funding and boasts over 20,000 "TaskRabbits." It is a unique player in the temporary hiring industry, which is estimated to be valued at $230 billion annually.

Source: http://techcrunch.com/2013/05/23/taskrabbit

9.2 MARKET ENTRY POSITIONING

Market entry positioning is not a fully developed strategy, but is a way a technology entrepreneur can get an initial foothold in a market. As is well known, the first sale and then the few sales to follow are the hardest to obtain. Yet it is necessary to get those initial sales as quickly as possible, so that more sales follow, with more and more revenue. In that way a positive cash flow is achieved as quickly as possible. This needs to be a, if not *the*, major focus of the technology entrepreneur. New technology ventures launch employing one or more of four major positioning strategies: a focus on aspects of the new product or service, parallel competitive parity, customer orientation, and/or government information.

Stressing the unique aspects of the new technology product or service is usually the best entry strategy because a new product or service employing a new technology having distinctive features will, at least, get the attention of potential customers. Such was the case of the RED camera, which used digital technology to record movie screen–quality pictures. Typically, new technology products have a lower failure rate than new technology services due to their higher barriers to entry. A winning combination is to offer the new technology product with its **unique selling propositions** and follow up with a new related service.[3] With this entry positioning, often a prominent and sometimes permanent leadership position can be created. The RED camera had this opportunity– if they had delivered on time at the price point originally offered. Having this position gives the new venture a head start, which can lead to significant public awareness and recognition.

A parallel **competitive parity** positioning strategy can also be used to effectively launch a new technology product or service.[4] This strategy can be employed when the features of the new technology product or service, while different, are not radically unique from the features of the products or services presently on the market. This is an attempt by the technology entrepreneur to fill a niche, a position in the market not presently being served. Such was the case for TaskRabbit: there were already service providers on the market filling the same need—hiring someone on a short-term basis to do a certain task.[5,6] A technology entrepreneur who targets the presently unhappy customers of a particular product or service of another firm, attempting to make them happier when buying his or her product or service, is employing a parallel competitive parity strategy. Of course, this entry strategy usually does not receive a strong market leadership position, because some other venture will employ a similar strategy later.

The third market entry positioning strategy—**customer orientation**—focuses on the customer and its changing attitudes and purchasing behavior.[3] The new Lincoln automobile is now geared toward a younger, more trend-setting customer.[7] Gone are the days when a significant segment of the population is satisfied with a steakhouse.

Sometimes customers can be obtained by a technology entrepreneur through becoming a second source of supply. When a customer is having difficulty with their present supplier, starting off as an alternative supply source, if done correctly, can lead to being the primary supplier. This strategy worked for Arnolite Pallet Company, who started off with its present customer base by being the alternative supplier of choice for the needed pallets used by several Fortune 500 companies. This can be an excellent entry strategy when there is not significant, radical uniqueness in the product or service being offered and there are delivery problems in the industry.

The final market-entry positioning strategy is **government information**. Often, information on new government rules or laws can provide the launch opportunity and entry strategy. The government can provide assistance, favored purchasing status, or rule changes. Such government agencies as the Small Business Administration,[8,9] NASA,[10] and Sandia National Labs[11] have programs that assist the launch of new technology ventures. Some of these are in conjunction with specific states. The U.S. federal government, as well as some state governments, has procurement policies that establish set-asides and quotas for small technology ventures.[12] Some of these procurement policies and practices have been incorporated into other government agencies and throughout large corporations.

Finally, as government regulations change and new laws are enacted, new launch opportunities often occur. Pressure and rule changes have mandated that utility companies encourage their customers to become more energy efficient and give credit for new appliances or the use of solar energy. One technology firm, R&H Safety Sales, launched by offering specific first-aid kits that met the requirements for specific industries. Similar firms started by supplying the mandated exit signage and safety devices.

9.3 MARKET PENETRATION STRATEGY

A **market penetration strategy** is a useful way for a technology entrepreneur to launch the technology product or service.[13,14] This strategy focuses on the company's existing product in one particular market. The goal is to penetrate that particular market with the existing product by encouraging customers to buy more, and, through satisfied customers and word of mouth, have additional customers in that market buy more. The marketing effort in the launch can encourage more frequent usage (purchasing) as well as reward customers who get others in the market to purchase through their referrals. Once a market is saturated, then follow-up products can be offered in the existing market or the same product can be introduced into a new market (market rollout). One company using regenerative medicine to cure arthritic hips in dogs such as golden retrievers is launching their medical product in San Diego through veterinarians, and then plans to roll out into the veterinary/dog owner market in Phoenix, followed by Dallas/Ft. Worth.

Another company successfully satisfied its customers in the Boston area with quality delivered bark mulch, then gave one free cubic yard of bark mulch to any customer referring another customer that purchased two or more cubic yards of quality delivered bark mulch. When market saturation was reached, the company then offered other delivered products, such as crushed clam shells to these same satisfied customers.

One company, Color Lines Clothing, started manufacturing and exporting high-quality children's clothing using such a strategy. Bela Katrak, a young woman with two young children, moved to Bengaluru, India in 1988. She could not find affordable, comfortable, durable ready-to-wear clothes for her kids. All the clothes available in the market were fussy, highly embroidered, and overadorned. Bela started her own clothing business with an investment of Rs. 40 Lakh (US$80,000). She first started her own retail outlet that sold high-quality kids clothes. But it was not long before she realized that the Indian market was not ready for her clothes. Indian mothers were not ready to buy "all cotton" clothes because maintaining vis-à-vis polyester was difficult. So she started exporting to other countries that were ready to accept these clothes and her business flourished.

Even though the international market was large, to start from one location and be successful was difficult. Bela started her journey in Australia. She made a lot of cold calls, but did not get any leads. She got her break when one large group was at a Target (U.S. chain) store. Her customer was focused more on quality clothes than quantity, and that was a perfect match for her. Today, Color Lines focuses on variety and styles. It prefers to have numerous orders with smaller runs than small orders with longer runs. It helps Color Lines to maintain unique product designs and styles. It is unique in India as it provides high fashion, high-quality, clothing for children sensitive to dyes. Color Lines has its own in-house studio and design team that works closely with buyers to provide garments that are high quality at an affordable price. They take high fashion adult wear from international designers and convert them into children's clothing. Color Lines remains the only export house in India that deals almost exclusively in children's wear.

Today, the exchange rate is a big concern for Bela. If the rupee appreciates, it will make India too expensive as a supplier for textiles. With an appreciating rupee, buyers may move to other countries for cheaper options. On the other hand, Bela is confident that the business, being based on mutual benefit and long relationships, will keep her buyers loyal to her and her company.

MINI-CASE
Apple Excels as First Mover

Apple has been incredibly successful at being the first mover in innovative technologies over the last decade. In 2001, they released the first iPod mp3 player, which sold over 100,000 units in its first year. In 2003, they launched the iTunes store, allowing consumers to purchase digital music online legally. They created the original smartphone, the iPhone, which was originally released in 2007 and has been a remarkable success. Building off of this technology, Apple launched the first tablet, the iPad in 2010. All of these products were the first of their kind on the market and have been able to maintain their advantage over the competition that quickly followed. Apple's proprietorship technology allows it to maintain greater entry barriers by increasing consumer switching costs and promoting brand loyalty.

9.3.1 First Mover Advantage

A well-planned launch of a new technology product or service provides the technology entrepreneur with the **first mover advantage**—being the first with the product or service on the market. Being first can result in a number of advantages that can enhance sales and profits of the new technology venture, such as developing a cost advantage, having less competition, developing better supply and distribution channels, gaining experience, and switching costs.[15] To be the first technology entrepreneur to launch a product in a market means that his or her company will be the first to start achieving a cost advantage through economies of scale and experience. The experience factor of producing more and more units enables the new technology venture to produce each unit more cost effectively. More units allows the new venture to spread its overhead costs, including the research and development costs, over more units and purchase in larger quantities, causing the cost of component parts and equipment and other supplies to decrease.

Also, being the first on the market (the first mover) means there is less competition in the market at launch. However, in a growing market, this advantage is lost as competitive technology products or services enter the market–unless exclusivity has been established by intellectual property protection of strategic features of the products or services. The first mover is usually more than compensated by the increasing sales from a growing market. These growing markets and increasing sales usually mean less, direct competitive actions such as price cutting.

Being the first mover provides the technology entrepreneur the opportunity to receive the best position in the supply chain both with suppliers and sellers. The best suppliers and distribution channel members can be selected and strong relationships developed. Exclusive relationships can sometimes be established through appropriate contracts. This may represent a barrier to entry for competing products.

First movers gain the experience of participation by being in the market. This can lead to improving the first generation of technology products or services, and to providing opportunities for developing new products for the satisfied customers obtained. The networks and satisfied customers provide knowledge and insight that can be very valuable to the technology entrepreneur.

Finally, the first mover may provide a barrier to entry by imposing switching costs on buyers. These switching costs can be established through marketing or contractual obligations. When customer satisfaction occurs, leading to brand loyalty, the high buyer learning and evolution costs that are inherent in purchasing a competitive product may inhibit a buyer from even considering another alternative. It is much easier, safer, and quicker to stick with the brand loyal product.

9.4 VALUE CHAIN ANALYSIS

To successfully enter a market, a technology entrepreneur needs to understand and untangle the value web in a particular market. The value risk is composed of many integrated value chains and is usually very interwoven and complex. A **value chain** is the sum of all the business activities that create value for the firm in an industry. These activities include engineering, design, marketing, production, shipping, warehousing, servicing, and support. The value chain activities are supplemented by the activities of their suppliers and channel members to create value for the consumer.[16]

To untangle the value risk, the technology entrepreneur should first identify the market leaders in the industry and determine which of their activities add value to their value chain, particularly those activities tend to be exclusive to their business. Is it their product features, their distribution, their warranty, or their quality of service?

Then the technology entrepreneur should identify the key customers, suppliers, channel members, and strategic partners in the market. These should be carefully examined to better understand the major costs and cost structures in the value chain. Hopefully, examining these cost structures will indicate a problem that can be solved by the technology entrepreneur, allowing him or her to develop a new value chain to solve the problem and provide value to the consumer. If this can be done with minimal impact on the way the customer presently does things, the new technology venture has a better chance for success.

9.5 DEVELOPING A CONTINGENCY PLAN

Regardless of the quality of the value chain established, and the launch strategy employed, there is no crystal ball for predicting success. A technology entrepreneur should consider many different scenarios and contingency plans in case changes are needed. Problems can occur with downturns and upturns in the economy, new products being introduced, and changes in consumer tastes and preferences. Any of these things can radically alter the market and industry and disrupt the planned strategy of the technology venture.[17]

To maintain growth in this turbulence, it is essential that the technology entrepreneur establishes and maintains product or service quality. A high-quality standard can be sustained and used as a competitive advantage through continuous improvement (the process of setting higher standards of performance with each interaction of the quality cycle); benchmarking (identifying and imitating the best in the world at various tasks and functions); and outsourcing (procuring the best quality from outside organizations).

MINI-CASE
Lincoln Motor Company Raises Quality in New Model

In 2006, Lincoln Motor Company released a new version of the Lincoln Zephyr, renamed the Lincoln MKZ, targeting the new younger generation of Luxury car buyers. Aiming for an entry-level luxury car market, the 2006 Zephyr came at a price of $29,995-$35,575. The car sold over 7,000 models in the first quarter of 2006 with an average buyer age of 56, whereas the typical age of a Lincoln Town Car buyer was above 70. The new purchasers were trading in nonLincoln Mercury model cars more than the company had seen with its sales of other models.[7] Lincoln continued to adapt the model, releasing updated versions in 2007 and 2010, as well as a hybrid version of the car in 2011. In 2013, the second generation MKZ hybrid was released and was declared the most fuel-efficient luxury vehicle in the U.S.

9.6 GROWING BEYOND THE STARTUP

Forces beyond the control of the technology entrepreneur always occur and push the venture in new directions. This requires new plans to be formulated and implemented in order for the venture to survive and grow.[18]

One of the corporate cultures that encourage change and innovation is corporate entrepreneurship. This concept is the focus of a book: Robert D. Hisrich and Claudine R. Kearney, *Corporate Entrepreneurship* (2012), New York: McGraw-Hill Books. **Corporate entrepreneurship**, which has also been called intrapreneurship or corporate venturing, is the process by which individuals inside an organization create and pursue opportunities that will better the organization. It usually involves pursuing opportunities independent of the current resources and involves formal or informal activities around creating something new of value to the organization in order to improve the company's competitive position and financial performance.

The basic aspects of corporate entrepreneurship, which can vary in intensity, or even presence, from organization to organization are indicated in the formula below:

$$L = I + O + C^2$$

where:

L = Level of corporate entrepreneurship
I = Innovation
O = Ownership
C = Creativity
C = Change

Each of these four aspects will be discussed in turn.

9.6.1 Innovation

Innovation is highly valued and necessary for the survival and growth of a technology venture.[19] Innovation can take three basic forms, as indicated in Exhibit 9.1—ordinary, technological, and breakthrough. As indicated, the majority of innovations (the innovation with the highest incident rate) are

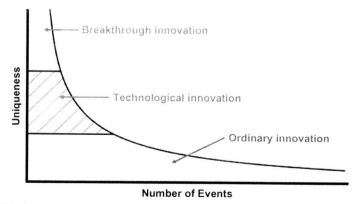

EXHIBIT 9.1
Three forms of innovation.

ordinary innovations. These are innovations that have the smallest change in the way things are presently done or in the features of the product or service being offered. It may be just a simple, but better change in the way the product tastes or performs, or a change in the packaging, or a new way of counting inventory.

The next infrequent type of innovation is the technological innovation. These are technological innovations that advance the process, product, or service beyond what is presently available. These, where possible, are protected by a patent, trade secret (confidentiality agreements) or trademark that covers any intellectual property involved.

The smallest number of innovative events fall under the category of break-through innovations. These cause a radical transformation from the way things are presently done and can impact lifestyles of the purchaser. These include such things as the personal computer, cell phone, Internet, social networks, and digital media. A technology company needs all these types of innovation constantly occurring so it can grow and prosper in its hypercompetitive, rapidly changing environment. Break-through innovations by successful companies are typically the subject of extensive intellectual property protection strategies.

9.6.2 Ownership

Ownership is also essential, and is actually a result of, corporate entrepreneurship because it reflects the overall organizational environment or culture of the technology venture. Ownership means that individuals in the company "own" or feel responsible for the job to such an extent that they want to make sure they do their jobs in the best way possible. They strive to perform in the most efficient and effective manner. They love to go to work. When this is part of the culture of a technology venture, it makes it very difficult for competition to enter.

9.6.3 Creativity

Creativity is the first C of the C^2 of the corporate entrepreneurship formula. This allows and even encourages individuals in the technology venture to bring into being from one's imagination something that is unique.[20] This often can be stimulated by the technological entrepreneur by using one of the creative problem-solving techniques indicated in Exhibit 9.2. Each of these will be briefly discussed below.

The first technique, **brainstorming**, the most widely used creative problem-solving technique, involves bringing a group of people together to focus on a defined problem under the guidance of a moderator. All the ideas mentioned in the meeting are recorded, and no criticism is allowed.

Given that it is easier for people to be positive rather than negative, reverse brainstorming is often used instead of brainstorming. Here, the group of

- Brainstorming
- Reverse brainstorming
- Brainwriting
- Gordon method
- Checklist method
- Free association
- Forced relationships
- Collective notebook method
- Attribute listing method
- Big-dream approach
- Parameter analysis

EXHIBIT 9.2

Creative problem-solving techniques. *Source: Robert D. Hisrich, Michael P. Peters, and Dean A. Shepherd, Entrepreneurship, 9th edition (2013), Burr Ridge, IL: McGraw-Hill/Irvin, p. 95.*

individuals first focuses on the negative aspects of a product, service, idea, or concept. After all are mentioned, the group then uses these to form the best possible solution. Again, these activities take place under the guidance of a moderator, and no criticism is allowed.

Brainwriting, a form of written brainstorming, is a silent, written generation of ideas by a group of people. Participants write down their ideas on special forms and then pass them around the group, which usually consists of 6-10 members. The process is moderated so that three ideas are generated by each individual during each 5-minute interval.

The **Gordon method**, unlike most other creative problem-solving techniques, starts with participants not knowing the exact nature of the problem. Someone will start by mentioning the general concept associated with the problem. The group then gives ideas to develop a concept. The actual problem is then revealed, with the group discussing and developing a final solution.

The **checklist method** develops a new idea through the use of a list of related issues or suggestions. A list of questions or statements can be used to guide the direction of the group in developing entirely new ideas or to focus on specific idea areas.

Free association, one of the simplest yet effective creative problem-solving techniques, is useful in developing an entirely new slant to a problem. A word or phrase related to the problem is written down, eliciting a new word or phrase in response, and then another and another. Each new word or phrase attempts to add to the process of creating a chain of new ideas.

Forced relationships is the process of forcing relationships among some product combinations. It asks questions about ideas in order to develop a new idea with each new combination, and eventually a new idea emerges.

In the **collective notebook method,** a small, easy-to-carry notebook, given to participating individuals, contains a statement of a problem with any needed background and blank pages. Participants consider solutions to the problem periodically throughout the day, recording the responses. At the end of the week, the list of ideas are gathered and summarized, and the group meets to select the best idea.

Attribute listing is a collective notebook technique where an item or problem is listed and then the group looks at it from a variety of viewpoints. Often originally unrelated objects are brought together to form a new combination or new uses.

The **big-dream approach** requires each individual to think big without constraints. Every possible idea needs to be recorded without worrying about any negative aspects or the resources required. Ideas are conceptualized without any restraints.

The final creative problem-solving technique, **parameter analysis,** focuses first on identifying the parameters that have to be considered in solving the problem. Once these have been identified, the relationships between them are examined developing the underlying issues. The solutions within these parameters are developed within these parameters—creative synthesis.

9.6.4 Change

In order for a technology venture to survive and grow, change must continuously be allowed and encouraged. This change should be constant, with continuous small steps to achieve greater organizational change.[21] Individuals in a technology venture tend to be more accepting of change when they can see, experience, and understand the change slowly and in small amounts. The change occurs incrementally and collectively, meaning that the technology venture should be continuously experimenting and modifying ideas and processes. The benefits of implementing corporate entrepreneurship in a technology venture are numerous, including:

- High performance culture and better morale established
- Employee turnover reduction
- Motivated workforce
- New business concepts
- New ways of doing things
- More flexible organizational structure
- Organizational learning
- Positive impact on revenues and profits

While generating new ideas for products or services and new ways of doing things, the venture has a motivated workforce that loves their jobs and likes coming to work. This of course, in turn significantly impacts the sales, profits, and returns of the company.

Employees benefit as well because they can obtain feelings of self-achievement and job satisfaction. They are able to be creative, can increase their skill sets in a learning organization, and can receive financial and nonfinancial rewards. These include:

- Feeling of self-achievement
- More job satisfaction
- Increased skills
- Financial and nonfinancial rewards
- Excitement toward their work
- Increased creativity

9.7 CHAPTER SUMMARY

This chapter focuses on the important tasks of launching the venture. In order to achieve a positive cash flow position where company revenues cover company costs, it is important to achieve sales and revenue as quickly as possible. The chapter starts by discussing four entry positioning strategies—focusing on aspects of the new product or service, parallel competitive parity, customer orientation, and government information. Then, aspects of a penetration strategy are discussed, followed by the advantages of being the first mover. The chapter concludes with a discussion of value chain analysis, building a contingency plan, and planning for change by implementing corporate venturing.

KEYTERMS

Unique selling propositions The unique aspects of the new technology product or service.
Competitive parity The features of the new technology product or service are not radically unique from the features of the products and services presently on the market.
Customer-or ientation The focus of the customer and their changing attitudes and purchasing behavior.
Government information Information provided by federal, state, or local governments to assist in the launch of new technology ventures.
Market penetration strategy A strategy to grow by encouraging existing customers to buy more of the firm's current products.
First mover advantage Being the first with the product or service on the market.
Value chain The sum of all the business activities that create value for the firm in an industry.
Corporate entrepreneurship Entrepreneurial action within an established organization.

ADDITIONAL READING

Akpoyomare, O.B., Adeosun, L.P.K., Ganiyu, R.A.. 2013. Approaches for generating and evaluating product positioning strategy. International Journal of Business Administration 4 (1), 46–52, This article explains how to develop a product positioning strategy. It investigates ways of evaluating the strategy using descriptive research, suggesting a positioning strategy that focuses on a niche, of which consumers do not normally consider.

Berte, E., Rodrigues, L.C., Almeida, M.I.. 2010. The lessons learned from the unique characteristics of small technology-based firms. International Management Review 6 (1), 62–70, The study tries to identify how aspects of small technology-based firms (STBFs) influence their formulation of strategy. The results suggest that the technical expertise of the founder shapes the mission and vision of the firm. It notes how STBFs focus their competitive advantage exclusively on market positioning strategy without considering other methods..

De Fan, T.P.C., 2010. De novo venture strategy: arch incumbency at inaugural entry. Strategic Management Journal 31 (1), 19–38. The article addresses market entry strategy for de novo ventures to avoid incumbent retaliation. It suggests avoiding large market overlaps with incumbents in the industry as well as creating an aggressive inaugural market entry or differentiated market positioning.

Kim, N., Im, S., Slater, S.F., 2013. Impact of knowledge type and strategic orientation on new product creativity and advantage in high-technology. Journal of Product Innovation Management 30 (1),136–53, This article reports on the results of a study that considers two dimensions of knowledge and two types of strategic orientation that are influential in determining positional advantages for new product development. The article explains how these variables may influence new product innovation and product advantages from the customer perspective. It suggests methods of knowledge sharing within an organization to maximize creativity with regard to new product development and market orientation.

Perren, L., O'Regan, N., Kling, G., Ghobadian, A., 2012. Strategic positioning and grand strategies for high-technology SMEs. Strategic Change 21 (5-6), 199–215, The article describes how firm size affects its strategic decisions. It also discusses how product innovation is stimulated by market stability and technology.

WEB RESOURCES

http://iipdigital.usembassy.gov/st/english/publication/2008/06/20080603214836eaifas5.498904e-02.html#axzz2hGXJmzwR: This is a U.S. government publication on entry strategy.

http://venturebeat.com/: This site is a leading publication for news and perspective on the most innovative technologies and funding for the latest tech ventures.

http://www.luxresearchinc.com/solutions/formulating-market-entry-strategies.html: This site provides entry strategy resources.

http://www.anzatechnet.com/tag/market-entry-strategy/: This is site is a blog on market entry.

ENDNOTES

1. Kotha, R., Y. Zheng and G. George. 2011. "Entry into New Niches: The Effects of Firm Age and the Expansion of Technological Capabilities on Innovative Output and Impact." *Strategic Management Journal*, 32(9): 1011-1024.
2. Sacks, D. 2013. "The Purpose-Driven Startup." *Fast Company*, 177: 92-106.
3. Dhamija, S., A. Agrawal and A. Kumar. 2011. "Place Marketing - Creating a Unique Proposition." *BVIMR Management Edge*, 4(2): 95-99.

4. Lien-Ti, B., C. Chia-Hsien and S. Yung-Cheng. 2011. "Positioning Brand Extensions in Comparative Advertising: An Assessment of the Roles of Comparative Brand Similarity, Comparative Claims, and Consumer Product Knowledge." *Journal of Marketing Communications*, 17(4): 229-244.

5. Geuss, M. 2012. "Got Too Much to Do? Hire a TaskRabbit." *PC World*, 30(3): 24.

6. Hortinha, P., C. Lages and L.F. Lages. 2011. "The Trade-Off between Customer and Technology Orientations: Impact on Innovation Capabilities and Export Performance." *Journal of International Marketing*, 19(3): 36-58.

7. Geist, L.C. and A. Wilson. 2006. "Lincoln Zephyr Attracts Younger Buyers." *Automotive News*, 80(6198): 62.

8. 2010. "SBA Expands Outreach to Small IT Business Owners." *Practicing CPA*, 34(7): 5-6.

9. Whittlesey, P. and S. Kulkarni. 2010. "Cable MSOs Ready to Support SMBs with Fiber." *Lightwave*, 27(8): 18-22.

10. Holmes, M. 2010. "SSTL CEO Predicts U.S. Breakthrough in 2011." *Satellite News*, 33(47): 4.

11. Stinnett, R. 2012. "Nanotechnology Policy and Education." *Journal of Business Ethics*, 109(4): 551-552.

12. 2012. "Report Says U.S. in Top Tier On Pro-Innovation Policies." *Telecommunications Reports*, 78(7): 36.

13. Mattare, M., M. Monahan and A. Shah. 2010. "Navigating Turbulent Times and Looking into the Future: What Do Micro-Entrepreneurs Have to Say?" *Journal of Marketing Development & Competitiveness*, 5(1): 79-94.

14. Schaltegger, S. and M. Wagner. 2011. "Sustainable Entrepreneurship and Sustainability Innovation: Categories and Interactions." *Business Strategy & the Environment*. (John Wiley & Sons) 20(4): 222-237.

15. Barnett, W.P., M. Feng and X. Luo. 2013. "Social Identity, Market Memory, and First-Mover Advantage." *Industrial & Corporate Change*, 22(3): 585-615.

16. Wachnik, B. 2013. "Analysis of IT Projects in the Models of Enterprise Value Building: A Summary of Research between 2010-2012." *Proceedings of the European Conference on Information Management & Evaluation*, 192-201.

17. 2013. "Business Development: Contingency Plans for a Difficult Economy." *Recruiter*, 6.

18. Czarnitzki, D. and J. Delanote. 2013. "Young Innovative Companies: The New High-Growth Firms?" *Industrial & Corporate Change*, 22(5): 1315-1340.

19. Finkle, T.A. 2012. "Corporate Entrepreneurship and Innovation in Silicon Valley: The Case of Google, Inc." *Entrepreneurship: Theory & Practice*, 36(4): 863-884.

20. Goddard, J. and T. Eccles. 2013. "Why Some Companies Consistently Outperform Their Rivals." *Business Strategy Review*, 24(4): 7.

21. Ford, Sv., E. Garnsey and D. Probert. 2010. "Evolving Corporate Entrepreneurship Strategy: Technology Incubation at Philips." *R&D Management*, 40(1): 81-90.

CHAPTER 10

Marketing and Selling Your Products

PERSISTENT MARKETING AND SELLING LAUNCHES SQUARE

The company, Square Inc., started in 2009 with a list created by Jack Dorsey, co-founder of Twitter, distributed to potential investors. The list had 140 reasons why the company would fail and counterpoints to each reason. Jack wanted to show these potential investors that he was aware of all the problems existing, but would nevertheless be successful in designing a product to allow anyone to accept credit card payments anywhere in the world—a particular frustration for small businesses throughout the world in the Internet age of online purchases. The company's first product was the Card Reader—a one inch square dongle that turns an iPhone, iPad, Android phone, or Android tablet into a credit card processor.

The idea for the product came from the frustration of friend Jim McKelvey on losing a $2000 sale of one of his hand-blown glass faucets because his company could not accept the potential buyer's credit card. McKelvey, Dorsey, and Tristan O'Tierney an iOS engineer, developed the prototype of Square to prevent this problem from occurring. In order to be able to work, the company would have to get the approval of major credit card companies. At that time, however, the major credit card companies allowed only entities with a merchant account to accept their credit cards, with the exception of one online aggregator—PayPal. It took 6 months and many demonstrations by Dorsey of the prototype for the major banks and credit card companies to change their rules and allow aggregators such as Square Inc. to act like a merchant or PayPal and accept credit card payments.

Using this first mover advantage, Card Reader successfully marketed the product, charging merchant users only 2.75% per swipe or a flat fee of $275 per month (for merchant users who swipe less than $250,000 per year). The Card Reader, which can be ordered free online on Square Inc.'s website, uses a creative point-of-sale app to replace the cash register. This allows a customer to swipe his/her credit card and sign on the screen with a finger.

The company continues to be on the cutting edge of technology for the new generation. It launched Square Wallet in 2011 – an app that allows customers to pay a merchant by just saying their names. Market acceptance has not been quite as fast as for the Card Reader, but by the end of first quarter 2013, 250 locations in the USA and Canada are allowing its use.

By combining quintessential technology with good marketing, Square Inc. has achieved significant success. It has raised more than $341 million in financing from venture capital funds and

Continued

other investors such as Starbucks. In spite of significant competition, (including a deal between Visa and Samsung by which Samsung's next generation of phones will have Visa's PayWave app already installed), the company had a valuation of $3.25 billion and 500 employees in 2013. The company plans to have Square available worldwide in the next years.

Sources: *https://squareup.com/; http://www.inc.com/audacious-companies/issie-lapowsky/square.html*

10.1 INTRODUCTION

Square Inc. like every successful company has one thing in common—customer focus. This, along with the commitment to solving problems through technology and to sound strategic marketing, allows these companies to successfully launch and grow. Being dedicated to understanding and satisfying the needs of customers in defined target markets, these companies motivate everyone in the organization to produce superior value leading to high levels of customer satisfaction.

Delivering customer satisfaction through marketing, the focus of this chapter, begins with understanding and selecting the target market(s) and then focusing. This area is followed by a discussion of the consumer purchasing (adopting) the product and the difficulty in disruptive technologies. The remainder of this chapter deals with a presentation of the marketing mix and building a marketing plan.

10.2 TARGET MARKET SELECTION

Selecting a **target market** allows technology entrepreneurs to be more efficient and effective by focusing on the segments identified (see Chapter 4) that can be best reached and satisfied profitably.[1] In target marketing, not only does the *who* need to be addressed, but also the *how* and *for what*, which are discussed later in this chapter. The target group of customers must be of sufficient size, able to be reached cost effectively, and be satisfied.

The degree of customer satisfaction depends on the perceived performance of the product or service in filling the customer's expectation. If performance matches or exceeds expectation, the customer is satisfied. Satisfied customers not only make repeat purchases and can become loyal customers, but also tell others about their good experience with the product or service—word of mouth advertising. The key is to match the performance of the company and its product or service with these expectations, making sure not to promise more than can be delivered.

How are customer expectations formed? Customers form their expectations of a product or service from past buying experiences, competition information, their experience of the performance of the product or service, and the opinion of others. A company satisfying these expectations creates value for the customer.[2] Customer value is basically the difference between the values gained from the product or service and the costs of its obtainment.

10.3 PURCHASING THE PRODUCT SERVICE

The purchasing process can be looked at in terms of five stages—need recognition, information search, information evaluation, purchase decision, and post-purchase behavior (see Exhibit 10.1). Sometimes these stages are collapsed, such as during a repeat purchase by a satisfied customer where there is no need for an information search or evaluation.

Customers buy things because they recognize a need and want to fulfill it. The stimulus of this need recognition can be from the inside (feeling hungry and having something to eat) or from the outside (seeing someone's computer and deciding to buy one).[3] This need recognition can be instantly satisfied with an immediate purchase or taking a longer period of time to fulfill, which often occurs in disruptive technologies.

In most situations, except when a need can be instantly gratified such as filling hunger by purchasing a McDonald's hamburger, information is needed. This information usually comes from one of four sources: personal sources, personal experience, public information, or commercial information.[4]

This information must then be evaluated, the most critical stage in the process, which forms the customer's expectation. The quality and characteristics of the product or service are evaluated based on both its characteristics and its utility. The goal is to obtain the best quality with the most utility at the best cost.

Based on this information evaluation, a customer selects a product or service and decides whether or not to purchase it. Social factors influence how long it will take to purchase, and if indeed a purchase will even be made. During the buying process, cusstomers are constantly assessing whether or not they will be satisfied with the product or service and receive the value expected.

EXHIBIT 10.1
Model of the customer buying process.

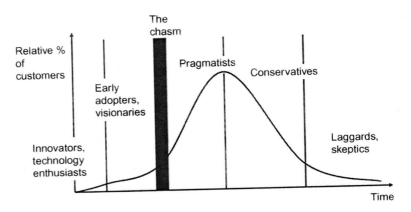

EXHIBIT 10.2
The product adoption curve.

The feelings of concern and perhaps not receiving satisfaction go on after the purchase—the postpurchase behavior. Sometimes cognitive dissonance occurs, which is an uncomfortable state occurring after the purchase due to: lack of product or service performance; new information; or a reappraisal of customer expectations. These feelings of dissatisfaction need to be successfully dealt with for customer satisfaction and so that repeat purchasing might occur.

The **adoption curve**[5] is a summation of customers making an initial purchase decision over time (see Exhibit 10.2). These customers can be classified into five groups: innovators, early adopters, late adopters, conservatives, and laggards. In the case of disruptive technologies, there is often a gap between the more innovative and early adopting groups purchasing the product and the remaining bulk of the market. This chasm, or gap of adoption, needs to be carefully bridged through a strategically constructed marketing plan in order to ensure success of the product or service.

10.4 MARKETING

The definition of marketing depends on the perspective of the individual and the discipline. Law, technology, finance, economics, operations, and the customer each view marketing from a distinct vantage point, or lens. In this hypercompetitive, rapidly changing technological environment, the following definition is most applicable. **Technology marketing** is the process of making decisions in a totally interrelated, changing business environment, or the activities that facilitate exchange satisfying the targeted customer while achieving the objectives of the company.[6] In order for an exchange (the process of obtaining a desired object, usually a product or service or experience, by offering something in return, usually money) to occur, sound management decisions need to be made on each controllable marketing activity. This exchange is best satisfied by the combination of marketing activities selected when everyone in the company focuses on the satisfaction of the target market. This is the essence of the marketing concept.

MINI-CASE
Whole Foods Differentiates on GMO Labeling

In March 2013, Whole Foods Market announced that all genetically modified organism (GMO) food products in their stores in the USA and Canada would require a GMO label by 2018. They are the first grocery store in the USA to enact this mandate. With increasing skepticism among consumers about GMO foods, particularly among health food consumers, they have created a valuable differentiated niche in the grocery store industry and perhaps have spearheaded a full movement toward GMO labeling. They currently sell more nonGMO Project verified products than any other retailer in North America, making their labeling strategy hard to replicate for grocery stores that may wish to follow their lead.

10.4.1 The Marketing Mix

The controllable marketing activities, or marking mix, indicated in Exhibit 10.3 are used in some combination to facilitate exchange.[7] They are best classified in the four areas indicated: product mix, price mix, placement (distribution) mix, and promotion mix. These are sometimes referred to as the 4 P's.

10.4.2 Product Mix

Product mix area relates to decisions by the technical entrepreneur regarding quality, breadth and depth of line, guarantee, service, and package.[8] A product or service is anything both tangible and intangible (including activities, benefits, or satisfaction) offered for consumption or use. If it is a technical product, the service component (in the form of installation, training, or repair) may be the most important element in the sale. Technical entrepreneurs need to think of their products at different levels. First, there is the core product which solves the problem of the customer: the basic bed of Jameson Inns for sleeping, the basic computer of Apple for analysis and processing, or the basic telephone of Samsung for communication. The core product needs to provide the benefits sought by the customer in solving the problem. These are the products' unique selling propositions, which, when matched with the benefits sought, help ensure the success of a technical product.

This core technical product must then be enhanced to deliver the actual product to be marketed. This enhancement usually includes some aspects of design, level of quality, features, brand name, and packaging. The Apple iPhone 5 is more than just a telephone: it provides access to email; it is a camera; it stores information;

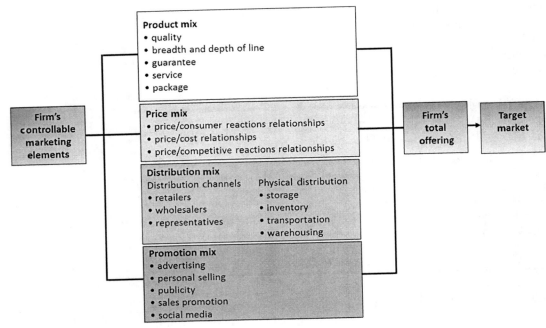

EXHIBIT 10.3
The marketing mix.

it has a design that is thin, light, and attractive; it has several colors as its package; and it carries the brand name Apple and all that this implies.[9] The transformation of the core product to the actual marketable product can be a challenge. One technical entrepreneur had a difficult time transforming the core product of a motorized wheelchair for children into a marketable one that would attract customers to purchase it. The final marketable product took one year to design at a cost of $500,000, which had the same operational and safety features as the core wheelchair product. Functionally, the wheelchairs were the same, but they had significant esthetic differences. Often, a technical product will have different levels of features so that different products can be offered at different price points to different target customers. This is a good way to develop a product line.

Branding is creating a name, design, logo, symbol, or a combination of these that identifies your company as the source of the product or service. One goal of a successful company is to turn that brand into a protectable trademark.[10] In choosing a brand name, the technical entrepreneur must make sure that it is not already registered in the market where the product will be offered. A brand name should usually avoid all geographic words because this may limit the expansion of the brand to other markets and, from a trademark perspective, is not inherently protectable. Some brands, however, use the word to take on some meaning associated with the brand. The city of Boston (USA) is known for its high standard of medical care and innovation, so a start-up company, Polymer Technology, named its first gas permeable contact lens the Boston Lens and its solutions the Boston Lens Cleaner and the Boston Lens Soaking Solution. The company was so successful in its introduction that it was purchased by Bosch & Lomb before the soft lenses were created and marketed.

A brand name needs to have the attributes of pronunciation (easy to say), connotation (related to the product area and not a negative one), and memorability (easily remembered).[11] Further, from a trademark perspective, the brand name should either be arbitrary (and unrelated to the product or service) or indirectly suggest the characteristics or nature of the product, rather than describe them directly. A brand-name that merely describes the product is typically not protectable as a trademark (see Chapter 5). An example of a logo that suggests the characteristics of the product is that of Whole Foods; the addition of a leaf over the "O" in Whole Foods' logo suggests the all-natural aspect of Whole Foods' product line while making it more appealing and approachable. The Whole Foods Market announced in early 2013 at the Natural Products Expo West, that all products that will be carried and sold in its stores in both the USA and Canada must have a label indicating whether the product contains GMOs.[12] This is the first national grocery chain to do so, but it may set a standard for all stores. Of course, not all brand names are "suggestive." "Apple", for example, is arbitrary as applied to electronic products; before Apple Computer, iPhones, and tablets, the term "Apple" was more closely associated with food and fruits than it was the computer area.

While packaging is often thought of as only applying to consumer technical products, it can be equally important for industrial technical products as well. Packaging involves designing and producing the wrapper or container for the package.[13] Sometimes there are two packages—one for the final product and one for shipping. A package needs to not only protect the product, but also be attractive, adhere to any applicable legal aspects and be adaptable to production line speeds. When a package is designed for sales, it needs to have the following features:

- Apparent size (given the perception of size without being deceptive)
- Attitude—drawing power (capture and hold the attention of a customer)
- Quality (convey the feeling of quality)
- Readability (the name and logo should be easily read)
- Aesthetically appealing (should have a good appearance)

For example, one small company producing a high-tech plastic extruding machine found that making the machine aesthetically appealing on the manufacturing floor was necessary for sales, even though the functional aspects of the machine did not change. For similar reasons, in 2013, Perrier water introduced a new slim aluminum 250 ml single-serve can in Canada when celebrating its 150th anniversary.[14] The cans contain the French carbonated water in several flavors—natural, lime, and pink grapefruit. Each flavor has no sugar and zero calories.

Package design has become very technologically advanced. Examples include:

- Appy Food & Drink Co. launched a new line of juice drinks for children in 2013. Each drink features a reality, technologically enhanced character from Nickelodeon. Each fruit drink is packaged in Tetra Pak cartons that allow consumers to scan the back of the carton with a smartphone to obtain props worn by a character. These props can then be stored on an iPhone and used to customize images.
- Marks & Spencer collaborated with Staeger Clear Packaging to create an intricately woven gold foiled tube for the company's dark chocolate–coated ginger. The packs, made from recycled plastic (polyethylene terephthalate (PET)), give a very classic, upmarket appeal to the product. The difficulty was being able to have the right heat and pressure to print a smooth detailed finish for the design.
- In 2009, Kimberly-Clark used computer screens outfitted with retina-tracking cameras to test the company's newest packaging for its Viva paper towels to determine which package design got noticed in the first 10 seconds a shopper looked at all the towels on the shelf.[15] This retina-tracking research was also used by Unilever for redesigning the bottle for Axe body wash. Using a virtual 3D environment, testers wearing specially equipped glasses outfitted with three balls with sensors, recorded, the eye movement of each consumer sideways and vertically within the virtual scene. The results made Unilever change the shape of the bottle from curvy to straight, increase the size of the product description, and make the brand more visible by embedding the black X in a blue background.

Additional features can then be added to create the final product for the market. These last additions to the product offering can include such things as guarantee, delivery, credit, installation, and after sales service. The guarantee is often very important, particularly for a disruptive technology that is new to the customer. It is also important to assure a new market that the new product is a good one to purchase. Swatch first entered the United States by guaranteeing its watches for life. Swatch also made it easy for U.S. customers to deal with a foreign company (Swiss) by opening an office in the United States. Cross Inc., a U.S. company founded in 1846, guarantees its pens and pencils for life. Usually a new pen or pencil is sent to the customer whose product ceases to work properly regardless of the date of purchase. Jameson Inns guarantees a good night's sleep or you do not have to pay for the room.

After sales, service can be the most important factor for the customer buying the product. Knowing that service is readily available or that it will be delivered in a prompt manner is important in certain purchase decisions. Discount Tire, a nationwide chain in the very competitive automobile tire business, will rotate the tires on a truck or car at no charge every 5,000–7,000 miles. And, this offer is even for tires not purchased at one of their stores. As one store manager explained to the author: "When rotating the tires, if one needs replacing, chances are the customer will do it then. And even if no tire needs replacing, the customer has now experienced our fast, friendly service." Minneapolis-based Best Buy created a better experience in its after-sale computer service business by creating the Geek Squad dressed in white pants and black ties and driving old cars to repair your computer.[16]

10.4.3 Pricing

Probably the area of the marketing mix most poorly done by entrepreneurs is pricing.[17] Two internal factors affect overall pricing decisions. One is the objectives for the product. These objectives can be: acquiring sales as quickly as possible; maximizing cash flow and profits; acquiring a strong market share and market position; indicating a level of product quality; and survival. The first objective—acquiring some sales as quickly as possible—is often a primary factor; it provides the technical entrepreneur with some needed cash and allows more funding to be raised, which is often needed in a technology venture.

The second internal factor—other elements of the marketing mix—should always be a consideration. Pricing is just one aspect of the offering of the technology venture. It needs to be carefully considered along with the aspects of the other three elements—product, distribution, and promotion—in order to develop a consistent, effective marketing mix. Some technical entrepreneurs have the price drive the entire marketing mix, and they price according to market conditions and competitive prices. Preferably however, a technology entrepreneur deemphasizes price and focuses instead on the product being offered and its unique features. This is usually a better strategy. It is almost always better when introducing a technological product not to have the lowest price, but rather have something different in the rest of the marketing mix that supports a higher price.

The three fundamental aspects of pricing (sometimes referred to as the 3 C's) as indicated in Exhibit 10.3 are cost, competition, and the consumer. Cost is the floor, the foundation, that the technology entrepreneur needs to determine the price. The price set for a product or service must cover all costs of producing, distributing, and selling the product or service and deliver an adequate rate of return. Costs take two forms: fixed and variable. Fixed costs, often referred to as overhead, are those costs that do not vary with the production or sale of the product. At the start of the venture, these should be very low except for any research and development costs for the product. Variable costs, sometimes referred to as "Cost of Goods Sold" (COGS), are those costs that do vary with the level of production. These costs should decrease as the company becomes more efficient and achieves some economies of scale of purchasing and production. One cost that needs to be taken into account is the cost of the channel of distribution used, which is considered under the distribution discussion of the market mix later in this chapter.

The second C—competition—establishes the boundary for the price. Unless there is something radically different about the product due to its technology,

technology entrepreneurs should price their products within these boundaries. This pricing strategy requires a thorough knowledge of all products presently filling the same market need, their features, and price. These competitive prices establish a benchmark for pricing and will aid in determining the differences in the product as well as possible prices to be considered. This analysis also helps the entrepreneur to gauge the competition's most likely reaction to your market entrance.

The final C, and the most important one, is the consumer. Persuading a consumer into becoming a customer is the ultimate goal of pricing and the entire marketing mix. The consumer, whether in the consumer, industrial, or government market, determines whether the price and/or product offering is the right one. The customer exchanges something of value (the price) for something of value (the benefits) when buying a product. Customers need to perceive value from the purchase and then realize that value once the purchase is made.

Two overall general pricing strategies are often used when introducing new products—market-skimming pricing and market-penetration pricing. Most technological products are introduced through **market-skimming**.[18] Here, a higher price is set so that more revenues can initially be skimmed from the market. Prices later are lowered as economies of scale occur and competition enters the market. This pricing strategy is most appropriate when: the image and quality of the technology product supports the price, such as, e.g., the Apple iPhone 5; enough customers will want the product at that price, such as the Apple iPad; high research and development costs and the high costs of the initial production runs exist, such as with a new Intel computer chip; or competitors are not able to enter the market quickly and easily, because of, e.g., market entry barriers or intellectual property protection.

The other general pricing strategy to introduce a new product is **market-penetration pricing**.[19] In this case, a new technology product is introduced at a low price in order to capture more quickly a larger market share. This strategy tends to work best when: a higher price would cause potential buyers not to purchase; volume production of the product will quickly and significantly lower the costs; any overhead costs are best allocated over more units; or the product has a very short life cycle and loses its distinctiveness very quickly.

10.4.4 Placement (Distribution)

The third part of the marketing mix is distribution, an increasingly important area in the rapidly changing global economy. It is composed of two rather distinct areas: channels of distribution; and physical distribution.

Many technology products are taken to market through intermediaries—members of a distribution channel. A distribution channel is a group of independent organizations involved in making a technology product or service available for purchase in the market. In moving technology products or services from the

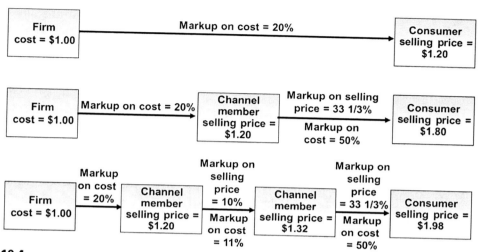

EXHIBIT 10.4
Channel members and price.

firm to the consumer, these independent organizations provide several functions: contact, marketing, promotion, and information. At times, channel members can provide financing, risk taking, and some negotiation. Of course, these services are provided at a cost which increases the selling price to the customer. As is indicated in Exhibit 10.4, the price can increase from $1.20 (direct) to $1.98 on a product costing $1.00, depending on the number of channel members used and their respective markups for handling the product. While manufacturers think of markups (margins) on a cost basis, channel members think of markups (margins) on a selling price basis. The conversion formula is:

% markup on cost = (% markup on selling price)/
(100% − % markup on selling price)

Of course, due to a larger denominator, the percentage markup on selling price is always lower than the percent markup or cost (see Exhibit 10.4). The two prominent channel members are retailers and wholesalers.

An interesting channel member that is widely used, particularly at start-up by technology entrepreneurs, is a manufacturer agent or broker. A manufacturer agent or broker can represent the technology company and sell its technology products on a commission (% of sales) basis. They do not take ownership (title) to the product or service, but market it to potential customers. They bring contacts and credibility, which are important for a new technology company. Significantly, they are only paid when a sale is made. This tends to mitigate potential cash flow problems for the emerging company.

The other aspect of distribution is physical distribution or the physical movement of the technology product or service to the customer. This can involve storage, inventory, transportation, and warehousing and has had increasing importance, particularly as the cost of transportation has significantly increased and will continue to do so. Physical distribution is directly related to customer service and can cause serious problems and costs, especially when the product is not available for purchase (out of stock conditions) or a product is not available for use when needed (late delivery).

The management of the distribution mix is called **supply chain management**. This is a management method for providing better customer service at a lower cost for the technology product or service through teamwork inside the company and among all the members of the channel system. Inside the technology company, each functional department needs to work closely together to maximize the logistics performance.[20] Outside, the company needs to integrate its logistics system with those of suppliers, channel members, and customers to maximize the performance of the entire distribution system.

MINI-CASE
Squishable Uses Facebook to Test Market New Products

The stuffed animal company Squishable.com found an innovative way to research which new stuffed animal it developed would be a success in the market. They use Facebook to survey over a quarter million fans to vote on their next creation. The top animals then enter the design phase. After designing the new animals, Squishable.com posts the sketches on Facebook for feedback. Facebook fans decide all final designing decisions on the stuffed animal creation, down to the colors. The new products are often back-ordered for months and Squishible.com continues to grow its Facebook fan base.

10.4.5 Promotion

The final part of the market mix is promotion. Promotion involves managing and integrating five areas:

- Advertising (paid-for communication of the technology product or service)
- Personal selling (personal presentation of the technology product or service)
- Publicity (nonpaid presentation of the technology product or service)
- Sales promotion (nonsecuring activities promoting the technology product or service)
- Social media (promotion in the Internet age)

Each of these five areas has several elements, as indicated in Exhibit 10.5.

Advertising is probably the most widely known, visible promotional activity. The primary components of advertising are media selection and message

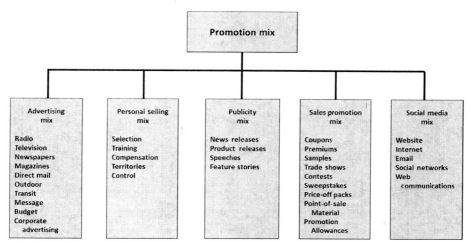

EXHIBIT 10.5
The promotion mix.

development. Such media as television, radio, magazines, newspapers, direct mail, outdoor, and transit are selected in light of the message to be communicated, the budget available, and the target audience. When possible, the technology entrepreneur should use cooperative advertising where the costs are shared with another entity, usually a channel member.

Unlike advertising, which is nonpersonal in nature, personal selling involves a personal presentation of the technology product or service to an individual or group.[21] Personal selling can be done face-to-face, on the telephone, or through the Internet. Personal selling involves such activities as selection, training, compensation, territories, and control.

Sales promotion is one of the least understood, diverse promotional areas, consisting of a variety of promotional activities that are not considered advertising or personal selling. These include such things as coupons, samples, trade shows, sweepstakes, price-off packs, point-of-sale material, and channel member promotional allowances. While not as understood, this promotional area is increasing in use and is a good method for the technology entrepreneur to develop a creative way to promote his or her technology product or service.

Publicity is a nonpersonal form of promotion that is not directly paid for by the technology company. Publicity involves a favorable presentation of the technology product or service, company, or entrepreneur, which tends to be viewed more favorably by a customer. It has more credibility because the source of the message is not perceived to be the company. It usually occurs in a print or broadcast media generated through news releases, product releases, or company events covered by the media. Speeches by employees, particularly the technology entrepreneur, are an effective form of sales promotion because customers more easily relate to people than products or services. Steve Jobs was an excellent spokesperson for Apple and had a significant influence on each new product introduced and the overall sales of the company. Word-of-mouth advertising or satisfied customers favorably talking about the technology product or service or referring the company to other potential customers is the best form of sales promotion because it has the highest level of source credibility.[22]

The final area of the promotion mix—social media—has become one of the most important areas of the promotion mix in this Internet age. Consisting of such diverse aspects as website, Internet, e-mail, web communications, and social networks, this area has now become the largest expenditure category in the promotion budget of any company. It is imperative that a technology entrepreneur have, in place at start-up, a website where information about the technology product or service and company can be obtained and the product or service purchased. All technology companies today need to establish an e-marketing presence that includes creating a website, placing ads and promotions online, creating or participating in web communities (social networks), and using e-mail and webcasting. An effective e-marketing presence is essential in today's global economy. Online marketing is a powerful tool for attracting customers, increasing sales, communicating company and product or service information, delivering products or services efficiently and effectively, and building customer relations.

The top five social media platforms for businesses are: Facebook, Twitter, LinkedIn, YouTube, and blogging, and, of these, Facebook and Twitter are the most popular. According to the *Social Media Examiner*, 97% of all businesses use social media, with 86% of them indicating that it is vital to the business.

Of increasing importance is using social media in marketing research, particularly since 80% of the businesses surveyed use social media sites to monitor and extract information. For example, when Walmart wanted to know whether to stock lollipop-shaped cake makers in its stores, it studied Twitter chatter. Estée Lauder's MAC cosmetics brand asked social media users to vote on which discontinued shades to bring back. The stuffed animal brand Squishable solicited Facebook feedback before settling on the final version of a new toy. And Samuel Adams asked users to vote on yeast, hops, color, and other qualities to create a crowd-sourced beer, an American red ale called B'Austin ale that got rave reviews. Also, Gilt asks customers to vote on which products to include in a sale and sets up Facebook chats between engineers and customers to help refine products. Kohls, which started asking its Facebook fans in July 2012 to pick products for inclusion in sales, said those fans were more heavily represented than its overall customer base in the 18-to-24-age demographic.

A unit, now called @WalmartLabs, looks at Twitter posts, public Facebook posts, and search terms on Walmart.com, among other cues, to help Walmart refine what it sells. Its technology can identify the context of words, distinguishing "Salt," the Angelina Jolie movie, from salt, the seasoning. @WalmartLabs noticed that enthusiasm for "The Avengers" and "The Dark Knight Rises" was surging before the movies were released and suggested that stores increase their orders of related merchandise. And, after Walmart started carrying a spicy chip called Takis, @WalmartLabs found that most of the positive chatter about it was coming from California and the Southwest.

MINI-CASE
Wal-Mart Labs Creates Seamless Shopping Experience

@WalmartLabs combines mobile, online, and stores to allow shoppers to buy whenever, wherever, and however they desire. To do this, they have brought in new technologies including Inkiru, OneOps, Tasty Labs, and Torbit. Using these advanced tools allows Walmart to obtain better analytics in order to enhance the performance of their website, allowing for a more flexible and customized online shopping experience. The @WalmartLabs team, made up of technology entrepreneurs, web developers, software engineers, and marketers, has created a state of the art online shopping experience for online consumers.

10.4.6 The Promotion Budget

One of the most difficult and yet important questions for the technology entrepreneur is to determine how much to spend on promotion. Total promotional spending varies greatly as a percent of sales from industry to industry. The technology entrepreneur needs to know the promotional expenditure by size of firm (small, medium, large) in his or her industry. Several key factors should enter into establishing the promotion budget: the product, its features, and stage in the life cycle; the channels of distribution; the promotion and company objectives; and most important, the target customer and characteristics.

Considering these factors, there are five methods that can be used for determining the promotion budget: arbitrary determination method, competitive parity method, objective and task method, percent-of-sales method, and affordable method.

In the **arbitrary determination method**, the promotion budget is determined in a seemingly arbitrary manner.[23] When using this method, technology entrepreneurs rely on their intuition and past experience in establishing the budget. This is a very unsophisticated method, but more firms use this method than any other one. A common benchmark used more frequently than others is 10% of last year's sales.

In the **competitive parity method**, a technology entrepreneur uses the promotion expenditures of competitors to establish their own budget. This information on the competitor's budget is also useful in evaluating the overall competitive environment and can be obtained through carefully analyzing a public company's financial statements or through a service that monitors all the advertisements in a given product or market area for a fee. The competitive parity method has some drawbacks because it employs historical data of competitive expenditures and assumes that all have very similar marketing situations.

Probably the best method for establishing a promotion budget is the third method—the **objective and task method**. When using this method, you establish definitive promotion objectives and then determine the amount and cost of the promotion to reach these objectives by costing each element of the promotion mix needed. Not only does this method allow for each element in the promotion mix to be effectively used, but it also makes sure that the amount of money to be spent on promotion is commensurate with the task at hand. A clearer basis for evaluation of the promotion expenditures also results.

The fourth method—the **percent-of-sales method**—is also widely used, mainly because it is easy to use. This method involves applying a fixed percentage to either past or future sales figures. If last year's sales were $200 million and the percentage used is 3%, the promotion budget for this coming year would be $6 million. The sales figure used can be the sales for the last year or an average of sales achieved over the past several years. By using an average sales figure, the impact of any erratic sales fluctuations on promotion expenditures is eliminated. You can also use a percentage of future sales, basing the promotions budget on future conditions, not on past events. When you use the percent-of-sales method for determining the promotion budget, you should compare it to the percents of leading competitors as well as industry norms.

The final method—the **affordable method**—is simplest to employ. It involves the technology entrepreneur setting the promotion budget at the level he or she thinks the company can afford. The level is usually determined by starting with total revenues, deducting operating expenses and other capital outlays, and devoting some part of the remaining funds to promotion. This is probably the most unsatisfactory method for establishing a promotion budget because it totally ignores the effects of promotion on sales and therefore revenues. It in actuality places promotion last in spending priorities, ignoring the fact that promotion is critical to the successful launch of the technology product or service and sales and growth of the company.

Regardless of the method used and the resulting size of the promotion budget, each promotion expenditure needs to be carefully evaluated to determine if it is effectively reaching and impacting the target customers, accomplishing the desired objectives, and achieving the desired sales.

10.5 CHAPTER SUMMARY

This chapter focused on the markets of the technology venture. Aspects of customer behavior and the impact on purchasing (adopting) the technology were discussed, along with the chasm occurring for disruptive technologies. The four aspects of the marketing mix were presented—product, price, placement (distribution), and promotion—as a way for the technology entrepreneur to inform the target customers about the nature and aspects of the technology product or service and company. The chapter concluded with a discussion of the promotion/marketing budget and methods by which it can be established.

KEYTERMS

Target market Specific group of potential customers toward which a business aims its marketing plan.
Adoption curve Summation of customers making a purchase decision over time.
Technology marketing The process of making decisions in a totally interrelated changing business environment or the activities that facilitate exchange satisfying the targeted customer while achieving the objectives of the company.
Market-skimming A process whereby a higher price is set so that more revenues can initially be skimmed from the market.
Market-penetration pricing A product introduced at a low price in order to capture more quickly a larger market share.
Supply chain management The management of the distribution channels.
Arbitrary determination method Determining the promotion budget in an arbitrary manner.
Competitive parity method A method a technology entrepreneur employs where the entrepreneur uses the promotion expenditures of competitors to establish his or her own budget.
Objective and task method A method where an entrepreneur uses where he or she establishes definitive promotion objectives and then determines the amount and cost of the promotion to reach these objectives by costing each element of the promotion mix needed.
Percent-of-sales method Involves applying a fixed percentage to either past or future sales figures.
Affordable method Involves setting the promotion budget at the level the technology entrepreneur thinks the company can afford.

ADDITIONAL READING

Ai, S., Du, R., Hu, Q., 2010. Dynamic optimal budget allocation for integrated marketing considering persistence. International Journal of Information Technology & Decision Making 9 (5), 715–36, This article details a model that attempts to create a dynamic optimal marketing policy. The model uses marginal analysis to determine the ideal budget application between different sales promotion strategies to maximize profits.

Dovleac, L., Bălășescu, M., 2012. Marketing mix for consumer high technology products. Bulletin of the Transylvania University of Brasov. Series V: Economic Sciences 5 (1), 41–8, The paper analyzes the marketing mix for high technology products. It details the fundamental marketing strategies used by companies in high technology for targeting consumer needs.

Jones, R., Suoranta, M., Rowley, J., 2013. Strategic network marketing in technology SMEs. Journal of Marketing Management 29 (5/6), 671–97, This article details a study that investigates strategic networks from the perspective of entrepreneurial marketing. The study develops the "Strategic Network Marketing Model," which classifies six types of strategic networks and their respective opportunities to create value.

Pullen, J.P., 2011. Dollars, sense and social media marketing. *Entrepreneur* 39 (6), 59–61, The article describes the use of social media marketing. It explains digital marketing and brand development strategies within online social networks Facebook and Twitter.

WEB RESOURCES

http://www.census.gov/: This site provides the latest data from the U.S. Census on demographic trends in the United States and around the world.

http://www.emarketer.com/: This is a for-profit market research site that is a leading destination for technology ventures to understand market trends.

http://www.gartner.com/: The Gartner Group is recognized around the world as a leader in the analysis of market and economic data to detect and describe market trends.

http://www.marketingpower.com/Pages/default.aspx: This is the website for the American Marketing Association, the premiere marketing trade and academic community. There are a number of useful resources on this site, including definitions of all the key marketing terms.

http://www.marketresearch.com/: This website has free and fee-based market reports on a wide range of industries.

ENDNOTES

1. Jiatao, L. and Weiping Liu. 2013. "Selecting a Target Segment: Market Structure and New Venture Entry Strategies." *Management Decision*, 51(7): 1402–1421
2. 2012. "Customers' Expectations Are Even Greater." *DIY Week*, 11.
3. Johnson, B. 2013. "Do Brands Really Understand the Needs of Customers? It Seems Not." *Marketing Week*, (01419285) 6.
4. Yoon, V.Y., R.E. Hostler, Z. Guo and T. Guimaraes. 2013. "Assessing the Moderating Effect of Consumer Product Knowledge and Online Shopping Experience on Using Recommendation Agents for Customer Loyalty." *Decision Support Systems*, 55(4): 883–893.
5. Van Ittersum, K. and F.M. Feinberg. 2010. "Cumulative Timed Intent: A New Predictive Tool for Technology Adoption." *Journal of Marketing Research (JMR)*, 47(5): 808–822.
6. Jones, R., M. Suoranta, and J. Rowley. 2013. "Strategic Network Marketing in Technology SMEs." *Journal of Marketing Management*, 29(5/6): 671–697.
7. Mintz, O. and I.S. Currim. 2013. "What Drives Managerial Use of Marketing and Financial Metrics and Does Metric Use Affect Performance of Marketing-Mix Activities?" *Journal of Marketing*, 77(2): 17–40.
8. Cannon, J.N., H.M. Cannon and J.T. Low. 2013. "Modeling Tactical Product-Mix Decisions: A Theory-of-Constraints Approach." *Simulation & Gaming*, 44(5): 624–644.
9. Lashinsky, A. 2013. "Room to Grow." *Fortune*, 168(2): 70.
10. Centeno, E., S. Hart and K. Dinnie. 2013. "The Five Phases of SME Brand-Building." *Journal of Brand Management*, 20(6): 445–457.
11. Schmitt, B. and S. Zhang. 2012. "Selecting the Right Brand Name: An Examination of Tacit and Explicit Linguistic Knowledge in Name Translations." *Journal of Brand Management*, 19(8): 655–665.

12   2013. "Mandatory GMO Labeling Coming to Whole Foods." *Label & Narrow Web*, 19(3): 14–16.

13   Lajante, M., O. Droulers, E. Jamet, S. Lacoste-Badie and M. Minvielle. 2013. "Influence of Packaging Visual Complexity upon Attention to the Brand." *Revue Des Sciences De Gestion*, 48(261/262): 10.

14   2013. "Perrier Reveals Slimmed Down Can." *Cantech International*, 20(7): 43.

15   2010. "Just Out." *Brand Packaging*, 14(4): 44–45.

16   Capozzi, M.M., R. Dye and A. Howe. 2011. "Sparking Creativity in Teams: An Executive's Guide." *Mckinsey Quarterly*, 2: 74–81

17   Huang, X. and G. Sošić. 2010. "Analysis of Industry Equilibria in Models with Sustaining and Disruptive Technology." *European Journal of Operational Research*, 207(1): 238–248.

18   Ingenbleek, P.T.M., R.T. Frambach and T.M.M. Verhallen. 2013. "Best Practices for New Product Pricing: Impact on Market Performance and Price Level Under Different Conditions Best Practices for New Product Pricing: Impact on Market Performance and Price Level Under Different Conditions." *Journal of Product Innovation Management*, 30(3): 560–573.

19   Vander Schee, B.A., T.W. Aurand, S. Suszek, A. Bastarrica, C. Asiegbu and B. Butler. 2010. "Integrated Pricing Strategy: A Corollary to Integrated Marketing Communications." *Proceedings of the Marketing Management Association*, 60.

20   Rezaei, J. and R. Ortt. 2012. "A Multi-variable Approach to Supplier Segmentation." *International Journal of Production Research*, 50(16): 4593–4611.

21   Atkinson, W. 2011. "The Role of Distributors and Dealers in New Technology." *Material Handling & Logistics*, 66(9): A-D.

22   Berger, J. and E.M. Schwartz. 2011. "What Drives Immediate and Ongoing Word of Mouth?" *Journal of Marketing Research (JMR)*, 48(5): 869–880.

23   Parsons, R. 2012. "Marketers Set Budgets High Despite Dip in Confidence." *Marketing Week*, (01419285) 35(4): 10.

CHAPTER 11

Contracts

INFORMAL AGREEMENTS CAN LEAD TO INTRACTABLE ARGUMENTS

A litigation involving Facebook.com founder Mark Zuckerberg, described by one judge as a "blood feud," illustrates the need for definitive written agreements. In December 2002, Harvard student Divya Narendra went to his fellow students Cameron and Tyler Winklevoss with an idea for an online social network for college students. They founded a venture, originally called "Harvard Connection" (later called ConnectU), to develop the idea.

Harvard Connection needed a programmer. Mark Zuckerberg, who later founded Facebook, had gained notoriety when he was placed on probation by Harvard University for hacking into the school's servers. Harvard Connection hired Zuckerberg and provided him their business plan and code. It wasn't long after that a dispute arose.

It is alleged that Zuckerberg understood that the information provided was proprietary. There was, however, no written agreement. Supposedly, Zuckerberg stalled Harvard Connection's launch while working on Facebook. In January 2004, 3 days after telling the Harvard Connection team that he had their coding done, Zuckerberg registered the domain name "TheFaceBook.com." E-mail correspondence indicates that the Facebook site was complete prior to a final meeting between Zuckerberg and Harvard Connection.

On September 2, 2004, ConnectU sued Zuckerberg for breach of contract, misappropriation of trade secrets, and copyright infringement. Zuckerberg countered that there was no contract, that he had never promised Harvard Connection anything, and that he was simply helping them out. The litigation lasted 3 years before it was ultimately settled prior to trial.

By February 2008, ConnectU had not been provided with its settlement terms. Nonetheless, the parties agreed to suspend proceedings and submit to mediation. After a two-day mediation, a one-and-a-half page "Term Sheet & Settlement Agreement" was signed. Facebook agreed to give the ConnectU principals a certain amount of cash and Facebook stock in return for their stock in ConnectU. However, ConnectU learned that its understanding of the value of the Facebook stock was in error, and the parties could not agree on the formal documents. ConnectU took the position that the mediation document lacked crucial terms and was insufficient to establish a meeting of the minds. The court disagreed. It held that the document was sufficient to create a contract and that ConnectU could have conditioned the agreement on a specific Facebook valuation or done its own due diligence. It had done none of those things.

Continued

The moral of the story is that complete and definitive agreements can prevent much heartache. Oral agreements and incomplete, imprecise written agreements are grist for the litigation mill.

Source: *ConnectU, Inc. versus Facebook, Inc. et al.*, District Of Massachusetts, case number 1:2007cv10593; *The Facebook, Inc. versus ConnectU, LLC et al.*, Northern District of California, case number 5:2007cv01389; O'Brien. Poking Facebook. November/December 2007.

11.1 INTRODUCTION

A contract is an agreement between two or more parties that is legally binding and enforceable in a court of law. Performance in accordance with the agreement is an obligation or duty under the law. A failure to perform is a breach for which the law offers a remedy.

Businesses are free to agree, within broad legal limits, upon whatever terms they see fit. Those terms define respective rights and obligations of the parties, and/or rules under which they operate, for the duration of the contract. A contract serves, in effect, as the private law of these businesses. That means that it is extremely important to understand when a contract is created and how it can be enforced. A failure to form a valid contract when intended, or inadvertently committing a business to obligations, can be catastrophic. (Just ask the Winklevoss'.)

In this chapter, we will consider the basic sources and concepts of contract law, the anatomy of a typical agreement, and introduce various types of agreements often encountered by technology entrepreneurs.

11.2 SOURCES OF CONTRACT LAW

When an issue of contract law arises, it is generally in connection with one of the following questions: Has a contract been formed? If so, what are the terms? And, if there is a breach, is there an applicable defense? If not, what are the remedies?

Answering these questions, however, is complicated by the fact that a different body of contract law applies depending upon the subject matter of the transaction, the location of the parties to the transaction, and the terms of a written agreement.

Two primary sources of law govern contracts between parties located within the United States. Historically, the common law of a state (precedential judicial decisions) was applied to contracts formed or performed within its boundaries.[1] Consequently, the law varied from state to state. As the frequency of interstate transactions increased, a **Uniform Commercial Code** (UCC) was developed for, among other things, the sale of **goods**.[2] The UCC, consisting of 10 articles, covers the rights of buyers and sellers in commercial transactions. Drafted in 1952, the UCC has been adopted in its entirety by every state except Louisiana (which has adopted about half of the code). Each state has unique adaptations of the code to fit its own common law findings. Generally, business owners can access a state's UCC through that state's Secretary of State Office. The UCC defines goods as all things that are movable at the time of identification to a contract for sale.[3] The UCC does not cover transactions relating solely to services, intangibles such as investment securities or intellectual property (IP), or real property.

The UCC is significantly different than the common law in a number of respects. Most notably, actions that would not create an enforceable contract under the common law may do so under the UCC, and the effective terms of an agreement may be different when interpreted under the UCC rather than under the common law.

In addition, since 1988, the United States has been a signatory to the 1980 United Nations Convention on **Contracts for the International Sale of Goods** (**CISG**). It applies to the sale of **goods** between a party in the United States and a party in another CISG country.[4] Effective October 2013, 80 countries are signatories of the CISG including: Canada, China, Mexico, Japan, Germany, South Korea, France, Venezuela, Netherlands, Italy, and Belgium, all of the top 10 trading partners with the United States, excepting the United Kingdom, and Saudi Arabia.[5]

The CISG covers agreements between parties in different CISG jurisdictions that would otherwise be covered by the UCC. It is, however, significantly different from the UCC in a number of respects, notably whether or not a contract is formed, the requirement for a written agreement,[6] and the availability of certain remedies. However, the parties to a transaction can opt out of the CISG by an express and explicit statement in a written agreement, such as:

> "The rights and obligations of the parties shall not be governed by the United Nations Convention on Contracts for the International Sale of Goods."

It is important to be able to determine the particular body of law that governs a transaction. Whether or not there is a contract, and even if so, the terms of the agreement may be different depending upon the applicable law. The process of determining the applicable law is illustrated in Exhibit 11.1.

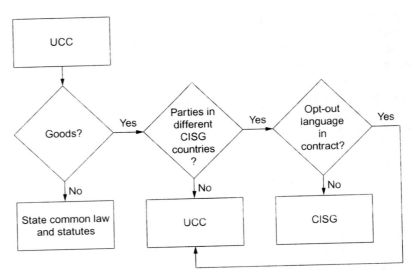

EXHIBIT 11.1
The CISG and UCC relationship.

11.3 CONTRACT FORMATION

To begin to understand the concept of contract formation, the first question to ask is: Has the interaction between two parties created an enforceable contract? Traditionally, the following must take place in order for an agreement between respective parties to create a legally enforceable obligation:

- The parties must have the legal capacity to enter into a contract.
- There must be an offer and an acceptance of that offer.
- There must be consideration to support the contract.
- The subject matter of the contract must be legal and not against public policy.
- There must not be any viable defense, such as mutual mistake, fraud, or duress.

In addition, under certain circumstances, the contract must be reflected by a written instrument.

11.3.1 The offer

The basic foundation of a contract is mutual assent or a meeting of the minds. In order for there to be a meeting of the minds, an offer must be made and that offer accepted. An **offer** is a proposal meeting certain requirements, made by an *offeror* to an *offeree*. In order for a proposal to constitute an offer, it must:

- Identify the person or class of persons (the offeree) to whom it is made
- Be unconditional
- Not require anything from the offeree other than an indication of acceptance

- Indicate an intent to create a contract with the offeree and not a mere invitation for offers (like an advertisement or request for proposals)
- Be objectively reasonable to believe that the person making it is prepared to be bound by the terms if it is accepted, that is, a reasonable person would not consider it a joke or facetious statement
- Provide a basis for determining the essential terms of the proposed agreement. Nonessential terms can be filled in by the court according to a "reasonableness" standard.[7]

The offeree's response to the offer determines whether or not a contract is formed. It can either reject or accept the offer. Rejection can be express (as when the offeree states verbally or in writing that the offer is rejected) or by lapse of time (i.e., the time for accepting the offer has passed). Once rejected, the offer need not be honored or renewed. Generally, only the person to whom an offer was directed may accept an offer. Under common law and the UCC, an acceptance is deemed to be effective when it was mailed, as opposed to when it was received.[8] This is sometimes referred to as the **mailbox rule**. Under the CISG, however, acceptance is not effective until it reaches the offeror.[9]

11.3.2 The Counteroffer

The response to an offer that proposes the parties go forward on different terms than proposed in the offer is a counteroffer. Whether or not a counteroffer constitutes acceptance and forms a contract is a function of the applicable body of law. Unless under the UCC, an acceptance must unequivocally meet and correlate with the offer in every respect in order to be effective. This is referred to as the **mirror image rule**.[10] Even if a response is labeled an acceptance, if it proposes any deviation from the terms of the original offer, it is considered a counteroffer, and a rejection of the original offer.[11] If the counteroffer is not acceptable, the original offeree cannot fall back on the original offer; the original offer need not be honored or renewed.

The mirror image rule does not apply to the sale of goods covered by the UCC.[12] The result is the sometimes difficult and confusing **battle of the forms**. When purchasing goods, a company often sends a preprinted purchase order form containing contract terms and conditions to a vendor. The vendor often responds with its preprinted order-acknowledgment form, containing terms and conditions that are almost always different from those contained in the order or offer.

Under the common law, the vendor's order-acknowledgment form is a counteroffer and the original offer is rejected. No contract would be formed. If, as is often the case, the parties fail to notice or ignore the inconsistencies in preprinted forms, and proceed with the transaction, and there is a dispute, contract law remedies would not be applicable. The parties would have to rely on the judicial doctrine of quasi-contract (discussed below) to prevent unjust enrichment.

Under the UCC, however, any definite and timely expression of acceptance will form a contract regardless of whether it contains different terms—unless acceptance is expressly made conditional on assent to the new terms.[13] Between merchants, the terms of the counteroffer become part of the contract unless:

- The offer expressly limits acceptance to the terms of the offer
- They materially alter the offer
- A notification of objection has already been given or is given within a reasonable time after notice of them is received[14]

A number of issues tend to arise: Which writing between the parties constitutes an offer and which constitutes the acceptance? Does a term proposed in a responsive communication "materially alter" the offer? And, even if the forms are so completely at odds that no written contract is formed, may conduct still form a contract[15]? The issue then becomes: What are the contract terms?

The UCC uniquely assists courts in determining provisions of contracts that are not explicitly spelled out between parties. If parties to a contract go to court to dispute something that was not explicitly spelled out in a written agreement, the court will rely upon the UCC to fill in the missing details. The UCC provides the court with what is known as gap fillers. These are reasonable provisions that are derived from the UCC and that help complete a contract. For example, if a party agrees to provide labor for $100, it may not be spelled out in the contract that U.S. currency is preferred. If a dispute arises because the second party wants to pay in Turkish Lira (a currency that devalues quickly),

MINI-CASE
A Battle of the Forms Can Result in Unwanted Contract Terms

Northrop Corporation sent out a request for proposals soliciting offers to sell Northrop, a particularly complex customized printed wire board. The request stated that any purchase would be made through a Northrop purchase order that would set forth terms and conditions that would override any inconsistent terms in the offer. Litronic had previously done business with Northrop and was familiar with Northrop's purchase order form, but nonetheless responded with an offer that was inconsistent with Northrop's purchase order form. Litronic mailed an offer to sell Northrop four boards for $19,000 apiece, to be delivered within 6 weeks. The offer contained a 90-day warranty stated to be in lieu of any other warranties and provided that the terms of the offer would take precedence over any terms proposed by the buyer. In contradistinction, the purchase order provided for a warranty that contained no time limit.

It was not until 5 or 6 months after delivery that Northrop completed testing of the boards. Finding the boards to be defective, Northrop attempted to return them. Litronic refused to accept the return of the boards, on the ground that its 90-day warranty had lapsed. Northrop's position of course was that it had an unlimited warranty, as stated in the purchase order.

Basically, the offer and acceptance contained materially different terms. There was clearly a contract (if for no other reason, by virtue of the conduct of the parties), but what were the terms of the contract? A lawsuit ensued, revolving around the applicable warranty provision. Did the contract incorporate a 90-day warranty as specified in Litronic's offer or an unlimited warranty as called for in Northrop's purchase order?

The answer was that the contract incorporated neither provision. The transaction was governed by the UCC, as enacted in the state of Illinois. In most states (including Illinois), where the offer and acceptance in a UCC transaction contain materially different terms, a contract is formed based on the terms common to both offer and acceptance, and the discrepant terms are ignored and replaced by a suitable UCC gap filler provision. Accordingly, the divergent warranty terms were both ignored and replaced with a gap filler provision to the effect that nonconforming goods may be rejected within a "reasonable" time from delivery. In this case, the court found that because of the complexity of the required testing, six months was a reasonable time in which to return the boards. Northrop prevailed, even though the warranty provision of its purchase order was discarded.

Significantly, some states (notably California) adopt a different view with respect to the treatment of terms in an acceptance that is materially different from the terms of the offer. According to that view, terms in the acceptance that are materially different from the terms of the offer are ignored. Had the California view been applicable, then Litronic would have prevailed. It was Northrop's good fortune that Illinois law, rather than California law, governed the transaction.

This is a perfect example of unintended (and perhaps serendipitous) consequences resulting from a "battle of the forms."

Source: *Northrop Corporation versus Litronic Industries*, 29 F.3d 1173 (7th Cir., 1994)

the first party may go to court, where the court will use gap fillers to determine the *intent* of the offer, acceptance, and consideration.

11.3.3 Acceptance

The **acceptance** is the offeree's assent to the terms of the offer. For example, if the offer to provide labor services was suitable, the party to receive the services would accept the terms. Signing a contract is one way a party may demonstrate assent. Alternatively, an offer consisting of a promise to pay someone if the latter performs certain acts that the latter would not otherwise do (such as repair a vehicle) may be accepted by the requested conduct instead of a promise to do the act. The performance of the requested act indicates objectively the party's assent to the terms of the offer.

The essential requirement is that there must be evidence that the parties had each, from an objective perspective, engaged in conduct indicating their assent. This requirement of an objective perspective is important in cases where a party claims that an offer was not accepted, taking advantage of the performance of the other party. In such cases, courts can apply the test of whether a reasonable bystander would have perceived that the party has impliedly accepted the offer by conduct.

11.3.4 Revocation of Offer or Acceptance

Occasionally, a party will attempt to revoke or retract an offer or an acceptance. If an offer is retracted before it is accepted, or acceptance retracted before it is effective, there is no contract. A revocation is effective when received.[16] An acceptance, under the mailbox rule of the common law and UCC, is effective

---

MINI-CASE
An Offer or Counteroffer Can Be Revoked Before It Is Accepted

The very day Ms. Miller put her property on the market, she received a written offer from Mr. Normile. The offer contained a provision to the effect that the offer "must be accepted on or before 5:00 p.m." the next day. Ms. Miller responded with a counteroffer, materially changing a number of the terms. Mr. Normile did not immediately respond to the counteroffer, but made statements to the effect that he was not sure that he could accept the modified terms and wanted to "wait a while before he decided what to do with it." In the meantime (at approximately 2:30 a.m. that night), Mr. Siegel made an offer that essentially replicated the terms of Ms. Miller's counteroffer to Mr. Normile. Later that day (at approximately 2:00 p.m.), Mr. Normile was informed that Ms. Miller's counteroffer was revoked and that the property had been sold to Mr. Siegel. After being informed of the revocation, Mr. Normile, but before the 5:00 p.m. deadline set in the original offer, delivered to Ms. Miller a written acceptance of her counteroffer, together with the earnest money required by the counteroffer. The timeline of events is shown in Exhibit 11.2.

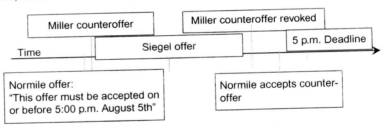

EXHIBIT 11.2
Timeline of Events.

Mr. Normile sued Ms. Miller and Mr. Siegel, seeking a court order requiring sale of the property to Mr. Normile. Mr. Normile argued that the "must be accepted on or before 5:00 p.m." the next day provision in the original offer was imported into Ms. Miller's counteroffer and, in effect, gave him an option on the property precluding revocation of the counteroffer until the 5:00 p.m. deadline. The court disagreed. In the first place, the "must be accepted on or before" language, is not the same as language to the effect that "the offer shall be held open until" language; it was not a promise or agreement that the offer (or counteroffer) would remain open for a specified period of time. Even more significantly, Ms. Miller's counteroffer was a rejection of Mr. Normile's original offer. Accordingly, no contract was formed on the basis of the original offer. Further, there could be no contract based on Ms. Miller's counteroffer until Mr. Normile accepted it. Accordingly, the acceptance deadline in the original offer, even if imported into the counteroffer, did not create any sort of contractual obligation on Ms. Miller. A counteroffer can be revoked at any time before it is accepted, and once revoked, it cannot be accepted. Accordingly, Ms. Miller prevailed.

on dispatch.[17] Under the CISG, it is effective when it reaches the offeror.[18] If the revocation of an offer reaches the offeree after acceptance is dispatched, but before it is received by the offeror, under the UCC (and common law), the revocation is too late and a contract is formed. Under the CISG, however, there is no contract.

11.3.5 Consideration

Consideration is the bargained-for exchange of an agreement. The adequacy of consideration is irrelevant to the formation of a contract. However, in the absence of consideration, there is generally no enforceable contract. Nevertheless, in some cases, it is reasonable for a person to rely upon a promise. And, if they do so to their detriment, a legal theory known as **promissory estoppel** provides a substitute for consideration and the promise is enforceable.

11.4 DEFENSES AGAINST CONTRACT ENFORCEMENT

In order for there to be an enforceable contract, there must be mutual assent as to the terms of the agreement. Assent, however, can be negated by a **mutual mistake** as to the subject matter of the agreement.[19] If both parties are mistaken as to a basic assumption regarding the contract, the contract may be avoided. For example, assume that an antique seller has agreed to sell what was thought to be an original, 1955 vintage, Berkeley Enterprises GENIAC computer. The purchaser was only interested in buying the item because it was one of the original devices. The seller was aware of this, and received a premium price. If it is later discovered that the item was not in fact an original GENIAC, this would be a mistake of fact material to the transaction—a mutual mistake—and the purchaser would have the right to cancel (rescind) the contract.

On the other hand, a mistake on the part of only one party to a contract, that is, a **unilateral mistake**, is generally not a defense to enforcement of the contract. However, if the nonmistaken party knew or should have known of the mistake, or if it was the result of a misstatement or misleading statement by the nonmistaken party regarding a material element of the contract, the contract may be voidable. In the example given above, if the purchaser mistakenly believed the item to be an original GENIAC and the seller had no reason to know of the mistaken belief and made no representations as to the item's nature, the unilateral mistake by the purchaser would not make the contract voidable.

A contract is unenforceable if the subject matter of the agreement is against public policy or illegal. Examples include:

- Agreements in restraint of trade; for example, an overreaching employee noncompetition agreement may be unenforceable as against public policy
- Gambling contracts (in some states)
- Usurious agreements
- Agreements to commit crimes or torts

An enforceable contract also requires true consent to the terms. If the agreement is entered into as a result of undue duress or coercion, there is no true consent. If the agreement is induced by fraud, e.g., the victim was not aware that the contract was being formed; such **fraud in the inducement** negates true consent and renders the contract voidable.

Under the **statute of frauds**, in the absence of a signed writing, certain contracts are voidable. This concept is incorporated into both common law and the UCC. Significantly, there is no statute of frauds under the CISG.[20] Under the statute of frauds, the following subjects must be in writing:

- Surety contracts that are agreements to pay debt of another
- Promises in consideration of marriage (e.g., "I'll marry you if you'll pay for that surgery I want"—where still enforceable)
- Long-term interests in real property such as the purchase of real estate, mortgages, easements, and leases with the term in excess of one year
- Contracts that by their terms cannot possibly be performed within one year (e.g., five-year employment contract)
- Promises by the executor of an estate to pay the estate debts out of the executor's private funds
- Sale of goods having a value of more than a specified amount, typically $500[21]
- Certain consumer purchases

The writing need not be a formal contract, or even a single document, but must reflect the existence of a contract. The common law requires that the writing includes the identity of parties, subject matter of the agreement, essential terms of the agreement, consideration, and signature of the party against which the contract is to be enforced. Under the UCC, the writing need only be signed by the party to be charged and specify a quantity term.[22] In addition, the UCC also relaxes the signature requirement for transactions between merchants; a timely written confirmation of an existing (oral) contract that has sufficient detail to make the contract enforceable against the merchant sending the confirmation satisfies the writing requirement against the merchant receiving confirmation unless the receiving merchant objects in writing within 10 days of receipt.[23]

11.5 PERFORMANCE AND BREACH

Contracts typically spell out performance requirements for the various parties. For example, a contract to purchase a vehicle from a car dealer specifies that the car will be delivered to the buyer in good quality and in working order. The dealer must perform to these standards to meet its obligations to avoid committing a **breach**. The buyer is obligated to meet payment terms, including a down payment and future monthly payments if the vehicle is financed. **Complete performance** of all of the obligations in the contract fulfills and **discharges** the contract. Anything less than complete performance is a breach.

Essential fulfillment of the obligations but with slight variances from the exact terms, unimportant omissions, and/or minor defects that do not defeat the purpose of the contract is **substantial performance**, but still a breach. (If, in the example above, the contract called for light brown upholstery, and the dealer delivered a car with dark brown upholstery, the dealer would have substantially performed under the agreement). A significant variance from the terms of the contract is considered **inadequate performance** or **nonperformance**. A failure to provide at least substantial performance, that is, inadequate performance or nonperformance, is a **material breach**.[24] (If, in the example above, the dealer failed to deliver a car it would be a material breach). But what are the consequences of less than complete performance? The answer depends upon how close the actual performance comes to being complete and the applicable law.

With respect to UCC transactions the gravity of a breach is no issue. Under the UCC **perfect tender rule**,[25] any deviation from the terms of the agreement (e.g., nonconforming goods) is a material breach and discharges the aggrieved party from performance.

This is not the case, however, as to common law and CISG contracts. A minor breach gives rise to damages, but the aggrieved party is typically still required to perform under the contract. On the other hand, if the breach is material, the aggrieved party can not only obtain damages, but can also stop performing under the contract.[26]

Unfortunately, distinguishing between substantial performance and a material breach can sometimes be difficult. There is no bright-line rule. If the aggrieved party is mistaken with respect to the materiality of a breach and stops performance, it is itself in breach. Fortunately, this issue can be minimized by a well-drafted agreement. The contract can expressly identify certain provisions as material or can provide a right to terminate in the event of any breach (material or not) that is not cured within a reasonable specified time period (Exhibit 11.3).

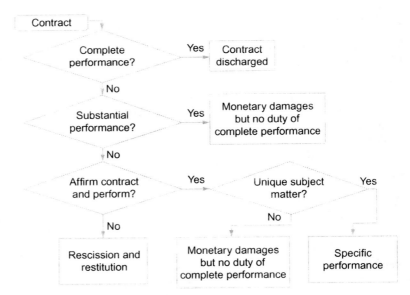

EXHIBIT 11.3
Alternative remedies available to aggrieved party in the event of breach.

11.5.1 Conditions

Performance under a contract can be made conditional on specified events or metrics. Satisfaction of a condition can be a prerequisite for a duty to arise, terminate an existing duty, or change the nature or extent of a duty. Failure to meet a condition is not a breach. Rather, the contract contemplates the possibility of the condition not being met and is carried out or terminated according to its terms.

11.5.2 Excusing Less Than Complete Performance

When one or both parties to a contract failed to provide complete performance, the parties can generally agree to discharge each other's obligations through any one of the following:

- A mutual rescission, whereby both parties are restored to the positions they were in before the contract was entered
- A substituted or amended contract, whereby the original duties are discharged and new duties imposed
- An accord and satisfaction, where a party to the contract agrees to accept a different performance in satisfaction of the existing duty
- A novation, where an amended contract substitutes a third party for one of the original parties to the contract

Occasionally, after an agreement has been entered into, something will occur that renders performance either improper or impossible. In such extraordinary circumstances, the duty to perform will be excused. For instance, if the subject matter of an agreement was legal when the contract was formed, but becomes illegal at a later time, failure to perform is excused.

Circumstances that make performance more difficult or burdensome, however, will typically not excuse contractual duties. However, under a **commercial impracticability test** adopted by many states and the UCC and CISG, performance by a party is excused if it is rendered impracticable by an event that is:

1. Beyond the control of the party
2. Not reasonably foreseeable by the party
3. Contradicts a mutual assumption on which the contract was based

Commercial impracticability requires more than just unexpected hardship or increased cost. It requires extreme and unreasonable difficulty or expense caused by an unforeseeable supervening event. The concept of commercial impracticability is often explicitly written into contracts. Such a provision is often referred to as a **force majeure** clause. Alternatively, an agreement can explicitly call for a party to assume the risk that a certain event may occur, in which case commercial impracticability would not excuse nonperformance.

11.5.3 Damages

In general, the aggrieved party is entitled to all damages that it can show were actually caused by a breach of the contract, as long as the damages were reasonably foreseeable at the time the parties entered into the contract, and are sufficiently certain so as not to be "speculative."[27] Damages arising out of a breach of contract are typically either in the form of compensatory damages or reliance damages. **Punitive damages** are typically not available in breach of contract cases.

Compensatory damages place the injured party in the position it would have been in had there been complete performance under the contract.[28] This includes:

- The loss of value (the difference between the value of the benefits that would have been received and those actually received) less any savings experienced from being excused from any obligations under the contract (the benefit of the bargain)
- Incidental damages directly resulting from the breach (e.g., costs of obtaining substitute performance)
- Consequential damages, also sometimes referred to as "special damages," which include any other damages or losses actually caused by the breach, as long as they are reasonably foreseeable; agreement terms can make clear that certain types of losses are foreseeable

Reliance damages are an alternative to compensatory damages, intended to place the nonbreaching party in the position it would have been in had the contract never been formed. Typical elements include expenses incurred in preparing for, and performing under, the agreement and lost opportunities.

In general, there is a duty to mitigate damages. Damages that could have been avoided with reasonable effort and without undue risk or burden cannot be recovered.[29] The particular acts required to mitigate damages depend upon the circumstances. For example, an aggrieved buyer of goods must **cover** by purchasing substitute goods at a reasonable rate and without unreasonable delay or is foreclosed from receiving any consequential damages that would have been prevented.[30]

Commercial contracts often include provisions purporting to limit liability. Contractual limitation of liability can allocate risk between the parties and be reflected in the consideration paid. If the party limiting liability assumes less risk, the price paid by the other party would be proportionately decreased. The provision thus permits lower charges. This is, of course, totally inapplicable to contracts of adhesion. In general, limitations on liability are not favored by the courts.

11.5.3.1 *Rescission and Restitution*

Rescission means, in effect, to undo a contract. **Restitution** means to return to the aggrieved party benefits that were unjustly conferred upon another. The aggrieved party is placed in the position it would have been in had the transaction never taken place. Rescission and/or restitution are available:

- As an alternative remedy for a party injured by material breach[31]
- To a party that breaches after partial performance for any benefit conferred in excess of the loss caused by the breach[32]
- When there has been a fault in the formation of a contract or it is otherwise unenforceable or voidable[33]

Restitution may also be pursued in situations where damages are difficult or impossible to prove and to recover specific property.

11.5.3.2 *Specific Performance*

When monetary damages are inadequate to compensate an aggrieved party, a court may order the breaching party to completely perform under the contract. This is called **specific performance**.[34] For example, money damages are inadequate where the subject matter of the contract is rare or unique (e.g., land or fine art). Under common law, and the UCC, specific performance is not available unless monetary damages are inadequate. However, the parties in international sales transactions under the CISG may obtain specific performance anytime there is a material (fundamental) breach, as long as they have not already resorted to an inconsistent remedy.[35]

11.5.3.3 Quasi-Contract

Even in the absence of an enforceable agreement, a court may construct a fictional contract—a **quasi-contract**—to prevent unjust enrichment. Quasi-contractual recovery is applicable when a party has partially performed in the mistaken belief that a contract had been formed or under a contract that is unenforceable. It permits the party to recover the reasonable value of the partial performance.

11.5.3.4 Reformation

In some cases, a written agreement will not accurately reflect the intentions of the parties. This most often occurs when there has been an error in documentation, a fraud, misrepresentation, or mutual mistake. In those cases, the equitable remedy of **reformation** is employed by the court to effectively rewrite the contract to reflect the actual intentions of the parties.

In other instances, a particular provision of an agreement may be void or unenforceable as **unconscionable**, against public policy, or illegal (due to a change in the law) if interpreted literally. When that occurs, the courts in various jurisdictions will take one of three approaches: (1) hold the entire agreement unenforceable; (2) delete the offending provision and enforce the remainder of the agreement; or (3) effectively rewrite the offending agreement, so that it is valid and enforceable.[36] In some instances, courts can be directed as to which approach to take by an appropriate provision in the contract.

11.6 ANATOMY OF A CONTRACT

Written contracts can take many forms and adopt many formats. As discussed earlier, an enforceable contract can be found even in the absence of a written agreement. However, a comprehensive written agreement can be a tool to ensure good relationships between participants. In this section, we will examine the general structure of a typical negotiated written agreement, and consider the significance of some of the typical provisions. For purposes of illustration, we will refer to a sample development agreement included as Appendix III.

11.6.1 Preamble

The preamble is a vehicle for identifying the parties, establishing the effective date of the agreement and, in many instances, for denominating the agreement for identification purposes.

11.6.2 Recitals

Recitals are a vehicle for ensuring a complete understanding of the party's respective expectations and memorializing any relevant background information and extrinsic considerations. Recitals preferably cover a range of considerations, including:

- Related agreements
- Any third party (3dP, in the example) that may play a role in the performance or that may be a beneficiary
- The relationship between the parties
- The basic premise of the agreement and expectations of the parties
- Any facts or circumstances relevant to the possibility of particular damages (and establishing that such damages were foreseeable)

In the example in Appendix III, the recitals establish that:

- Developer knew one of the reasons that the company engaged it to do the development work under the agreement was that the company understood developer to have special expertise in the subject matter of the agreement. (If the developer did not have that special expertise, it could not be a unilateral mistake on the part of the company).
- Developer's failure to perform under the agreement would cause the company to breach various third party agreements. (Ensuring that the damages resulting from reaching the third party agreements are foreseen and thus recoverable, consequential damages in the event of a breach by developer.)
- A confidentiality agreement exists between the company and 3dP pertaining to some of the information that will be provided to the developer.
- Developer is aware that confidentiality is a significant concern to the company and that an obligation of confidentiality on the part of developer was a prerequisite to the engagement.

11.6.3 Definitions

A well-written agreement will include a definition for each term used that could conceivably be misunderstood. In complex agreements, these definitions are typically included as a separate section either at the beginning or end of the agreement. However, in shorter, simpler agreements, the definitions may be provided in the body of the agreement, as the terms are introduced in context. By convention, defined terms are typically designated within the text of the agreement by capitalization. Explicit definitions are particularly important in international transactions where there are cultural and language differences.

11.6.4 Performance

The provisions of an agreement relating to performance by the parties vary greatly depending upon the subject matter of the agreement. In any event, it is extremely important that all expectations be explicitly reflected in the provisions. A simple rule of thumb: if a detail is not explicitly recited in the agreement, do not expect it to happen.

All deliverables should be precisely and completely defined, including all details that are significant to the parties. Deliverables can be described integrally with the performance events, or in separate provisions. Such descriptions are often provided by reference to a specification or documentation. This is the case in the agreement of Appendix III (see Paragraph 1.1, and Schedule 1, although, examples of the appended documentation are not included in the Exhibit). In fact, as illustrated in Appendix III, one of the initial deliverables can be a specification and development schedule for subsequent deliverables (See Schedule 2 Performance Schedule, Phase 1).

Conditions and contingencies are frequently built into the performance provisions. Use of relative timing (e.g., "Developer will deliver a prototype within 30 days of receiving the specification") rather than absolute dates can be used to expressly make performance conditional on the performance of the other party.

There are myriad ways to specify the performance of the parties. Each performance event can be described with particularity in successive provisions. In the case of the contract of Appendix III performance is defined using a "Performance Schedule" (Paragraph 2.1 and Schedule 2). In a situation where there is considerable interaction between the parties, a performance matrix can be a useful organizational tool, particularly when used in conjunction with a separate detailed description of deliverables. The "Performance Schedule" in Appendix III illustrates a skeletal generic performance matrix. Respective columns are provided for event timing and actions by each of the parties. Each sequential performance event is represented by a successive row. Details with respect to particular performance events can be particularized in the text of the agreement if not clear from the matrix. Please note that Appendix III omits a number of provisions that are often in development agreements: many of the particulars with respect to the cooperation and communications between the parties; specifics as to the expected features of the deliverables, and so on.

11.6.5 Intellectual Property

If performance under the agreement is likely to result in the development of IP, the issue should be dealt with explicitly in the agreement. Ownership of IP resulting from performance under an agreement tends to be highly negotiated. Some considerations include:

- Is the purpose of the contract to develop the IP or is the development ancillary to performance?
- Is the "preexisting" IP of one party being used as the basis for development of the new IP?

- Was the preexisting IP maintained as a trade secret to which the other party would not have access but for the agreement?
- Is the development effort entirely by one party or is it a cooperative effort?
- Is the originator of the new IP being fully compensated for its contribution to the development effort?

It should be recalled that in the absence of an express contractual provision, where independent parties are involved, irrespective of which party pays for development, whichever party actually creates the IP will own it. The circumstances of the agreement may give rise to an implied **license** to use the IP, but it is not good practice to leave the existence and scope of a license to the courts.

Transfers of rights in IP are typically in the form of an **assignment** or a grant of a license. Such transfers are encountered in, for example, agreements setting up a business venture, agreements between a company and its employees, and development/consulting, manufacturing, licensing, technical assistance, joint venture, distributorship, and marketing agreements.

The sample development agreement in Appendix III specifically contemplates the creation of IP during the development process. The developer is required to assign new technology or IP resulting from performance under the agreement ("resultant IP," Paragraph 4.1) to the company, but retains ownership to its preexisting technology. No third party technology or developer's preexisting technology (or any unassigned technology or IP) can be incorporated into the deliverables, unless it is specifically identified (Paragraphs 4.2.1 and 4.2.2) and the company granted a license (of clearly defined scope) to use it (Paragraph 4.3).

It is particularly important that any third party technology, especially software, to be incorporated into the deliverables is identified during the negotiation process. A determination can then be made whether use of the third party technology is acceptable, and if it is, whether or not any necessary licenses from the third parties can be obtained under acceptable terms. Acceptable terms may not always be available, and substitution of other third party technology may sometimes be required. For example, the terms of certain of the prevalent licenses for open source software can have significant repercussions with respect to the ability to maintain exclusivity of certain IP and are not appropriate in some situations.

11.6.6 Consideration

The provisions in an agreement relating to consideration can take a wide variety of forms. In many cases, the consideration is simply a single lump sum payment. In other cases, payment is tied to a milestone. Where a performance matrix is employed, as in Appendix III, such payments are simply reflected as performance events. Other agreements, such as license agreements and distribution agreements, may call for periodic payments, the amounts of which vary on the basis of the marketplace performance, for example, sales of a product. Such agreements might call for quarterly royalty payments or commission payments, in an amount equal to a percentage of gross sales. In those cases, provisions particularizing the computation of the payments, as well as maintenance of accounting records, reporting, and audits of the records, are typically included in the agreement.

11.6.7 Representations and Warranties

Representations are, in essence, a statement of facts relied upon by the parties in entering the agreement. **Warranties** are, in effect, a promise that a product, service, or aspect of performance will meet certain standards or metrics, or do certain things. In many agreements, representations and warranties will be combined ("Developer represents and warrants that …") so that the statement is both a representation upon which the other party relied and a promise.

Representations and warranties are a vehicle for ensuring that the expectations of the parties are understood and met. If something is important to a party, inclusion of a representation or warranty on that point can ensure that there is no misunderstanding as to the party's expectations. The breach of a representation or warranty is also readily made an event that would trigger a right to terminate. Representations are often provided with respect to:

- Authority to enter into the agreement without any third party consents
- The expertise of the parties
- Staffing level to be dedicated to the project
- Level of efforts to go into performance (e.g., "best efforts" or "reasonable commercial efforts")
- Novelty, confidentiality, and/or originality of deliverables
- Absence of pending or threatened claims or proceedings relating to the subject matter of the contract
- Absence of conflicting agreements
- Trade secret status of aspects of the licensed subject matter
- Quality, functionality, and performance characteristics of deliverables
- Completeness and/or accuracy of information provided to the other party

In Appendix III, representations and warranties are provided in Section 5. Representations and warranties may be absolute or qualified. When an exhaustive investigation would be required in order to make a representation, a party can qualify the representation by limiting it to their personal knowledge (e.g., "To the best of Developer's knowledge, the deliverables do not infringe any third party patents").

Warranties that are explicitly stated in the agreement are referred to as **express warranties**. The terms of express warranties are totally in control of the parties.

However, certain other warranties may be implied by law unless explicitly and conspicuously disclaimed in the agreement.[37] In contracts involving goods, both the UCC and the CISG provide for a number of implied warranties, including a:

- Warranty of noninfringement: that the goods sold do not infringe the IP rights of any third party[38]
- Warranty of merchantability: that the product is fit for the ordinary purposes for which it is used[39]
- Warranty of fitness for a particular purpose: that the product is fit for the particular purpose of the buyer[40]

The scope of these **implied warranties** is left open to interpretation of the courts. The better practice is to disclaim the implied warranties and provide express warranties that clearly specify the rights and obligations of the parties.

11.6.8 Indemnity

An **indemnity**, also sometimes referred to as a hold-harmless, is another mechanism for allocating risk between the parties. The indemnitor is obligated to pay for loss or damages incurred by the indemnitee if certain specified events occur. A typical indemnity provision shifts liability to the party who is likely to be more actively or primarily responsible for the events giving rise to the liability. However, negligence or fault is not necessarily a prerequisite for coverage.

For example, indemnifications by the party supplying technology in connection with an agreement against infringement of third-party IP rights are common. The scope of IP indemnifications can vary greatly depending upon the terms of the provision. Frequent areas of negotiation include:

- Choice and participation of counsel and control of litigation
- Reimbursement for attorneys' fees
- Infringement not solely due to the deliverables from the indemnitor, that is, by virtue of additions, modifications, or combinations not provided by the indemnitor
- Remedies in the event of an injunction

An indemnification provision is often used instead of a representation or warranty when the subject matter of the indemnity is beyond the knowledge of the parties. If an event renders a representation false, there is a breach giving rise to liability and perhaps excusing further performance by the nonbreaching party. If, on the other hand, the same event triggered an indemnification, essentially the same liability would accrue, but there would be no breach; the contract simply proceeds according to its terms.

11.6.9 Term and Termination

The **term** of the contract refers to the length of time it remains actively applicable to the parties. When performance involves a single discrete transaction, such as the purchase of goods or the assignment of IP, the duties under the agreement are discharged when the transaction is completed. However, many contracts involve a continuing right or duty, or a series of transactions. In those cases, the term of the agreement should be explicitly stated.

> **KEY POINT**
>
> **The Term of an Agreement**
>
> There are a number of different approaches to establishing the term of an agreement. The term of an agreement can be:
>
> - For a stated period of time (e.g., "the term shall be for a period of 1 year from the date of this agreement")
> - For consecutive periods of stated duration, automatically renewed unless terminated (e.g., "If not earlier terminated, this agreement shall have an initial term of 5 years after the effective date and shall be automatically renewed for successive one-year periods thereafter. Either party may terminate this agreement at or after the end of such initial term upon 90 days written notice.")
> - For an indefinite period, but terminable by one or both parties upon the occurrence of specified events ("The term of this agreement shall commence upon the effective date and shall continue until terminated pursuant to this paragraph. This agreement may be terminated: (i) by either party, upon 30 days written notice, at any time after the second anniversary of the effective date; …")

When a contract applies to an indefinite period, the termination trigger events are often crafted to establish an initial period during which the agreement can be terminated only in the event of a breach or failure of condition. Thereafter, the agreement can be terminated by one or both parties upon notice to the other. Termination is typically effected by a written notice, although termination is often not effective until the lapse of a specified time period after the notice.

Certain types of rights and obligations would not be particularly effective if they were terminated or discharged at the same time as the other aspects of an agreement. Confidentiality and indemnification provisions often fall within this category. It is not uncommon for an agreement to expressly state that designated provisions will survive termination of the agreement. (See Appendix III, Paragraph 8)

11.6.10 Miscellaneous Provisions

So-called **boilerplate** provisions typically address various issues of law that are ancillary to actual performance of the agreement. These issues tend to recur in many types of agreements, and over the years, standardized language, the boilerplate, covering the issues has been developed. The issues include:

- Notices: The manner in which formal notices under the contract are to be made
- Waiver: The failure to enforce a right in one instance not constituting a waiver preventing enforcement emulator instances
- Applicable law: Specifying the body of law under which the contract will be interpreted
- Dispute resolution processes: Admission of jurisdiction/exclusive jurisdiction, forum selection,[41] arbitration, mediation
- Integration: Precluding reference to communications not incorporated into the written agreement
- Severability/reformation: Treatment of an enforceable provision
- Assignment: Whether or not assignment of rights and/or delegation of duties are permitted, and any limitations; in the absence of contractual provisions to the contrary, with the exception of personal services, contractual rights can be assigned and duties delegated[42]
- Force majeure: Performance excused if made impracticable by forces outside of the control of the party
- Relationship of the parties: Agreement not creating partnership

11.7 TYPES OF AGREEMENTS ENCOUNTERED

Over its life, a technology venture is likely to encounter a number of different types of agreements. This section will introduce some of the commonly encountered agreements.

11.7.1 Operating Agreements

When a business entity is formed, an operating agreement between the principals is desirable, if not necessary. This is particularly true when the business entity is formed to implement a joint venture between separate business entities. A partnership agreement, Limited Liability Company (LLC) operating agreement, or corporate shareholder's agreement permits the principals to define the internal rules under which the venture will operate and directly or indirectly assign management responsibilities. We discussed these topics in more detail in Chapter 6.

There are limits to the extent to which management can be separated from stock ownership in a corporation, particularly S-corporations. However, a shareholder's agreement can require individual shareholders to vote their stock for particular board members, thus effectively defining the management structure. Partnership agreements and LLC operating agreements are also vehicles for defining how profits and losses will be allocated among the principals. Other considerations often addressed in these agreements include:

- Any special powers to be given to the management
- Contributions of IP by the principals, such as licenses to the venture
- Ownership and use of IP
- Confidentiality
- Services to be rendered on behalf of the venture by individual participants
- Compensation, if any, to principals for services provided to or work done on behalf of the venture
- Transfer restrictions on ownership interests
- Vesting schedules
- Exit procedures, such as repurchase and buy-sell arrangements and processes for dissolution and winding up
- Rights and obligations with respect to further financings and issuance of additional equity interests, such as antidilution provisions and capital contribution provisions
- Indemnifications of the individual principals for actions taken on behalf of the venture
- Special tax allocations

11.7.2 Employment Agreements

Agreements with employees are often essential to obtaining and maintaining rights in IP. In general, such agreements typically require each employee to assign to the company all rights to any IP that relates to the company's business created by the employee during the term of employment. In the absence of provisions in a written agreement to the contrary, the allocation of rights to any inventions made by the employee will depend upon whether the employee was "hired to invent" and when and where the invention was made. The agreement typically also defines an obligation of confidentiality. In some cases, where the employee has access to particularly sensitive information or it is particularly likely that the employee will use proprietary information in connection with other employment, a noncompete provision (discussed below) may be included. Appendix IV provides an example of an employment agreement.

11.7.3 Noncompete Agreements

In the sale of a business, or where particularly critical confidential information is involved, or when an employee is in a crucial position, noncompetition provisions may be appropriate. Noncompetition provisions tend to be very strictly construed and are typically unenforceable if they are not reasonable in scope as to geography, time, and precluded activities. Paragraphs 10 and 11 in

Appendix IV are examples of noncompete provisions premised on maintaining confidentiality.

In other instances where the primary purpose of a noncompetition provision is to prevent the appropriation of the goodwill of the company, by, for example, the sellers of a business or an employee that has been put in a position to establish relationships with customers of the business, the precluded activities tend to be limited to any number of the following:

- A specific geographic area and specific types of positions
- Contact with customers of the employer/sold business
- Certain channels of trade

11.7.4 Confidentiality Agreements

A confidentiality agreement, sometimes also referred to as a nondisclosure agreement (NDA), should be in place before any third party is given access to proprietary know-how outside of the context of a broader agreement already including confidentiality provisions. The NDA precludes disclosure and also limits the use of the information by the recipient to a specific purpose in connection with the business relationship between the parties. The need for written confidentiality agreements in order to establish an obligation of confidentiality to maintain trade secret status of information was discussed in Chapter 5. An example of a generic confidentiality agreement is provided in Appendix I.

In some situations, it is difficult to obtain an executed NDA. For example, in instances where a proposal is being pitched to, a prospective investor who is unfamiliar with the subject matter of the proposal or not yet sure that they are interested in pursuing the proposal, they are likely to be reluctant to take on an obligation of confidentiality.

> **KEY POINT**
>
> *Sanitizing the Disclosure as an Alternative to an NDA*
>
> Rather than relying on an NDA, technology entrepreneurs can excise sensitive information from the documents they share with potential investors. According to former venture capitalist George Lipper, "The perfect business plan is one that gets mouths watering without giving up the secret sauce."[43]

Some confidentiality agreements impose an obligation of confidentiality for a specified time period. The confidential status of anything disclosed under the agreement is lost at the end of that time period. The time period is typically triggered by disclosure of the confidential information. However, in some NDAs, the obligation runs from the effective date of the agreement irrespective of when the information was disclosed during the course of the agreement. In either case, the time period is arbitrary as it relates to the actual confidentiality of the information. A much better approach is to require confidentiality to be observed with respect to information until such time as the information falls within one of the exclusions to confidentiality status (see Appendix I, Paragraph 4, Appendix III, Paragraph 1.3 and Appendix IV, paragraph 6). This avoids placing an artificial lifetime on confidential status.

11.7.5 Consulting and Development Agreements

In a **consulting agreement**, a consultant (non-employee) is engaged to apply a particular expertise on behalf of the hiring company. The consulting agreement may relate to a particular project or may be more open ended, with the consultant available for any issue involving the consultant's expertise.

A **development agreement** is sometimes considered a species of consulting agreement that specifically contemplates that the consultant will create IP. In many cases, the developer is not intimately involved with the operation of the hiring company and may work on a largely independent basis. A development agreement calling for all IP resulting from the agreement to belong to the hiring company is sometimes referred to as a work-for-hire agreement.

Depending upon the nature and subject matter of the engagement, the agreement may also include provisions precluding the consultant from providing similar services to competitors during the term of the engagement and for a reasonable period after the termination of the agreement. A sample development agreement is provided in Appendix III.

11.7.6 Maintenance and Support Agreements

Maintenance and support agreements cover a wide variety of circumstances, but are most commonly encountered in connection with software products. The agreements often involve the provision of updates and error correction of licensed software. Some of the issues that are commonly addressed in maintenance and support agreements are:

- Ownership of, and the rights of the customer to, new developments
- Availability of updates or error corrections
- Representations by the service persons pertaining to the software
- Criticality of "bugs" and required response times
- The amount and nature of support to be provided

11.7.7 Manufacturing Agreements

Under a **manufacturing agreement**, a company hires a another entity (manufacturer) to manufacture an item according to certain specifications. The arrangement often involves communication of confidential information (e.g., the specification). Appropriate confidentiality provisions and, in some cases, restricted licenses under other types of IP (e.g., patents and know-how) are then included in the agreement. Other areas often addressed in manufacturing agreements include:

- Ordering and delivery procedures
- Quality control and approval procedures
- Ownership of molds and tooling
- Ownership of IP arising out of the agreement
- Representations and warranties as to quality
- Whether or not subcontracting is permitted
- Noncompetition provisions that restrict the manufacturer from manufacturing similar items for competitors or that restricts itself from becoming a competitor

11.7.8 Assignment Agreements

An **assignment agreement** is a document that transfers title to specified IP rights. In most instances, to be effective with respect to a patent, copyright, or trademark, an assignment must be recorded with the patent and trademark or copyright office. In general, there should be a recorded assignment with respect to each patent owned by a company and with respect to each copyright or trademark acquired from outside of the company.

11.7.9 License Agreements

In a **license agreement**, the licensor grants to the licensee certain rights with respect to IP of the licensor. A license is to be distinguished from a sale or an assignment in which substantially all commercial rights and title to the IP are transferred to the assignee. In the case of a license, the licensor retains title to the IP.

Exclusive rights to use IP are the primary sources of competitive advantage in the marketplace (see Chapter 5). Yet, there are a number of reasons why you, as a business venture, might permit others to use your IP. One obvious reason is obtaining a royalty income in consideration for use of the IP. This is particularly so when you are not presently using the technology or are unable or unwilling to meet the demand for products using the IP. For example, you may lack the capital to expand your production facilities or may have other priorities. It may not be practical or economical for you to export product into a particular geographical area or to establish your own manufacturing facilities in that area. Or, you may not have the resources or contacts to develop a necessary distribution system in the area.

Other reasons for licensing IP are not necessarily obvious. For example, having an additional source for a product may increase its market acceptance. In other instances, you might be using the licensed product as a part in another product or sell the licensed product as part of an overall line of products, but find it uneconomical or impractical to manufacture the licensed product yourself. The fact that the product might also be available to your competitors is offset by the compensation paid to you by the licensee.

In other instances, licensing a local entity in a given geographical area to manufacture one product may create a market in that geographical area, or at least increase market acceptance, for others of your products.

The consideration for a license is often a license fee, typically referred to as a **royalty**. The fee can be a single up-front payment, but is often in the form of periodic royalty payments. The periodic royalty payments can be determined upon any number of bases, such as:

- A percentage of a designated royalty base (e.g., gross sales of products employing licensed IP)
- A fixed per-unit fee
- A percentage of savings through use of the licensed IP

The license may also be made conditional upon meeting specified milestones or performance requirements. Milestones can include such things as resources devoted, promotional expenditures, staffing, facilities, and product introduction. Performance requirements could relate to such things as sales level and market penetration. Generally, the conditions are reflected as termination provisions in the agreement.

License agreements often grant rights with respect to more than one type of IP. In many cases, a licensee may be granted rights with respect to both patented technology and unpatented know-how. Many of such licenses also will include a license to use (and sometimes a requirement to use) the licensor's trademark.

11.7.9.1 *Patent Licenses*

Under a **patent license**, the licensor agrees not to enforce the patent (or certain claims of the patent) against the licensee with respect to certain acts that would otherwise be precluded by the licensor's patent, typically making, using, or selling processes or products covered by the patent. It may be no more than a covenant not to enforce the patent, or it may be coupled with a transfer of know-how or technical assistance. Essentially all countries that have patent systems in place recognize patent licenses.

11.7.9.2 *Know-How Licenses*

Under a **know-how license**, the licensor permits the licensee to have access to, and use, the licensed know-how. The know-how may be proprietary or nonproprietary. If proprietary know-how is involved, the agreement would include nondisclosure provisions and, typically, provisions restricting the manner in which the know-how can be used. Such provisions are necessary to protect the licensor's proprietary interest in the know-how. Other subjects that tend to be addressed in know-how licenses include:

- The particular mechanism for transferring the know-how
- The nature of deliverables
- A time schedule for the transfer
- The amount and type of technical assistance

11.7.9.3 Trademark Licenses

Under a **trademark license**, the trademark owner permits the licensee to employ the licensed trademark in connection with the licensee's goods or services. It permits the licensee to take advantage of the goodwill associated with the trademark. At the same time, presupposing that the agreement includes the appropriate provisions, the licensor expands its goodwill through, and generates revenue from, the licensee's efforts. Trademark licensing is recognized under the laws of most, but not all, countries. In most instances, however, in order for the trademark owner to retain rights in the mark, the trademark owner must maintain careful quality control over the products of the licensee with which the mark is used.

11.7.9.4 Franchise Agreements

A **franchise agreement** is a specie of license agreement that establishes a close, ongoing relationship between the franchisor and franchisee. Very often, the franchise agreement relates to a turn-key business, including trademark know-how and, sometimes, patent licenses and technical assistance. The franchisor grants the franchisee a license to replicate the business and provides them with the know-how and often other resources to do so. The franchisee then builds a business—that they own—modeled on the successful characteristics of the franchise. The net result is that, as far as the consuming public is concerned, the franchisee is the alter ego of the franchisor.

While franchises tend to be more prevalent for low technology ventures, a number of technology entrepreneurs have successfully leveraged their ventures through franchising. These include franchises in a number of different technologies, including technology consulting; computer products, wireless electronics, Internet marketing, video games, and ink and toner printer cartridges; photo, video, and web services; IT services; digital document storage; cell phone and wireless device accessories; technology education; technology consulting; computer and electronics repairs, services, and sales; and Internet services.

On the other hand, entrepreneurs tend not to be happy as franchisees; a franchisee is working to advance the franchisor's idea, not their own, and must abide by a set of rules standards and controls set by the franchisor. Franchise agreements are closely regulated throughout the United States and in many countries.

11.7.9.5 Technical Services Agreements

Technical assistance or technical services agreements are, in general, hybrid consulting–know-how agreements relating to provisions of know-how, instruction, and training to the receiving party. Proprietary or nonproprietary know-how, or both, may be involved. For example, an entity with special expertise may be engaged to assist in implementing or installing an equipment plant or process and to instruct and train the contracting party in the operation and management of the plant or process.

11.7.9.6 Distribution Agreements

In a **distribution agreement**, the owner of a product (manufacturer and developer) engages a distributor to market the product in substantially unmodified form. In some instances, the distributor may install, service, or customize the product for an end user. Issues often arise with respect to:

- The extent to which, if at all, the distributor is licensed to use the owner's IP internally
- The extent to which, if at all, the distributor is permitted to use the manufacturer/developer's trademarks
- The allocation of risk of liability for infringement of third-party IP rights

In addition, when the product is predominantly software, the distributor agreement typically specifies the manner in which the software is to be marketed by the distributor to the end user. Alternatives include: selling a copy of the software to the end user; acting as an agent for the developer company; acting as a broker for an end-user license agreement between the developer company and the end user; or entering into a sublicense agreement (with specified terms) with the end user. The agreement also typically specifies the origin of copies of software that will be delivered to the end user, for example, provided by the proprietor on an as-ordered basis, provided from an inventory maintained by the distributor, or made by the distributor on an as-needed basis from a "master copy" provided by the proprietor.

11.7.9.7 Value-Added Reseller and Original Equipment Manufacturer Agreements

Value-added reseller (VAR) and **original equipment manufacturer** (OEM) agreements are utilized to permit the VAR or OEM to market the technology or product as part of an overall system or product. OEM and VAR agreements are often distinguished by the extent to which the supplied technology or product retains its separate identity. Where the supplied product is a separate module identified to, or identifiable by, the end user, the agreement is typically characterized as a VAR agreement. Conversely, under an OEM agreement, the supplied technology or product tends to lose its separate identity and becomes an integral part of a product marketed under the OEM's name.

Many of the same IP and compliance issues arise with OEM and VAR agreements that arise with distribution agreements. The agreement, if competently drafted, will also include provisions allocating risks of liability with respect to defects in the products and provisions relating to the infringement of third-party IP rights. Such provisions are particularly significant in VAR and OEM agreements, and even more so where the VAR or OEM is permitted to modify the supplied product. Indemnity provisions often make the owner liable to infringement claims as to unmodified product, and run to his or her benefit to the extent that infringement claims relate to modifications of the product by the VAR or OEM.

11.7.9.8 Purchase Agreements

In general, a **purchase agreement** is an agreement under which a vendor (supplier) transfers the title to a product to a purchaser (customer). The UCC and/or CISG are typically applicable to purchase agreements, and terms of the purchase agreement are often established by preprinted forms: purchase order and order-acknowledgement or confirmation forms. Issues tend to arise as to the precise terms of the agreement when the terms in the purchase order conflict with those of the confirmation or acknowledgment form. This is the "battle of the forms" discussed earlier.

11.8 CHAPTER SUMMARY

This chapter has explored in detail both the legal underpinning for contracts and some of the many different types of contracts that technology ventures may encounter. In general, a contract between respective parties serves to define the rights and obligations of the parties and/or the rules under which they operate for the duration of the contract. Within broad legal limits, the parties are free to agree upon whatever terms they see fit. Inadvertently creating a contract, or failing to form a valid contract when a contract is intended, can be catastrophic to a business.

There are various sources of contract law, and it is important to identify which body of law will control the transaction. The common law—the multifarious precedential decisions that have been handed down by judges at all levels for decades—provides the basic foundation, and, in general, still governs contracts other than those for the sale of goods. Where the sale of goods is involved, either the UCC—one form or another of which has been adopted in essentially all of the United States—or the United Nations Convention on CISG will control. In many cases the particular law that applies can make a difference in whether or not a contract exists, and if so, the terms of that agreement.

In general, contract formation requires the basic ingredients of offer, acceptance, and consideration. The offer is a proposal, an attempt, by an offeror directed to an offeree, to initiate a meeting of minds. In order for a proposal to qualify as an offer it must meet certain requirements, for example, identify the offeree, be unconditional, and establish the essential terms of the proposed agreement.

The offeree's response to an offer determines whether or not a contract is formed. If the offeree unequivocally accepts all of the terms of the offer a contract is formed on those terms. Conversely, if the offeree unequivocally and totally rejects the offer, no contract is formed. However, the response to an offer is often a counteroffer, a proposal that the parties go forward on terms different than those of the original offer. Whether or not the counteroffer results in formation of a contract, and if so, the terms of the contract, depend upon the applicable law. Under the common law, and the CISG, no contract is formed unless the acceptance mirrors the offer—unequivocal acceptance of all of the terms of the offer, without any additional terms. However, that is not the case under the UCC, often resulting in the so-called battle of the forms where a contract is formed, but the particular terms of the contract can require considerable analysis. The other necessary ingredient for a contract is consideration—an exchange of something of value—that signifies a transaction has occurred.

This chapter also explored various defenses against contract enforcement. For example, if there was a mutual mistake—both parties were mistaken—as to a basic assumption regarding the contract, the contract can be avoided. Likewise, if it can be shown that one party caused the other party to be mistaken, or knew the other party was mistaken, the mistaken party can avoid the contract. However, a unilateral mistake, strictly by one party, where the other party is not at fault, does not make a contract voidable. A contract is also unenforceable if its subject matter is illegal or against public policy, consent was provided under duress or as a result of fraud, or, with respect to certain types of agreements, it is not in writing.

Once a contract is formed, the parties must perform according to the terms of the contract. Failure to completely perform is a breach. The consequences of a breach (and whether or not the nonbreaching party can cease performance under the contract) depend upon the terms of the contract (which can dictate the consequences), the severity of the breach, and the applicable law.

A comprehensive written agreement can be a tool to avoid misunderstandings and ensure good relationships between participants. Terminology should be carefully and precisely defined, and all items of importance to the parties should be explicitly and unambiguously stated in the agreement. An attempt should be made to deal with all contingencies. Addressing future problems does not imply any particular expectation that the problems will arise, nor does a complete and unambiguous contract imply lack of trust. A comprehensive written agreement goes a long way to ensure that the parties have a complete understanding of each other's expectations. A minimalist agreement, on the other hand, leaves too much room for misunderstanding.

While contracts can take many forms and adopt many formats, there is a common anatomy to most contracts. We examined the general structure of a typical negotiated written agreement and considered the significance of some of the typical provisions, with context provided by the sample agreements of Appendix III and IV. We also reviewed some of the types of agreements commonly encountered by technology entrepreneurs. These discussions should give you the basic grounding in contracts necessary to be a successful technology entrepreneur.

KEYTERMS

Acceptance An unqualified expression of assent to the terms of an offer.

Assignment Transfer of all rights, essentially a sale.

Assignment agreement An agreement transferring title to specified intellectual property rights from the assignor to an assignee.

Boilerplate Standard provisions that address issues of law ancillary to actual performance of the agreement.

Breach An unexcused failure to perform in accordance with the terms of a contract.

Contracts for the International Sale of Goods (CISG) A UN-sponsored multilateral convention establishing rules applying to the sale of goods between parties in different CISG countries.

Commercial impracticability The legal doctrine whereby performance under a contract is excused where an unexpected event, for which risk is not assigned by the contract (or custom), makes performance extremely and unreasonably difficult or expensive.

Compensatory damages A sum of money intended to place the aggrieved party in the position it would have been in had there been complete performance under the contract.

Complete performance Completely fulfilling all obligations under the contract in accordance with its terms.

Consideration The bargained-for exchange of an agreement.

Consulting agreement An agreement where a consultant is engaged to apply a particular expertise on behalf of the hiring company.

Contract An agreement between two or more parties that is legally binding and enforceable in a court of law.

Counteroffer A response to an offer that proposes the parties go forward on different terms than proposed in the offer.

Cover Under the UCC as a prerequisite to consequential damages for an aggrieved buyer, the good faith purchase of substitute goods, at a reasonable rate and without unreasonable delay.

Development agreement An agreement where a developer, typically having particular expertise or resources, is engaged to create technology and/or intellectual property for the hiring company.

Discharge of a contract The termination of a contractual duty.

Distribution agreement An agreement through which the owner of a product (manufacturer, developer) engages a distributor to market the product in substantially unmodified form.

Express warranty A warranty expressly stated in a contract.

Force majeure clause A contractual provision which excuses (for suspends) performance under the contract in the event performance is rendered impracticable by events outside of the control of the parties.

Franchise agreement A species of license agreement that establishes a close, ongoing relationship between the franchisor and franchisee, with the franchisee granted a license to, in effect, replicate a franchise business. Typically trademarks and know-how, and sometimes patents, are licensed.

Fraud in the inducement What occurs when the victim is not aware that the contract was being formed.

Goods Under the UCC, anything that is movable at the time of identification to a contract for a sale.

Implied warranty A warranty that is implied by law in certain contracts unless explicitly disclaimed in the contract.

Inadequate performance A significant variance from the terms of the contract.

Indemnity A promise by an indemnitor to pay for the liability, loss, and/or costs incurred by the indemnitee if specified events occur.

Know-how license An agreement whereby the licensor permits the licensee to have access to, and use, specified know-how.

License A grant of a right to use the licensed subject matter for a specified period of time, analogous to a rental of real property.

License agreement An agreement whereby a licensor having certain intellectual property grants certain rights with respect to the intellectual property to a licensee.

Manufacturing agreement An agreement whereby a company hires another entity to manufacture an item according to certain specifications.

Material breach Failure to provide at least substantial performance under a contract and generally depriving the aggrieved party of reasonably expected benefits from the contract.

Mirror image rule The concept that in order for a response to an offer to be an acceptance, the response must accept every aspect of the offer, without modification, addition, or condition.

Offer A proposal made by an offeror to an offeree.

Original equipment manufacturer (OEM) agreement An agreement by which the owner of a technology or product licenses an OEM to market the technology or product as part of an overall system or product, where the supplied product is fully integrated into the system and is not identifiable as being a product of the technology owner.

Patent license An agreement whereby the licensor agrees not to enforce a patent (or certain claims of the patent) against the licensee with respect to certain acts that would otherwise be precluded by the licensor's patent.

Perfect tender rule A UCC provision (UCC §2-601 et seq) such that any deviation from the terms of the agreement is a material breach and discharges the aggrieved party from performance.

Promissory estoppel An equitable doctrine that makes a promise enforceable by law, even though there is no consideration for the promise when the entity to which the promise is made, reasonably relying on the promise, changes its position to its detriment.

Punitive damages A sum of money not directly corresponding to losses incurred by the aggrieved party, but intended to punish and act as a deterrent to egregious activities.

Purchase agreement An agreement under which a vendor (supplier) transfers the title to a product to a purchaser (customer).

Quasi-contract An equitable doctrine where in the absence of a legally enforceable contract the court creates a fictional contract to prevent one person from being unjustly enriched through retaining money or benefits that in fairness belong to another.

Reliance damages A sum of money intended to place the aggrieved party in the position it would have been had the contract never been formed.

Representation A statement of facts relied upon by parties in entering an agreement.

Rescission Abrogation of a contract, effective from its inception, restoring the parties to the positions that they would have been in had no contract been formed.

Restitution Returning the benefits unjustly conferred upon another to the aggrieved party.

Royalties Consideration paid for in return for the grant of a license.

Specific performance Where a court orders complete performance under a contract.

Statute of frauds A law requiring certain types of contracts to be evidenced in writing.

Substantial performance Essential fulfillment of the obligations but with slight variances from the exact terms, unimportant omissions, and/or minor defects that do not defeat the purpose of the contract.

Term of a contract The length of time the contract remains actively applicable to the parties.

Trademark license An agreement whereby the trademark owner (licensor) permits a licensee to employ a specified licensed trademark in connection with certain goods or services of the licensee.

Unconscionable A contract provision so egregiously one-sided that a court will refuse to enforce it.

Uniform Commercial Code A relatively comprehensive set of laws relating to business and commercial transactions in the United States.

Value-added-reseller (VAR) agreement An agreement by which the owner of a technology or product licenses an OEM to market the technology or product as part of an overall system, but the supplied product is a separate module identified to, or identifiable by, the end user as being that of the technology owner.

Warranty A promise that a product, service, or aspect of performance will meet certain standards or metrics, or do certain things.

ADDITIONAL READING

Bloom, B., Contracts, Wolters Kluwer Law & Business, New York, New York.

Emerson, R., 2009. Business Law, fifth ed. Barron's Educational Series, Inc., Hauppauge, NY.

Ferriell, J., 2009. Understanding Contracts. second ed. LexisNexis, New York, NY.

Lechter, M., 2014. Protecting Your #1 Asset: Leveraging Intellectual Property, second ed. TechPress Inc., Phoenix Arizona.

WEB RESOURCES

http://www.cisg.law.pace.edu: This website is a case law database for the United Nations Convention on Contracts for the International Sale of Goods (CISG).

http://www.expertlaw.com/library/business/contract_law.html This site is an excellent introduction to contract law, covering many topics relevant to the launch and growth of technology ventures.

http://www.law.cornell.edu/ This website provides a searchable database of case law, statutes, and articles on law.

http://smallbusiness.findlaw.com/business-contracts-forms.html This website provides an overview of contract law and the different types of contracts and forms commonly used by businesses.

http://www.law.cornell.edu/ucc This site provides a searchable copy of the Uniform Commercial Code (UCC).

ENDNOTES

1. Louisiana is an exception. Its law is based on a civil code, rather than the precedent of judicial decisions.
2. UCC Article 2. The UCC also includes articles covering, among other things, Negotiable Instruments (Article 3), Bank Deposit (Article 4), Letters of Credit (Article 5), Bulk Transfers (Article 6), Documents of Title (Article 7), Investment Securities (Article 8), and Secured Transactions (Article 9).
3. UCC § 2–103.(1) (k).
4. There are a number of exclusions from the CISG, including "goods bought for personal, family, or household use." CISG Art. 2(a).
5. http://treaties.un.org/pages/ViewDetails.aspx?src=TREATY&mtdsg_no=X-10&chapter=10&lang=en
6. CISG Article 11. But see, Forestal Guarani, S.A. v. Daros International, Inc. 2008 U.S. Dist. LEXIS 79734, (D. NJ, 7 October 2008), Zhejiang Shaoxing Yongli Printing & Dyeing Co., v. Microflock Textile Group Corp., 2008 U.S. Dist. LEXIS 40418, 2008 WL 2098062 (S.D. Fla. May 19, 2008), Chateau des Charmes Wines Ltd. v. Sabaté USA, Sabaté S.A 328 F.3d 528 (9th Cir 2003).
7. See UCC §204C. For example, price is not an essential term. UCC §305 See also CISG Art 14(1).
8. See Restatement (Second) of Contracts §63, UCC §206(1).
9. CISG Art 18(2).
10. Restatement (2d) Contracts §59. CISG Art 19. CISG Art 19(2) indicates that additional or different terms that do not materially alter the terms of the constitutes an acceptance and, unless objected to without undue delay, become part of the agreement. However, CISG Art 19(3) denominates most relevant terms as material, including: "terms relating, among other things, to the price, payment, quality and quantity of the goods, place and time of delivery, extent of one party's liability to the other or the settlement of disputes."
11. See Janky v. Batistatos, 86 USPQ2d 1585 (N.D. Ind. 2008).
12. UCC §2-207 See *Architectural Metal Systems, Inc.*, 58 F.3d 1227, 1230 (7th Cir. 1995), Janky v. Batistatos, 86 USPQ2d 1585 (N.D. Ind. 2008), Uniroyal, Inc. v. Chambers Gasket and Mfg. Co., 380 N.E.2d 571, 575 (Ind. Ct. App. 1978).
13. UCC §2-207(A).
14. UCC §2-207(B).
15. UCC §2-207(C).
16. See Restatement (Second) of Contracts §42, CISG Art 22.
17. See Restatement (Second) of Contracts §63, UCC §206(1).
18. CISG Art 18(2).
19. See Restatement (Second) of Contracts §§17, 20.
20. CISG Art 11.
21. UCC §2-201.
22. UCC §2-201.
23. UCC §2-201(2).
24. The CISG defines a breach as "fundamental" if it "results in such detriment to the other party as substantially to deprive him of what he is entitled to expect under the contract, unless the party in breach did not foresee and a reasonable person of the same kind in the same circumstances would not have foreseen such a result." CISG Art 25, Art 49

25 See UCC §2-601 et seq.
26 See Restatement (Second) of Contracts §§237, 373, CISG Art 49
27 Restatement (Second) of Contracts §352.
28 See Restatement (Second) of Contracts §344, UCC 1–106, CISG Art 74.
29 See Restatement (Second) of Contracts §350, UCC §2-715(2)(a), CISG Art 77.
30 UCC §2-715(2)(a).
31 See Restatement (Second) of Contracts §373.
32 See Restatement (Second) of Contracts §374.
33 See Restatement (Second) of Contracts §376.
34 See Restatement (Second) of Contracts §359, UCC §2-716.
35 See CISG Art 46, Art 62. See also CISG Art 28.
36 See Restatement (Second) of Contracts §184.
37 See UCC §§2-312, 2–314, and 2–315, CISG Art 35(2). Under the UCC, the disclaimer must be conspicuous. UCC §2-316.
38 UCC §2-312, CISG Articles 42 – 44.
39 UCC §2-2-314, CISG Articles 35(2)(a).
40 UCC §2-2-315, CISG Articles 35(2)(b).
41 See, Atlantic Marine Construction Co. v. United States District Court for the Western District of Texas, 571 U. S. ____ (2013), Slip Opinion No. 12–929 decided December 3, 2013
42 see, e.g., UCC §2-210
43 Herbert M. Confidentially Speaking. *Inc.* April 2005; 27 (4): 50.

PART 4

Growth and Exit

CHAPTER 12

Venture Management and Leadership

ENTREPRENEURIAL LEADER LEAVES COMFORTABLE JOB TO LAUNCH TECH VENTURE

Todd Basche wasn't always an entrepreneur. In fact, he spent many years working in big companies before he decided to strike out on his own. As Apple Computer's vice president of application software, Basche's responsibilities were always entrepreneurial in nature. In addition to his creative days at Apple, Basche had also worked in product development at Hewlett-Packard. Despite the creativity of his day job, Basche dreamed of starting his own venture one day.

When he turned 50, Basche decided it was time to create something that was completely his own. His breakthrough idea came to him one day when he was trying to think of an intuitive number for the lock to his pool. What this helped him discover was a $1 billion consumer market with revolutionary potential.

What captured his attention was the combination lock industry. While not exactly a high-tech market, Basche believed that memorizing numbers for various locks is not intuitive. He thought about his experience in the computer industry, and the prevalence of User IDs and passwords, most of which are letter-based codes that are easy to recall.

Basche and his wife, Rahn, developed and licensed the initial product to Staples for a trial run. Quickly realizing its potential, they both quit their jobs in January 2007 to focus on building their new company, "Wordlock."

The Santa Clara, Calif.–based company was self-funded and home-based during the first year, but experienced strong growth in 2008. By the first quarter of 2008, Wordlock was in 900 retail locations; by the end of the second quarter, it was in 12,000 stores.

Basche sees potential in other markets, as well. "There are other consumer product areas where the market giants have been sleepy," he says. "The learning curves and challenges for an entrepreneur are considerable," but Basche says the experience has been profound. "When the product is your own, versus a corporation's, it's an amazing feeling to watch it go from your head to the store shelf."

Source: Adapted from Julia Boorstin, "Staples Lets Customers Do the Designing," Fortune, April 18, 2005; Jennifer Wang, "A Step Down That's a Step Up," Entrepreneur, November 14, 2008.

CHAPTER 12 Venture Management and Leadership

12.1 INTRODUCTION

A common problem that arises for many technology entrepreneurs is the lack of preparation as the venture grows, and the founder must make the transition from lone entrepreneur to manager of a growing enterprise.[1] Many technology entrepreneurs have not had formal training in managing or leading others, and some are happiest and at their best when working in isolation on challenging technical problems. That is not to say that such individuals are incapable of developing and even mastering the skills of effective management.[2] We are simply saying that many technology entrepreneurs have spent their careers learning and working within highly specific technical niches and have not focused on developing their managerial talents and skills.

The good news is that it does not require an MBA to be an effective manager or leader of a technology start-up. Many technology entrepreneurs simply take the "on-the-job-training" approach, but that can be hazardous. Start-up ventures often have very little room for failure and must operate effectively from the beginning to succeed. Not only can the venture suffer if the entrepreneur is learning to manage and lead for the first time "on the job," but investors and other stakeholders may have little patience for the learning process.[3]

Although it is imperative that entrepreneurs learn at least some of the art and craft of managing and leading from experience, a few lessons from books and advisors can also be helpful. In this chapter, we provide some fundamental ideas about managing and leading a technology venture. The basic concepts and leadership techniques discussed here are intended to provide a framework for understanding the variety of issues that inevitably arise in start-up ventures. These ideas should be applied to leading and managing, and the results of applying them should be monitored and assessed. Most likely, each of the skills, roles, and techniques described in this chapter will need to be adapted and modified for unique situations.

We begin our discussion of management and leadership by reviewing several of the higher order skills that most entrepreneur managers will need to organize and control the venture work environment.

12.2 ENTREPRENEURIAL LEADERSHIP

Leadership is an important and necessary skill for achieving individual, group, and venture performance. The entrepreneur influences the attitudes and expectations that encourage or discourage performance, secure or alienate employee commitment, reward or penalize achievement. Entrepreneurial leaders must be able to influence others to work as hard as they do in order to create a valuable company. Some scholars have noted that the technical skills that make entrepreneurs successful in the early stages of a company actually hinder them as the company grows and greater emphasis on managing and leading are required.[4] Research has determined that many entrepreneurs are effective in creating and starting a company on their own. However, when growth requires additional employees the entrepreneur does not have the necessary skills to motivate and inspire followers.

Recall that we have characterized the start-up company as a "temporary organization" that should be focused on searching for a repeatable, scalable business model. This phase of the venture will require less formal leadership and management, but the entrepreneur must still ensure that everyone is operating effectively. The search process in the start-up phase is full of ambiguity, chaos, and uncertainty. The entrepreneur-leader must ensure that everyone is focused on running the kinds of experiments that will result in validated learning. The build-measure-learn feedback loop (Exhibit 2.1), that we discussed in Chapter 2, requires the entrepreneurial leader to focus the team on learning from experience.

Once a repeatable, scalable business model has been found, the venture goes through a transition from searching to executing. The entrepreneurial leader must now shift focus to building a company capable of executing the business model that was discovered during the search phase. Exhibit 12.1 highlights the transition from "search" to "execution."

Technology entrepreneurs who lack leadership skills hang on to their "independent" actor status for as long as they can. However, their growing companies are often characterized by high employee turnover and general employee dissatisfaction. Usually, the entrepreneur who lacks leadership skills either eventually leaves the company or finds someone else who can lead employees day-to-day while the entrepreneur takes on a different role within the company—perhaps as chairman, as an externally oriented CEO, or Chief Technology Officer (CTO).

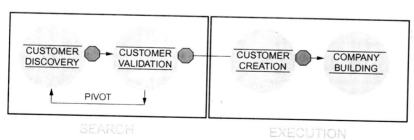

EXHIBIT 12.1
The transition from search to execution.

In this chapter, we define leadership as "the ability to influence, through communication, the activities of others, individually or as a group, toward the accomplishment of worthwhile, meaningful, and challenging goals." First, this definition indicates that one cannot be a leader unless there are people (e.g., co-workers and followers) to be led. Second, leadership involves the application of influence skills. The use of these skills has a purpose; to accomplish goals. Finally, an objective of leadership is to bring about influence so that important goals are achieved. This influence is brought about not only directly through authority or motivation, but also indirectly through role-modeling. Research has shown that employees generally have higher expectations of leaders as models or exemplars of the organization.[5]

Leadership is a general concept that applies in many different social contexts: sports, politics, organizations, and entrepreneurial ventures. You probably will agree that the traits and behaviors that make a person an effective leader in politics may be quite different from those that make a person an effective leader of a sports team. The same is true of leaders in business. The traits and behaviors that are necessary for leaders of large organizations are different from those that are required for leaders of entrepreneurial ventures.

12.2.1 Influence

The exercise of **influence** is the essence of leadership behavior. Leaders use influence as their primary tool to move the venture toward its goals. Of course, the entrepreneur must also use such tools as compensation, employee feedback and evaluation, and organizational structure to move a venture forward. Seven influence strategies have been proposed as vital for entrepreneurial leadership roles:[6]

1. Reason—Using facts and data to develop a logically sound argument
2. Friendliness—Using supportiveness, praise, and the creation of goodwill
3. Coalition—Mobilizing others in the organization
4. Bargaining—Negotiating through the use of benefits or favors
5. Assertiveness—Using a direct and forceful approach
6. Higher authority—Gaining the support of higher levels in the hierarchy to add weight to the request
7. Sanctions—Using rewards and punishment

Entrepreneurial leaders need to learn a variety of influence strategies. As their business grows, they cannot rely solely on exercising the power they possess by virtue of their position as founder and owner. Research indicates that employees demonstrate higher levels of motivation if they are allowed to influence the way the organization works. Scholar Noel M. Tichy said, "The ultimate test for a leader is not whether he or she makes smart decisions and takes decisive action, but whether he or she teaches others to be leaders and builds an organization that can sustain success even when he or she is not around."[7]

Entrepreneurial leaders must be able to exercise their influence without developing arrogance or an air of superiority. This is difficult for novice entrepreneurs who may be unfamiliar with power and authority. Suffice it to say that tyrants rarely inspire great effort among their people.[8] In fact, people respond to, and are influenced by, leaders who can demonstrate concern for their personal growth. Of course, in business a leader cannot be concerned about personal growth as it pertains to nonbusiness areas, such as a person's romantic life or personal financial acumen. Rather, people respond to leaders who are able to provide them with a continuing stream of challenging projects that are appropriate to their current skills, abilities, and temperament.[9]

12.3 LEADERSHIP SKILLS

Performance in a start-up or growing technology venture does not just happen. Committed and skillful entrepreneurs carrying out specific roles as leaders or managers make it happen. The entrepreneur as a leader of a growing venture influences performance by defining objectives, recognizing and minimizing obstacles to the achievement of those objectives, and effectively planning, organizing, and controlling resources to attain high levels of venture performance. This section focuses on the leadership skills that must be applied to everyday situations experienced in start-up technology ventures.

Learning to lead effectively as the venture grows can be a difficult challenge for many technology entrepreneurs. Most first-time technology entrepreneurs have never received formal training in management. Their only point of reference may be the individuals who led them at some point in their careers. And *those* individuals may also never have been exposed to formal leadership training.

It is not necessary to have had formal leadership training to be an effective leader, but the concepts and tools provided via formal training can lead to more effective performance. Entrepreneurs who do not adapt to leadership roles often have to cede control of their venture to more experienced people. This can work, and many technology entrepreneurs have done so. But for those entrepreneurs who want to stay engaged at the forefront of their venture over the long haul, developing personal leadership skills will be essential.[10]

Most investors and experienced entrepreneurs will readily attest to the importance of leadership in start-up ventures.[11] When it comes to judging the success of a start-up technology venture, actual performance is all that matters. Good intentions, promises, and wishes will not matter to the entrepreneur or to investors if solid execution and performance is lacking. Fortunately, most of the key skills and abilities necessary to execute are learnable, and the technology entrepreneur who is willing to learn has the best chance to succeed.[12] For most entrepreneurs, learning to lead people should occur both via reading and understanding, and through on-the-job-training.

Regardless of the venture type, entrepreneurial leaders must possess and seek further to develop many critical skills. A **skill** is an ability or proficiency in performing a particular task. Leadership skills can be learned and developed. In general, all technology entrepreneurs as *leaders* should seek to develop skills in the following areas:

- Analytical
- Decision making
- Communication
- Conceptual
- Resilience
- Team building
- Self-awareness

12.3.1 Analytical Skills

Analytical skills involve using repeatable approaches or techniques to solve organizational problems. In essence, **analytical skills** are concerned with the ability to identify key factors affecting venture performance, to understand how they interrelate and how they can be managed to achieve venture goals. Analytical skills include the ability to diagnose and evaluate the issues that face the venture on a daily basis. They are needed to understand problems and to develop action plans for their resolution.

Analytical skills also include the ability to discern and understand how multiple complex variables interact, and to conceive of ways to make them interact in a desirable manner. These skills include the ability to analyze one's own talent, as well as the talent of others associated with the venture. The entrepreneur who is able to analyze and accept his or her own strengths and weaknesses will be in a better position to achieve performance goals. The entrepreneur who has accurately analyzed his or her own capabilities will hire those who complement strengths and compensate weaknesses.

Most technology entrepreneurs already possess strong analytical skills, but they are mostly focused on *technology* issues rather than *business* issues. Running a venture requires shifting focus from analyzing the venture's key technology as a technology to analyzing it as a business. Technically, a venture's offerings may be the best in the world. From a purely business perspective, however, the entrepreneur must assess and analyze the potential market, the costs associated with bringing the product to market, the scalability of the enterprise over time, the financial resources that will be required to build and grow the venture, and many other things. We have discussed many of these analytic tools throughout this book. Successful technology entrepreneurs learn to be adaptable in their thinking, developing an ability to shift their analytic focus between technical and business issues as required.[13]

12.3.2 Decision-Making Skills

All entrepreneurs must make decisions, and the quality of these decisions determines their degree of effectiveness. An entrepreneur's **decision-making skill** in selecting a course of action is greatly influenced by his or her ability to focus and guide the venture despite ambiguous information and conditions. One of the hallmarks of start-up ventures is their ambiguity.[14] They are often ambiguous with respect to their market, their value proposition, their competition, and even their ability to persist. Entrepreneurs must learn to resist pressures for a quick fix when problems and issues arise. They must learn to live with uncertainty and ambiguity, and to recognize often subtle differences between actions that work and those that do not work.[15]

Decision making in start-up ventures is almost always done in the face of irresolvable ambiguity. This is something that the technology entrepreneur must learn to accept. Yet, the ambiguous nature of many of the issues the entrepreneur faces must not lead to inaction or paralysis. The entrepreneur must have a predilection toward action and must be able to make decisions in the face of incomplete information. The term that is often used to describe this situation is **satisficing**. This means that the entrepreneur as manager must choose the best solution to a problem despite incomplete information about both the problem and the likely outcome of the decision that is taken.[16] Choosing the *best* alternative is a vastly different decision than choosing the *correct* alternative. Making a decision among alternatives in running a start-up venture is different than selecting among alternatives on a multiple choice exam or in crafting a solution to a problem posed in a textbook. In the world of the start-up, there usually is no *correct* course of action, but there often is a *best* one. That is the one the technology entrepreneur must learn how to identify, select, and act upon.

The analysis of decision alternatives likely will alternate between technology and business issues. Technology entrepreneurs must make choices about the technology development road map and the business development road map. This will require flexible analytical skills, as we discussed above. But being flexible should not lead to an inability to decide firmly and move ahead. Most technology entrepreneurs are very good at analysis. To be entrepreneurial, however, requires an equal ability to take action once a decision has been taken.

12.3.3 Communication Skills

Since entrepreneurs must accomplish much of their work through other people, their ability to work with, communicate with, and understand others is important.[17] Effective communication, written and verbal, is vital for venture performance. The skill is critical to success in every venture, but it is crucial to entrepreneurs who must achieve results through the efforts of others. **Communication skills** involve the ability to communicate in ways that other people understand, and to seek and use feedback from others to ensure that one is understood.

The entrepreneur's communication skills will also be tested among investors, shareholders, and other stakeholders. One of the primary ongoing tasks for the technology entrepreneur is fundraising. Fundraising requires that the entrepreneur be able to tell a succinct story about the venture's offerings and market, its value proposition, and its business model. This short story about the venture is sometimes referred to as its **elevator pitch**. The elevator pitch is so named because it is a short and clear overview of the venture that might be told to a potential investor during a chance meeting on an elevator. Generally, an elevator pitch should take no more than one minute and should be clear and even a bit exciting.

In addition to the elevator pitch, the entrepreneur must be able to communicate the intent and direction of the venture via a written business plan. The business plan is a required document when speaking to potential investors, lenders, or other key stakeholders. Business plans should be supplemented by a one- to two-page executive summary.[18] Substantial writing skills are required to boil an entire business plan down to an executive summary. There are firms that provide entrepreneurs with fee-based support in writing and editing business plans and executive summaries, but research has indicated that it is better for the entrepreneur to write these documents.[19] Not only is there potential value to the entrepreneur in the research and learning that goes into business plan writing, there is also a need to constantly update and/or adjust the plan.

Finally, the entrepreneur must be skilled in presenting the plan to investors, lenders, and others. The verbal ability to articulate the intent and goals of the business, as written in the plan, requires the ability to understand and adapt communication style to the audience. For example, many technology venture entrepreneurs must present their business plans to others who are not as technically adept as they are. Speaking in technical jargon or using complex graphics to convey the business idea will likely not produce the desired results in such cases. The technology entrepreneur must adapt communications to match the ability of the other party to the discussion.

12.3.4 Conceptual Skills

Conceptual skills consist of the ability to see the big picture, the complexities of the overall organization, and how the various parts fit together. Entrepreneurs use their conceptual skills to develop long-range visions for their companies. Conceptual skills enable the entrepreneur to project how prospective actions may affect a company one, three, or even five years in the future. The single most important foundation of entrepreneurial success today is leadership—especially visionary leadership. It is critical that every venture develop a vision and mission to guide the many choices that it must make in the present and into the future.

Visionary companies possess values based on a core ideology that is unchanging and transcends immediate customer demands and market conditions. The unifying ideology of visionary companies guides and inspires people. Coupled with an intense "cult-like" culture, a unifying ideology creates enormous solidarity and *esprit de corps*. Lastly, visionary companies subscribe to what some have called "big, hairy, audacious goals" **(BHAGs)** that galvanize people to come together, team, create, and stretch themselves and their companies to achieve greatness over the long haul.[20]

In their book *Built to Last*, Jim Collins and Jerry Porras advocated that entrepreneurial leaders should develop **BHAGs**. The authors elaborate that "a true BHAG is clear and compelling, serves as a unifying focal point of effort, and acts as a catalyst for team spirit. It has a clear finish line, so the organization can know when it has achieved the goal; people like to shoot for finish lines. A BHAG engages people—it reaches out and grabs them. It is tangible, energizing, highly focused. People get it right away; it takes little or no explanation."[21] Exhibit 12.2 provides an illustration of how a BHAG combines passion, expertise, and economic reality.

Another key skill the entrepreneurial leader needs to develop is tolerance of ambiguity and uncertainty. In fact, the lead entrepreneur should develop the ability to absorb ambiguity, enabling the rest of the team to focus on the BHAG. The uncertainty that all new ventures experience can only be relieved over time via market exploration and discovery. The entrepreneurial leader must decide on which experiments the venture will run and how to interpret the results to enable forward progress.

Technology ventures that have achieved the highest levels of success have often been launched by visionary founders. Steve Jobs of Apple, Next, and Pixar is perhaps the most visible of these visionaries. His overarching vision for Apple was to create "insanely great products." This simple vision elicited incredible loyalty and effort from employees and provided the world with a stream of novel and path-breaking products for more than two decades.

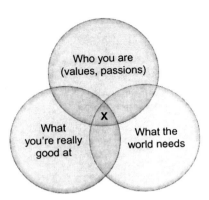

EXHIBIT 12.2
Creating your BHAG: "X" marks the spot.

Other visionary leaders have used their focused visions to revolutionize industries as diverse as package delivery, grocery retailing, and airlines. Fred Smith founded FedEx on the vision of delivering packages overnight across the entire United States. His vision was that people would call FedEx for packages that "absolutely, positively had to get there overnight." Whole Foods founder John Mackey created a new concept in grocery retailing based on his vision of "whole foods, whole people, whole planet." This vision permeates the culture, operations, and merchandise of Whole Foods. Herb Kelleher founded Southwest Airlines on his vision that an airline could provide great "no frills" service at a low price. Southwest subverted the traditional wisdom of the industry and continues to be one of the few profitable airlines in the United States. Even its stock trading ticker symbol—LUV—expresses its customer-focused mission and vision.

12.3.5 Resilience Skills

Resilience is one of the more central characteristics required of the entrepreneurial leader. The turbulence that most start-ups endure is often difficult to manage emotionally, financially, and physically. The entrepreneurial leader must be the one who maintains the equilibrium of the venture through the turbulence. Everyone possesses some ability to remain strong in the face of ambiguity, uncertainty, and stress. The entrepreneurial leader should be exceedingly aware of his or her personal capacities in this area and strive to improve through experience, reflection, and personal growth. Increased resilience will happen naturally through experience for most entrepreneurs as they struggle, fail, and recover. But there are things the entrepreneurial leader can and must do to enhance personal resilience. For example, we discussed in Chapter 3 the need for emotional intelligence. This means that the entrepreneurial leader is aware of personal emotions and uses proven and tested strategies to manage them. For example, if the venture is experiencing difficult times financially, the entrepreneurial leader must control personal feelings about those difficulties to prevent those feelings from affecting the performance of the venture team. If the leader walks around with a "woe is me" look or grumbles aloud about the state of the venture's finances, it is likely to have a negative effect on team performance. At a time when performance is needed more than ever, the entrepreneurial leader's lack of emotional control could subvert that performance. The Mini-Case below highlights one young entrepreneur who preserved through several failed ventures to achieve eventual success.

MINI-CASE
Resilience Proves Vital for One Tech Entrepreneur

Jesse Schwarz was always trying new things to make money. In college, he ran a small used car lot with his friend. Later, he opened a chicken restaurant that lasted a total of 30 days before going under. Dusting himself off from that failure, Jesse launched a new business providing high-end coffee makers to professional offices. This lasted until his foreign suppliers proved too unreliable to build a scalable business around. With a dwindling savings account, a wife, and two kids, Jesse knew his next venture had to be a hit. So, rather than investing in some new product, restaurant, or other scheme, he decided to invest in himself. Even though he had no technical background, Jesse decided to teach himself how to succeed in e-commerce. He learned everything he could from free resources on the Internet. Eventually, he decided to launch a service that would provide consumers with an easy to understand comparison shopping site for high speed Internet service providers. Using Commission Junction as his primary service for managing click-through transactions, Jesse built a thriving business. Within months he was making more than $3,000/month, and within a couple of years his venture was cash flowing more than $50,000/month. Jesse's resilience through several failed ventures led to his ultimate success.

Source: Duening, T.N. 2011. *Experience Starting an Internet Marketing Company.* (Colorado Springs, CO: Businesses2Learn Publishing.)

12.3.6 Team-Building Skills

Rare is the lone-wolf entrepreneur. Most technology ventures are too complicated for a single person to operate without help from others. Successful entrepreneurs are usually talented team builders. They are able to attract other people to their vision and then build them into a coherent team, all focused on pursuing the same goals. **Team building** is based on identifying gaps in talent and skills that are required for the venture to succeed, then finding people with these necessary traits. Successful entrepreneurs are able to attract employees, advisors, and investors who provide the venture with missing talent that is required to achieve venture goals.

Building successful teams requires each of the skills mentioned above and also a strong helping of humility. Successful technology entrepreneurs attest to the importance of hiring and motivating people more talented than themselves. In fact, the strongest performing entrepreneur-leaders are not afraid to hire people who are more talented than they are. This statement may seem obvious, but it is not uncommon for technology entrepreneurs to feel intimidated by people who are more talented than they are. After all, throughout their formal technical education they were compared with, and competed against, other technically talented people. That competitiveness does not just disappear; an entrepreneur has to reorient his or her thinking to overcome the tendency to compete. Technology entrepreneurs must replace their feelings of competitiveness among their peers with competitiveness as a business manager, which means swallowing one's pride and hiring talented people that will help the venture compete in its market and achieve its goals.

The foundation of successful teams is simple: clarity of goals and responsibilities. Highly talented people are usually self-motivated to a high degree. They want to do a good job and enjoy working on goals that are clearly defined and are measurable. Technology entrepreneurs have learned that they can expect a high degree of commitment from talented people without a lot of managerial intervention. That is, talented people usually perform at their peak when their leaders provide them with goals and then get out of their way. Teams organized around specific projects and goals will often self-organize. Google, for example, allows its engineers to choose the projects that they work on, with very little day-to-day oversight on their work. This relaxed approach can only work in an environment where the vision and goals are clear and understood by all and where rewards and incentives are directly linked to performance that helps the venture achieve its objectives.

12.3.7 Self-Awareness Skills

Improving yourself in any role you are in, such as being a student, a parent, a friend, or a leader, requires taking control of determining who you are by conducting your own personal analysis. You need to think about situations in which you succeeded. What were the factors that led to that success? Hard work? Good planning? Problem solving skill? Patience? Examining previous experiences involves self-talk, reflection, and analysis.

Unfortunately, memories often leave out data, information, and critical incidents. Creating a daily journal or diary is a good method to keep track of vital components of your successes. Keeping a journal for 10–20 days about your interactions with other people—friends, colleagues, teachers, bosses, and vendors—requires discipline. The journal can be simply structured by using a date and summary comments about:

- What I did today
- Whom I interacted with today

- What kind of thoughts I had today
- What I used to solve any problems or new situations I was faced with

Simple, concise journal entries will help you conduct a review and analysis. About 10 days of entries will provide, at a minimum, a picture of how you typically behave and think.

A second method of data gathering for self-awareness is to complete self-assessment surveys. They can be a confidential source of information that only you control. To improve any skill, there is a need for feedback. Online assessments about your personality, skills, abilities, and attitudes can provide eye-opening feedback for you.

Friends and colleagues are another source of feedback. Some colleagues and friends will be reluctant to provide honest, accurate, and revealing feedback. You may have to create a framework and presentation that explains clearly why you need their feedback. Asking for their help to improve your understanding and knowledge of "who you are" can be convincing when presented as a request for assistance.

Matching the feedback of others with your own self-awareness analysis is educational and can be invaluable. You learn firsthand how others see you and it enables you to match this information with your self-concept. Discrepancies can occur and need to be evaluated carefully. The evaluation should consider the information—your own and that of others. Once a thorough evaluation is completed and analyzed, it is then time to consider what skills you need to work on to improve. What skills do you need to be successful in your life, career, and job? The feedback process will definitely provide you with insight into areas that need to be sharpened and refined. Overall, your self-assessment should focus on three major areas as a means of improving your personal leadership skills: personality, values, and attitudes.

12.3.8 Personality

Personality is a set of physical and psychological variables that create individual uniqueness. A key personality factor is self-concept; the view you hold of yourself as a physical, social, and spiritual person. Two related aspects of self-concept are self-esteem and self-efficacy. Self-esteem is a person's belief about his or her own worth. People with high self-esteem see themselves as worthwhile, vital, and important. Self-efficacy is a person's belief that he or she can successfully accomplish a job, a task, or an activity.

12.3.9 Values

Values are a person's preferences concerning appropriate courses of action. Values provide a picture of a person's sense of what is correct, fair, or right.

Exhibit 12.3 Rokeach's Values Classification.

Terminal Values	Instrumental Values
A comfortable life	Ambition
An exciting life	Broad-minded
A sense of accomplishment	Capable
A world at peace	Cheerful
A world of beauty	Clean
Equality	Courageous
Family security	Forgiving
Freedom	Helpful
Happiness	Honest

Parents, mentors, friends, teachers, and role models, in general, impact a person's values. Since learning and experiences differ from one person to another, values are also different.

The noted psychologist Milton Rokeach developed two broad value categories. Terminal values reflect a person's preferences concerning the "ends" to be achieved. Instrumental values reflect a person's preferences for the means to be used in achieving end states.[22] Exhibit 12.3 presents Rokeach's classification of his values system.

Empirical research has determined that both terminal and instrumental values differ across groups (e.g., leaders, union members) and individuals.

12.3.10 Attitudes

Attitudes can be characterized in various ways. First, they tend to persist unless something is done to change them. Second, attitudes can be favorable or unfavorable. Third, attitudes are directed toward some object and reflect a person's feelings and beliefs. Thus, attitudes are a persistent tendency to feel and behave in a particular way toward some object.

Attitudes can be depicted as consisting of three components: emotion, information, and behavior. The emotional component consists of a person's feelings or affect. The informational component consists of the beliefs and information you have about an object. It does not matter whether the information is correct or incorrect. The behavioral component consists of your tendencies to behave in a certain way. Only the behavioral component can be observed by others. One cannot see emotions unfold or the beliefs a person has about something. These two components must be inferred.

Attitudes serve a number of functions. First, attitudes help people adjust to their work environment.[23] When employees are treated well, fairly, and honestly, they tend to develop positive attitudes. When employees are embarrassed, intimidated, and threatened by leaders, they are likely to adopt negative attitudes toward others, the job, and the employer. Second, attitudes help you defend your self-images. By changing and altering attitudes, a person is able to keep a balanced self-concept and ego. Keeping one's self-image positive and in balance are important for being motivated to perform well. Third, attitudes provide the basis for expressing your values. For example, a person who has a strong work ethic will tend to voice attitudes about the importance of providing "a good day's work for a good day's pay." The person's core values are articulated to others and this provides a view of what the person represents.

12.4 ENTREPRENEURIAL LEADERSHIP AND ETHICS

Ethics is often a difficult topic to address in a business-oriented textbook. For the most part, people who excel in business are not deeply familiar with the terminology and concepts that comprise a standard course in ethics. Many businesspeople also regard the topic as "fuzzy" and difficult to comprehend because there are so many differing perspectives on the topic. In fact, many businesspeople live daily by the adage "if you can't measure it, it doesn't exist."

Ethics can be simple or hard. It is simple when individuals decide on a few basic principles that will guide them in their lives and then stick with these principles, come what may. It is hard when people believe that ethics can be nuanced and that situations must be judged independently to know what the right course of action might be. This textbook does not have enough space to take the nuanced approach to understanding entrepreneurial ethics. As such, it takes a more straightforward, principles-based approach to understanding this important topic.

Fortunately, it does not require that you have a master's degree in philosophy to understand some basic principles of entrepreneurial ethics. One very successful media entrepreneur, Karl Eller, summed up entrepreneurial ethics in the title of a book he wrote about his own successful ventures: "Integrity Is All You've Got."[24] Integrity is defined, basically, as doing what you say and saying what you do.

Integrity is a good starting point for a set of ethical principles. It would be difficult to argue the opposing perspective that integrity is not important, or that a lack of integrity is important to business success. In fact, most successful businesspeople will attest to the role that integrity plays in their ability to make things happen. Most business is transacted within a framework of trust. Businesspeople trust that those with whom they are associated will follow through on commitments and contracts to which they are counterparties.

Although contracts and commitments are normally also governed by legal rules, businesspeople do not want to have to resort to lawyers, lawsuits, and the court system every time they want to achieve a business goal. Rather, they want to work with people who will deliver on promises. A single failure to fulfill a contractual obligation, or to honor one's word, or to deliver on a promise can ruin a reputation.

Another principle that should be closely aligned to integrity is honesty. In business, the virtue of honesty is also sometimes referred to as "transparency." This simply means that the venture is operating each day in a manner that would pass a formal financial audit, and it is operating generally according to what is referred to as "good faith." This term means that counterparties to a contract or business relationship are using their best efforts to deliver their end of the contract. Clearly, not all business transactions or arrangements live up to their expectations. However, parties that act in good faith and in a transparent manner generally will not suffer any negative legal consequences as a result of a failure. The free-market system is often described as a "profit and loss system." That is, business transactions occasionally fulfill and even exceed profit expectations. On the other hand, oftentimes they do not, and parties to the transaction suffer financial loss. This is normal, and parties can recover to work together again in the future if they believe each has operated honestly and in good faith, and will likely act similarly in the future.

A final ethical principle that we will discuss in this chapter and that seems essential for business success is humility. Humility has often been described as a character trait, but it can also be expressed as an ethical principle. An individual that expresses humility is one that recognizes that many of the good and bad things that happen in life and in business are often a function of chance events. Consider the founders of Google, Sergey Brin and Larry Page. Certainly these are two shining stars among the technology entrepreneurs of our time. They not only have created a singularly impressive company, but they have achieved incredible levels of personal wealth. Their success, no doubt, can be attributed to their respective talents in computer programming and website design. These talents are clearly relevant to the technology age in which we live. What if Page and Brin had come of age in 1890 rather than 1990? Would their unique talents have been as applicable in that long-ago time period? Clearly, they would not have founded Google. Not only was there no Internet in 1890, there were no computers, databases, or even electric power. The point is, despite the great success of Google, Page and Brin have to admit they were lucky to have been born during an age when their unique talents are highly prized and rewarded. That would not have been the case had they come into the world 100 years earlier.

Most success is attributable to a complex mix of personal talent, fortunate circumstance, and chance events. Entrepreneurs who recognize this will be able

EXHIBIT 12.4

A definition of humility. *(Source: David Marcum and Steven Smith.* Egonomics: What Makes Ego Our Greatest Asset (or Most Expensive Liability)*. (Fireside Publishing, 2007))*

to maintain an authentic humility during times of success and will also be more balanced personally during difficult times. In fact, several authors have coined the term "egonomics" to refer to the importance of keeping one's ego in check and developing humility.[25] These authors list four warning signs that indicate when an individual's ego has taken over:

Being comparative When we're comparative, we tend to either pit our strengths against another's weaknesses, which may lead us to an exaggerated sense of confidence, or we compare our weaknesses to their strengths, which can cause negative self-pressure.

Being defensive "When we can't 'lose,' we defend our positions as if we're defending who we are, and the debate shifts from a we-centered battle of ideas to a me-centered war of wills."

Showcasing brilliance "The more we want or expect people to recognize, appreciate or be dazzled by how smart we are, the less they listen, even if we do have better ideas."

Seeking acceptance "When we equate acceptance or rejection of our ideas with acceptance or rejection of who we are, we 'play it safe.' We tend to swim with the current and find a slightly different way of saying what's already been said as long as acceptance is the outcome. That not only makes us a bland follower, but an uninspiring leader."

It is important to recognize how humility fits within the spectrum of possible character orientations, from a completely empty ego to egotism. Exhibit 12.4 provides a useful illustration that humility is in the middle of these two ends of the spectrum. According to this illustration, humility represents a healthy and intelligent understanding of one's unique talent and skills, but avoids the destructive potential of overconfidence and egotism.

12.5 CHAPTER SUMMARY

This chapter has explored entrepreneurial leadership. As we mentioned throughout this chapter leadership is developed over a lifetime of practice, reflection, and adjustment. There are no absolutes in leadership—no one is a leader in every circumstance. Each entrepreneurial manager and leader must find the practices that work best for him or her according to the circumstances encountered.

Despite the lack of absolutes, however, we have explored principles that have stood the test of time and are likely to be highly applicable to whatever technology venture you may be considering or already are operating. For example, the ethical principles of integrity, honesty, and humility likely will apply across any type of venture based on the fundamental rules of good business practice.

Additionally, the central skills of entrepreneurial leadership should form a basis for continued learning. As we stated, many technology entrepreneurs have not been exposed to formal training in the management or leadership sciences. As such, many learn the art and craft of managing and leading via "on the job-training." While this is laudable and acceptable, in many cases the venture environment is too complex to allow this relatively slow and inefficient learning process. While it is not necessary for entrepreneurs to have had formal training in leadership, it does help to have a conceptual framework available to interpret the events and situations that are experienced when operating a venture. Conceptual frameworks provide some structure to the chaos of leading. The central skills discussed in this chapter are meant only as a starting point to further reading and exploration of the concepts and principles that define effective managing and leading.

The major takeaway from this chapter is to continue to study and learn about leadership. No single book or management guru knows everything about each of the important topics in this book. It is imperative that you do not become discouraged by the fact that there is no single, correct answer to management and leadership questions. Rather, this knowledge should lead to a constant pursuit of excellence in managing and leading the technology venture. This constant pursuit will lessen the chance that mistakes will be repeated and enhance the likelihood of success.

KEYTERMS

Leadership The ability to influence through communication of the activities of others, individually or as a group, toward the accomplishment of worthwhile, meaningful, and challenging goals.
Influence The ability to achieve goals and objectives through other people.
Skill An ability or proficiency in performing a particular task.
Analytical skill The technology entrepreneur will use analytical skills to make sense of all the data the venture uses to manage its performance.
Decision making skill Technology entrepreneurs must learn to be decisive, often in the face of great uncertainty and/or ambiguity.
Satisficing Choosing the best, but not necessarily the optimum, option among various courses of action.
Communication skill The ability to communicate effectively with the venture's various stakeholders.
Elevator pitch A short, clear overview of the venture that might be told to a potential investor in a short elevator ride.
Conceptual skill The ability to see the big picture and how the various parts of the venture fit together.

Team-building skill The ability to identify the gaps in talent the venture needs to succeed and then bring in people to fill those gaps and get them to perform as a team.

ADDITIONAL READING

Bennis, W., 2009. On Becoming a Leader. Basic Books, New York.

Covey, S.R., 2013. The 7 Habits of Highly Effective People, Deluxe ed. Simon & Schuster, New York.

Greenberg, D., McKone-Sweet, K., James Wilson, H., 2011. The New Entrepreneurial Leader: Developing Leaders Who Shape Social and Economic Opportunity. Berrett-Koehler Publishers, San Francisco.

WEB RESOURCES

http://www.1000ventures.com/business_guide/crosscuttings/leadership_entrepreneurial.html: This site covers the topic of entrepreneurial leadership exclusively. While not the most visibly appealing site ever posted to the web, it does have a lot of resources, ideas, and leadership tips.

http://www.entrepreneur.com/management/index.html: This website is hosted by *Entrepreneur* magazine. This portion of the site provides a lot of information about management and leadership challenges that most entrepreneurs face. There are plenty of articles, tips, and case studies for learning.

http://www.inc.com/resources/leadership/: This site is hosted by *Inc.* magazine. This portion of the site focuses on leadership issues that entrepreneurs face.

ENDNOTES

1. Richard G. Hamermesh, James L. Heskett, and Michael J. Roberts. Note on Managing the Growing Venture. *Harvard Business Review*, January 2005, p. 1–8.
2. George C. Rubenson and Anil K. Gupta. "Replacing the Founder: The Myth of the Entrepreneur's Disease". *Business Horizons* 35(6): November/December 1992; p. 53–57.
3. Daniel P. Forbes. "Managerial Determinants of Decision Speed in New Ventures". *Strategic Management Journal*, 26(4): April 2005; p. 355–366.
4. John Hamm. "Why Entrepreneur's Don't Scale". *Harvard Business Review*, December 2002, pp. 110–115.
5. J. Clifton Williams. "Self-Control". *Baylor Business Review*, Fall 1997, pp. 9, 32.
6. David Kipnis, Stuart M. Schmidt, Chris Swaffin-Smith, and Ian Wilkinson. "Patterns of Managerial Influence: Shotgun Managers, Tacticians, and Bystanders". *Organizational Dynamics*, Winter 1984, pp. 58–67.
7. Cited in Vivian Pospisil. "Nurturing Leaders". *Industry Week*, November 17, 1997, p. 35.
8. Robert I. Sutton. 2007. The No Asshole Rule: Building a Civilized Workplace and Surviving One that Isn't. (New York: Business Plus).
9. Mihaly Csikszentmihalyi. 1991. Flow: The Psychology of Optimal Experience. (New York: Harper Publishing).
10. William K. Todorovic and Francine K. Schlosser. An Entrepreneur and a Leader! A Framework Conceptualizing the Influence of a Firm's Leadership Style on a Firm's Entrepreneurial Orientation—Performance Relationship". *Journal of Small Business & Entrepreneurship*, 20(3): 2007; p. 289–307.

11. Guy Kawasaki. "The Art of Execution". *Entrepreneur*, 36(4): April 2008; p. 48.
12. Sheetal Singh. "Practical Intelligence of High Potential Entrepreneurs: Antecedents and Links to New Venture Growth". *Academy of Management Proceedings*, 2008; p. 1–6.
13. Dev K. Dutta and Stewart Thornhill. "The Evolution of Growth Intentions: Toward a Cognition-Based Model". *Journal of Business Venturing*, 23(3): May 2008; p. 307–332.
14. Yosem E. Companys and Jeffrey S. McMullen. "Strategic Entrepreneurs at Work: The Nature, Discovery, and Exploitation of Entrepreneurial Opportunities". *Small Business Economics*, 28(4): April 2007; p. 301–322.
15. Deniz Ucbasaran. "The Fine 'Science' of Entrepreneurial Decision Making". *Journal of Management Studies*, 45(1): January 2008; p. 221–237.
16. S.G. Winter. "The Satisficing Principle in Capability Learning". *Strategic Management Journal*, 21(10/11): 2000; 981–996.
17. John R. Darling and Steven A. Beebe. "Enhancing Entrepreneurial Leadership: A Focus on Key Communication Priorities". *Journal of Small Business & Entrepreneurship*, 20(2): 2007; p. 151–167.
18. Nicole L. Torres. "Sounds Like a Plan". *Entrepreneur*, 33(3): March 2005; p. 102–104.
19. Julian E. Lange, Aleksandar Mollov, Michael Peralmutter, Sunil Singh, and William D. Bygrave. "Pre-start-up Formal Business Plans and Post-start-up Performance: A Study of 116 New Ventures". *Venture Capital*, 9(4): October-December 2007; p. 237–256.
20. Jim Collins and Jerry Porras. "Building Your Company's Vision". *Harvard Business Review*, 74(5): May 1996; p. 65–77.
21. Collins, J.C. and J.I. Porras. 1994. Built to Last: Successful Habits of Visionary Companies. (New York: HarperBusiness.)
22. Rokeach, M. 1973. The Nature of Human Values. (New York: Free Press.)
23. Katz, R.L. 1974. "Skills of An Effective Administrator." *Harvard Business Review*, 33(1): 90–102.
24. Karl Eller. 2004. Integrity is All You've Got. (New York: McGraw-Hill.)
25. David Marcum and Steven Smith. 2007. Egonomics: What Makes Ego Our Greatest Asset (or Most Expensive Liability). (Fireside Publishing.)

CHAPTER 13

Valuing and Exiting Your Venture

NESS COMPUTING ACQUIRED BY OPEN TABLE

Ness Computing, makers of a popular restaurant recommendations app was acquired in February 2014 by the restaurant reservation platform OpenTable. The acquisition was an all cash deal worth $17.3 million. Terms of the acquisition include the provision that the Ness Computing team will work at OpenTable's San Francisco headquarters. The Ness app and site will be discontinued as the technology will be integrated into OpenTable.

Although the deal seems to be positive for Ness and its investors, the reality is somewhat different. Ness, launched in 2011, has raised more than $20 million in investor capital. Ness started out as a personalized search engine technology for mobile. Ness pivoted from that model to specific restaurant recommendations—but always had its sights set on eventually developing beyond just restaurants. That growth model was behind its $15 million Series B investment that it concluded in 2012.

With this transaction, Ness's technology, and talent, will be tasked with enhancing OpenTable's restaurant recommendations. The Ness technology should help OpenTable add more features to attract more restaurants to its platform and potentially take home more returns on the commission it charges them.

Source: Lunden, I. 2014. "OpenTable Buys Ness for $17.3M to Beef Up Mobile and Restaurant Recommendations." TechCrunch, February 6.

13.1 INTRODUCTION

Perhaps the most ironic part of entrepreneurship is the need to build your venture every day with some idea of how you will exit. The reason this is important is that there are different options for exiting your venture and each option requires that you build your venture in different ways. For example, if you intend to pass your venture on to your heirs, there are things that you should do to prepare them for a smooth succession. On the other hand, if you are planning on exiting your venture via a sale of stock on a public stock exchange you will be focusing on different types of things than if you were handing the keys to your heirs.

A good bit of advice that entrepreneurs learn to follow is to "manage your business every day as if you are going to sell it tomorrow." The reason this is good advice is because you never really know when the right offer may come along that enables you to sell the business that you've diligently been building. If you've been practicing good business techniques along the way, the path to completing the sale will be far smoother. For example, no one wants to purchase a business at top dollar if it has been doing a terrible job with its accounts or if it is dealing with multiple ongoing lawsuits. Expert entrepreneurs know that it is important to pay attention to the details of the business. And this needs to be done whether it is the entrepreneur or someone else who actually keeps track of the details. Ultimately, of course, the entrepreneur is responsible for creating the business culture that leads to a well-run business venture.

In the case that you intend to sell your venture to another party, whether it is another entrepreneur, another venture, or a venture capital firm, the other party is going to perform what is called **due diligence** on your venture. The due diligence process usually includes a thorough on-site visit to your venture's headquarters, review of the accounts, and many other things. In this chapter, we'll introduce you to the due diligence process and discuss how you can be prepared for that in the event that you will one day desire to sell a venture you've created.

The due diligence process is generally undertaken for two purposes. One is to evaluate the long-term viability of the venture, as we've mentioned. The other reason to conduct due diligence is to help establish a value for the venture. **Valuation** is the term used to refer to the process and calculations used to establish a dollar value for your venture. You will need to establish a value for your venture when you exit. You'll also need to establish a value for your venture anytime you sell equity to others. We discussed valuation very briefly in Chapter 8 when we explored the venture fund-raising process. The same techniques that you use to value your venture for an exit are used to value your venture for a fund-raising transaction. We will provide you with some simple techniques that can be used to establish a reasonable value for your venture.

The primary purpose of a venture's **exit strategy** is to develop a roadmap by which early-stage investors can realize a tangible return on the capital they invested. The entrepreneur spells out a reasonable scenario—the exit strategy—by which a later round of funds will be raised. This later round should provide the original investors with an opportunity to sell their shares and realize a capital gain.[1] Second, the intent of an exit strategy is to suggest a proposed time window that investors can tentatively target as their **investment horizon,** placing a limit on their involvement in the early-stage funding deal. The founding team wants to assure investors that the venture will grow to a point whereby the initial investors will receive all of their original investment back, plus a sizable return for the risks they took by investing in a start-up venture. To investors and other shareholders of a private venture the exit is a **liquidity event.** It is the event which enables them to convert their shares into cash.

The problem in projecting an exit strategy is that it's based on several major assumptions. The speed of market penetration, the ability to sell at expected price levels, the costs of doing business, the margins on sales, the management team's ability to arrange consistent deals, and the impact of competition and other economic factors collectively affect the new venture's projected market share and bottom line. Certainly, when these factors are all aligned in the most favorable way the firm will experience significant market share, consistent high-growth sales rates, strong profit margins, and positive earnings. And such a scenario in the first two to three years is typically the story told by firms that eventually go public.

In this chapter, we provide an overview of the various ways that you may elect to exit your venture. The options that we will review include passing on the venture to heirs (succession), selling the venture to another party (acquisition), merging the venture with another firm, or taking the venture public through an initial public offering (IPO). Each of these exit strategies provides you with different challenges and opportunities as the entrepreneur and owner. We will examine each exit strategy, look at how the venture should be positioned during its growth stages to best pursue each of the various exit strategies, and review the advantages and disadvantages of each.

Let's begin by examining the due diligence process that precedes most exit strategies.

13.2 DUE DILIGENCE

If the exit process includes bringing in new buyers or investors it will also include in-depth analysis of the venture, its operating history, its future prospects, and many other things. This analysis is referred to as *due diligence* and is normally conducted by the parties that are interested in acquiring or investing in the venture.[2] Due diligence involves examining a number of key elements of the target venture, including financial health, status of the venture's product line, the potential for synergy with the acquiring firm, market position and future potential of the venture, research and development history and roadmap for the venture, legal considerations, and plans for managing the acquired entity.[3] Let's examine each of these factors in turn with an eye toward common analyses used in the due diligence process and how the entrepreneur can prepare for a due diligence review.

13.2.1 Finances

The due diligence process usually begins with a financial analysis—analyzing the profit and loss figures, operating statements, and balance sheets for the years of the company's operation, concentrating on the more recent years. Past operating results, particularly those occurring in the preceding three years, indicate the potential for future performance of the company. Key ratios and operating figures indicate whether the company is financially healthy and has been well managed. Areas of weakness, such as too much debt, too little financial control, dated and slow turning inventory, poor credit ratings, and bad debts are also carefully evaluated. The entrepreneur can prepare the venture for this portion of the due diligence process by ensuring that accurate, timely, and regular accounting reports have been prepared and filed. Here's where the entrepreneur can take advantage of having invested in hiring an accountant to ensure all the books have been well managed and stand up to auditing review.

13.2.2 Product/Service Line

The past, present, and future of a firm's product lines will also be examined. The strengths and weaknesses of the firm's products are evaluated in terms of design features, quality, reliability, unique differential advantage, and proprietary position. The life cycle and present market share of each of the firm's products are verified. One method for evaluating the product line is to plot sales and margins for each product over time.[4] Known as S or life-cycle curves, these indicate the life expectancy of the product and any developing gaps. The S-curve analysis could reveal that all products of this firm are at or near their period of peak profitability. We discussed the product life cycle in Chapter 10. Exhibit 13.1 is a representation of the product life cycle. It's important to know the life cycle status of the venture's product lines to understand whether there is a fit with the acquiring party's product portfolio and operating expertise. For example, a venture that is good at introducing new products would not be a good fit with a company whose primary product line is mature and stable.

13.2.3 Synergy

The phrase, "the whole is greater than the sum of its parts" is a good definition of the concept of **synergy**. Entrepreneurs who seek to exit their companies via acquisition should consider whether the company provides any synergies to potential acquirers.[5] The synergy should occur in both the business concept—the acquisition functioning to move the acquiring company toward its overall goals—and the financial performance. Acquisitions should positively impact the bottom line of the acquiring firm, affecting long-term gains and future

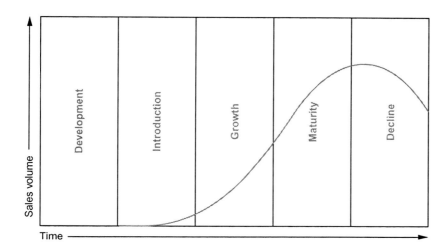

EXHIBIT 13.1
The product lifecycle.

growth. Lack of synergy with the existing business is one of the most frequent causes of an acquisition failing to meet its objectives.

13.2.4 Markets and Customers

The due diligence process includes evaluation of the entire marketing program and capabilities of the venture. Although all areas of marketing are assessed, particular care is normally taken in evaluating the quality and capability of the established distribution system, sales force, and manufacturers' representatives. For example, one company may acquire a venture primarily because of the quality of its sales force. Another may acquire a venture to obtain its established distribution system, which allows access to new markets. Entrepreneurs should be aware of which of these elements of enterprise value will be of interest to potential acquirers.

Acquiring companies can gain insight into the market orientation and sensitivity of target ventures by looking at their marketing research efforts. Does the venture have facts about customer satisfaction, trends in the market, and state-of-the-art technology of the industry? Ventures that are setting up for an exit by acquisition should be collecting, storing, and actively mining their customer data.

13.2.5 Research and Development and Intellectual Property

The future of the venture's products and market position is affected by its research and development (R & D) and the extent to which the venture has established intellectual property rights in the results of the R & D. The due diligence process normally probes the nature and depth of the venture's research and development and its ability to adjust to changing market conditions. The due diligence process will assess the strengths and weaknesses of the venture's innovation capabilities and its IP portfolio. Although the total amount of dollars spent on research and development is usually examined, it is more important to determine if these expenditures and programs are directed by the venture's long-range plans and whether or not the venture has been successful in introducing new offerings to customers. The due diligence process will scrutinize the intellectual property protection held by the venture, and assess its efficacy to sustain competitive advantage, generate royalty income from third parties, and/or secure freedom to operate for the venture through cross-licensing with competitors.

13.2.6 Operations

The nature of the venture's business processes, for example, the facilities and skills available, its efficiency and work flow, and its productivity, are also examined as part of the due diligence process. Are the facilities obsolete? Are they flexible, and can they produce output at a quality and a price that will compete over the coming years? Are there procedures in place to systematically review products and operations for potential intellectual property assets, and to establish and maximize protection of those assets? Acquiring firms do not want to build this infrastructure themselves. Typically, an acquiring company focuses on the growth prospects of the target venture and looks carefully at its overall operation to determine if it is poised to deliver on its growth potential.[6]

13.2.7 Management and Key Personnel

Finally, the due diligence process evaluates the management and key personnel of the venture. The individuals who have contributed positively to past success in sales and profits of the firm should be identified. If a company is conducting due diligence as part of a potential acquisition, they want to know if the key personnel will stay once acquisition occurs? Have they established sound objectives and then implemented plans to successfully reach those objectives?

Additionally, acquiring companies examine whether any of the venture's personnel are indispensable. Generally, it is not good for the acquiring company to have such individuals. There is too much risk associated with those key people leaving, becoming disabled, or worse. As such, technology ventures strive to ensure that knowledge and other key assets are not dependent on any single individual or groups of individuals.

13.3 VALUATION

Investors expect entrepreneurs in whom they invest to use their money wisely and carefully to build **enterprise value**. Enterprise value refers to the value of the enterprise as a whole. Enterprise value is easy to determine for companies that trade on the many public stock markets worldwide. The value of an enterprise whose shares trade on public markets is simply the price per share times the number of outstanding shares. This is also referred to as the firm's **market capitalization** or **market cap**. In the case of private ventures, the efficiency and effectiveness of operations and demonstrative intellectual property assets (such as patents) are the essence of its enterprise value. Efficiency refers to the cost-effectiveness of the venture's systems. Effectiveness refers to how well the systems produce value that customers demand.

13.3.1 Valuation Techniques

There are a number of techniques that can be used to determine the value of a venture.[7] Some of these techniques employ sophisticated mathematical formulas and statistics; others use primarily qualitative judgments and educated hunches. We'll look at a subset of these techniques, including the multiples technique and the discounted cash flow method.

13.3.2 Multiples Technique

The multiples technique is the least mathematical of the methods for determining venture valuation and perhaps the one most often used. It is a straightforward technique that relies on identifying recent comparable transactions, i.e., recent acquisitions of, or sale of equity in, comparable entities, where definitive valuations were established by the transaction. The valuations of those "comps" are then used to establish the valuation of the venture. This is analogous to the process used by appraisers to establish the value of real estate. Where technology ventures are involved, however, is not always easy to identify comparable entities; what metrics should be considered to in identifying "comps" for a venture? In some cases, key metrics for valuation have been established within an industry. For example, in the Internet industry a common key metric is "registered users." Instagram was acquired by Facebook in 2012 for $1 billion. At the time of the acquisition, Instagram had only a dozen or so employees, no revenue, and no real business model. However, it did have something that Facebook valued highly: registered users. At the time of the acquisition, Instagram boasted 33 million registered users, and it had only launched 18 months prior. In essence Facebook calculated that the value of each registered user was approximately $33. To arrive at the $1 billion valuation, then, is simple arithmetic:

$$30,000,000 \text{ users} \times \$33/\text{user} = \$1,000,000,000$$

Companies in similar industries can now use the same multiple (i.e., $33 per registered user) to develop a reasonable valuation based on the registered user metric. Note that the valuations vary depending on which variable a venture chooses to use in generating a reasonable valuation. In general, the owners of a venture will want to choose the variable that gives them the greatest value, even though potential investors will argue for the variable that gives the venture the least current value. There is no such thing as an "absolute" or "true" venture value. It all depends on the argument that can be made for one or another metric and a comparable company to serve as the baseline for calculating valuation.

13.3.3 Discounted Cash Flow Technique

For most start-up ventures the old mantra clearly applies: "Cash is king." What that simple phrase means is that start-up ventures must organize their growth and operations to ensure an ample supply of cash on hand to meet current obligations, including operating expenses and debt liabilities. One of the primary ways that entrepreneurs track and manage cash in the venture is via the cash flow statement. Unlike the income statement, the cash flow statement does not track operating performance according to rules of accounting. The cash flow statement tracks the actual movement of cash into and out of the firm. The technical term "discounted cash flow" is a refined measure of cash that subtracts one-time capital expenses and dividend obligations from the projected cash position at some future date.

To determine valuation using the discounted cash flow method requires understanding the concept of **present value**. Present value is defined as the value in the present of some future cash flow. A simple technique has been developed to determine the present value of a future cash flow. Let's say that you are promised some cash right now or $100 dollars one year from now. How much cash would you need to receive right now to decide to forgo the $100 future cash flow? Important to this consideration is how much a person could have earned by investing cash in hand over the one-year period. If the investment would return more than the $100, then you should take the cash in the present and invest it. If it would return less than the $100, then you should take the money in the future. The interest rate used to determine the return on current cash is called the **discount rate**. Here's how the calculation works:

$$PV = FV/(1 + r)$$

In this equation, PV is the present value of the future value (FV) divided by the discount rate, r. If the discount rate was 8% (.08) with a FV of $100, someone would need to give you at least $92.59 in the present to make it worth your while to forgo the future cash. Applying this logic to a venture requires identifying a future cash flow that will be put into the calculation above. As we've noted previously, your business plan should include a financial forecast that provides interested investors with cash projections for at least three years into the future. A simple way to generate the FV variable in this equation is by using the projected cash position of the venture at the end of the three-year period. This would be the last line of the cash flow statement in month 36 (if you need a refresher on the cash flow statement, see Chapter 6).

Investor expectations of future returns are based on their estimate of risk. One element of that risk is the alternative investments they could have made. One alternative investment would be to purchase fixed-income securities, such as U.S. treasury bills. Such notes are backed by the U.S. government and are assumed to be risk-free. Thus, the prevailing interest rate paid on U.S. treasuries is known as the risk-free rate of return. At minimum, investors in start-up ventures would expect to meet and exceed this risk-free rate. If the entrepreneur cannot demonstrate how that is possible, then investors would purchase the bonds instead.

In addition to the alternative investments that comprise one element of the risk associated with investing in any particular venture is the risk of the venture itself. The risk-free rate of return is relatively straightforward, and can be known with a high degree of accuracy. However, the risks that are associated with any particular venture are more difficult to quantify. One technique that is used to arrive at a discount rate that includes both the alternative investment risk and the venture risk is known as the **risk-adjusted discount rate** method, or RADR. This approach is summarized in the equation below.

$$r_{vt} = r_{ft} + RP_{vt}$$

This equation states that, for a particular venture, v, that yields an uncertain cash flow at some future time, t, the discount rate is expressed as rvt. In the equation above, rft is the so-called risk-free rate of return, and RPvt is the risk premium that is associated with the venture. The discount rate that is used to determine the present value of future cash flows, then, is a combination of the risk-free rate and the risk premium. This is the minimum rate of return that investors would expect to receive on their exit from the venture.[8]

The important point to remember is that, using this method, valuation is based on the likely future cash flows that will be generated by the venture. The various assets of the organization, such as intellectual property, products and services, customer lists, and brand value are not considered. Many entrepreneurs don't have a lot of experience in predicting future cash flows, and they are at a disadvantage when the time comes to place a value on the venture. This lack of experience can be offset by consulting with independent firms that specialize in placing a value on ventures. Paying for such a service prior to seeking capital, especially from experienced venture capital firms who are usually very good at determining the value of a venture, can be a worthwhile expense. Arming oneself with rational and defensible arguments about the future value of a venture can help when the time comes to negotiate with others about the value of a venture.

13.4 EXIT VIA SUCCESSION

Succession is a common exit strategy used in family-owned businesses. Such ventures often are handed down from generation to generation. Some of the largest private companies in America, such as Koch Industries, and Milliken & Company are family owned and continue to be run primarily by members of the Koch and Milliken families, respectively. These companies have developed detailed succession planning guides that help them groom and prepare the next generation to take over the company. Exhibit 13.2 indicates the primary industry categories for family businesses in the United States.

Many family-owned businesses aspire to keep the business in the family via succession, but they don't plan very well for it. As a result, many family-owned businesses suffer setbacks and occasionally bankruptcy because of poorly designed succession. For example, many fail to help the next generation come to terms with the difficulties of running a company, assuming instead that a college degree or MBA will prepare them for leadership. By way of contrast, family-owned ventures that manage the succession process effectively realize that there's no better way for the next generation to learn the business

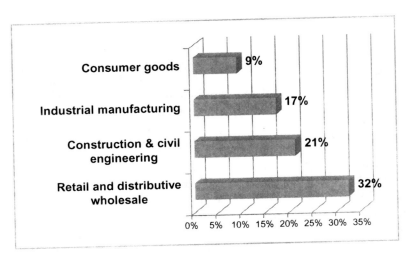

EXHIBIT 13.2
Industry types for family businesses in the United States. Source: "Choosing Your Next Big Bet." PriceWaterhouseCoopers family business survey 2010-2011; p. 6.

than the way the first generation did. Effective succession planning generally exposes the upcoming generation to all facets of the business. Family members who aspire to lead the venture someday will spend time "in the trenches" with other employees learning about the business from the ground up. Such lessons can be invaluable as the older generation steps aside and the next generation takes over.

There has been extensive research into the processes that can be used to better manage the succession process. One of the first things that should be considered in succession planning is to ensure that the transfer process doesn't incur burdensome taxation. A good accountant will be essential to help ease the tax burden associated with transferring ownership of a major asset (the business) to heirs.[9] Many family businesses have had to be liquidated on succession because the heirs could not afford the taxes associated with ownership transference. An estate planning accountant and attorney can be valuable in avoiding that problem.[10]

Of course, in order to use a succession strategy there must be someone in the family willing and able to take over the business. If no one has a passion for taking over, it would be wiser to exit the venture via acquisition—selling to someone outside the family. If there is someone in the family who passionately wants to run the business and is prepared to take the reins, you have no problem. But, what if multiple family members are interested in the business? If that circumstance arises, there are several things that can be done to avoid hard feelings. The most common strategy in such a situation is to bring everyone together and determine who is best suited for which role in the venture. This is best handled after the family members have each had an opportunity to work in various roles within the venture to determine the best fit for their own talents and experience. Allocating roles within the venture is part of the challenge when multiple family members are interested in running the venture; the other challenge is ownership percentages. A good way to manage that is to transfer ownership based in part on performance goals. In that way, no one can claim a large ownership stake without having demonstrated the ability to manage and lead the venture effectively.

Ultimately, the decision must be made concerning who will be in charge of running the company on succession of the current CEO. A firm decision must be made on this important part of the succession plan. Some family businesses opt for leadership by committee, where a committee of family members makes most of the major decisions for the venture. This can work but it can also be unwieldy and political as the venture grows or experiences challenging times. Not many large companies are run by committee, and it is not recommended that family businesses operate under that style. Most management and leadership scholars will point to the value of having a single individual who ultimately is responsible for the strategic and operational effectiveness of the venture.

Finally, the most effective succession planning eases the new generation into their respective leadership roles rather than all-at-once immersion. We mentioned before that it is useful for family members aspiring to leadership to work in various roles within the venture prior to assuming day-to-day leadership. It is also very useful to have the previous generation stay on during the early days of succession to help the new generation of leaders understand the context of their decisions. Although it is not advisable for the senior generation to intervene too soon, as some of the greatest lessons for the new generation will come via mistakes, it is advisable for the senior generation to help the new generation understand the lessons being learned.[11] This is not to say that the outgoing leaders should sit idly by if disastrous decisions are being made. The point of the nonintervention strategy is to allow the new generation to make mistakes that don't threaten the health of the business. Gradually stepping aside and allowing the incoming generation to assume larger and larger responsibilities is the preferred succession management technique. In fact, succession planning can often be a 10-year process.[12]

13.4.1 Advantages of Exit Via Succession

There are a few advantages to exit via succession. Perhaps a primary advantage is that the current business owner can decide when, where, and how he or she will exit. Family successions can be managed over a long period of time, as the incoming generation is brought up to speed on managing the venture. In addition, for business owners who have a difficult time "letting go" of the venture the exit via succession option can allow them to ease out over a longer time. Family members are generally going to be more tolerant of a former owner having some responsibility than is a total stranger.

Another advantage of exit via succession is the opportunity to pass on a healthy and viable business to one's family members. Many entrepreneurs work long and hard hours in large part for the benefit of their family. This motivation is strong for many entrepreneurs and small business owners and is rewarded when the next generation is groomed effectively to take over the business.

13.4.2 Disadvantages of Exit Via Succession

There are some disadvantages to the exit via succession strategy. Obviously, one of the disadvantages is the potential for family squabbles to arise. For example, if one family member is passed over to lead a business in favor of another family member, hard feelings could arise. In that sense, it is important for a succession plan to be put in place early so that people are allowed to "grow into" their future roles in the business.

Another disadvantage of exit via succession is that the current owner may not receive full compensation for the hard work he or she has put into the venture. An outright sale of the venture to a nonfamily member would be negotiated more strenuously than would the transition to next-generation ownership. As such, the current owner must take as part of his or her reward for years of hard work the satisfaction of passing on the venture to family members.

Where there are investors in the entity that are not family members, exit of the founders by a succession can be problematical; it does not necessarily provide a liquidity event for those investors. Accordingly, the organizational documents (e.g., operating agreement, or shareholders agreement) may include provisions that would effectively preclude and exit via succession strategy or make it very difficult and costly.

As we mentioned, the first thing that is required for the succession strategy to be used to exit the venture is to have a willing and able family member who is ready to take over. If one is not present, the business owner may be better positioned to exit the venture via selling to another party. This is known as "acquisition." Let's explore exit via acquisition next.

13.5 EXIT VIA ACQUISITION

Before the entrepreneur can build value attractive to a potential acquirer, he or she must understand the buyer's perspective.[13] Most **acquisitions** are undertaken because the acquiring company sees one or more strategic advantages that can be obtained by buying another business. These advantages, or **value drivers**, vary from industry to industry. They also change over time in specific industries because of evolving product life cycles and external developments that affect the industry. Generic categories of strategic value drivers include:

- **Broadening product lines:** The buyer adds complementary products to increase revenues. This is a common strategy for both product and service companies.
- **Expanding the technology base:** The buyer adds technology skills or intellectual property that enhances or complements the company's current base.[14]
- **Adding markets and distribution channels:** The acquirer obtains channels it doesn't currently serve. Companies that start out with a **vertical strategy** can add new industry expertise by widening its distribution capabilities. A vertical strategy is one where a venture develops expertise in a given industry that can be expanded to other industries with only slight modifications.
- **Increasing the customer base:** The buyer adds a company that is similar in product offerings or in its business model, yet focuses on a different customer segment. This strategy is enhanced if the target company has a good reputation or strong brand. It also can expand the acquirer's geographic coverage.

- **Creating economies of scale:** The combined company can offer a more efficient use of physical assets or overhead—a critical need in consolidating industries.
- **Extending internal skills:** The buyer can add new capabilities such as consulting or service offerings, international management skills, or various types of management and business skills. These skills can be offered as independent revenue producers or enhance a company's competitive edge.[15]

Many acquirers hope to leverage several value drivers in a single deal. Google, for example, has regularly used acquisitions as a strategic tool to bring in new skills, new technology, and, sometimes, new customers. Google has developed expertise for integrating these newly acquired technologies into their complete product offerings.

When the perspective of potential buyers is factored into the development of a new venture's strategy, the strategic planning process becomes very important. Internal characteristics of the company will influence the acquirer's final valuation.[16] These factors are related to fundamental business management, strong cash flow, and accurate books and records. Although they are obviously necessary for closing a deal, they must be augmented by significant value

KEY POINT

Preparing Your Venture for Acquisition

As you begin to study your industry, you will need to look at both industry trends and the strategies of competing companies in handling those trends. Some of the questions that need answers include:

- Why are your customers presently buying from you?
- What are the customers' sourcing alternatives, including direct competitors, internal competition within the customer company, inaction or not buying at all, and innovative alternate solutions?
- How will customers' needs change?
- What alternatives will customers have tomorrow?
- What is happening in related markets to influence buying patterns?
- What will the market look like tomorrow?
- What will other players be looking for in executing acquisitions?

The critical conclusion to this process is in the last two questions about the future. Doing your homework through the earlier points should help you develop a vision of future trends and value drivers in the industry.

Source: *Adams, M. 2004. "Exit Pay-Offs for the Entrepreneur." Mergers & Acquisitions: The Dealmaker's Journal, 39(3): 24-28.*

drivers to motivate the buyer to pay a premium price.[17] The Key Point box below highlights some actions entrepreneurs can take to prepare their venture for an eventual acquisition.

13.5.1 The Acquisition Deal

Once the venture has been identified as a good candidate for acquisition, an appropriate deal must be structured. Many techniques are available for acquiring a firm, each having a distinct set of advantages to both the buyer and seller. The deal structure involves the parties, the assets, the payment form, and the timing of the payment. For example, all or part of the assets of one firm can be acquired by another for some combination of cash, notes, stock, and employment contracts. This payment can be made at the time of acquisition, throughout the first year, or over several years.

The two most common means of acquisition are the **direct purchase** of the target venture's entire stock or assets, and the **bootstrap purchase** of these assets. In the direct purchase of the firm, the acquiring company often obtains funds from an outside lender or the seller of the company being purchased. The money is repaid over time from the cash flow generated from the operations. Although this is a relatively simple and clear transaction, it usually results in a long-term capital gain to the seller and double taxation on the funds used to repay the money borrowed to acquire the company.

In order to avoid these problems, the acquiring company can make a bootstrap purchase, acquiring a small amount of the firm, such as 20% to 30%, for cash. The acquiring company then purchases the remainder of the target venture by a long-term note that is paid off over time out of the acquired company's earnings. This type of deal often results in more favorable tax advantages to both the buyer and seller.

13.5.2 Advantages of Exit Via Acquisition

There are numerous potential advantages of the acquisition strategy. The most obvious is that the entrepreneur and any other owners of the venture's equity have the opportunity to convert their interest into cash. This is the ultimate goal for a lot of entrepreneurs, and it is certainly the goal of every one of the investors. The amount of cash that will be involved in the acquisition transaction depends upon the valuation of the venture. The more enterprise value the entrepreneur has created, the greater will be the cash required to purchase the venture. The cash that is used to purchase the venture is then distributed to each of the investors according to each investor's percentage of ownership interest.

Another advantage of acquisition is that the entrepreneur now has the cash needed to launch a new venture. Many so-called **serial entrepreneurs** continue to build new ventures with the cash they have obtained through past deals.

You may have heard of the boom-and-bust nature of entrepreneurship. Many entrepreneurs obtain a lot of cash early in their careers and erroneously conclude that entrepreneurship is easy or that they are uniquely talented. Some end up bankrupt because they used all of their cash to try to achieve the thrill of success once again. Expert entrepreneurs follow the affordable loss principle throughout their entrepreneurial careers. That is, even though they may have had a big payday from a successful exit via acquisition, they don't risk all of their gains on their next ventures. Instead, they put away enough money to live on, if they can, and continue to leverage other investors to help them reduce their personal financial risk.

Other advantages of the acquisition strategy include the opportunity to exit a venture that has grown too large for the entrepreneur. By getting out of a growing venture that exceeds the entrepreneur's leadership and management skills, the entrepreneur can avoid the problems associated with decline and, potentially, failure. Handing over the reins of the venture to better qualified leaders is common for expert entrepreneurs who prefer starting ventures over managing large and growing ones. Finally, the entrepreneur gains more respect and credibility among investors and others potential stakeholders if he or she can demonstrate a track record of building and selling companies. It will be far easier in future ventures for the entrepreneur to raise capital and acquire needed resources with such a track record of success.

13.5.3 Disadvantages of Exit Via Acquisition

One of the major disadvantages of exit via acquisition is that the entrepreneur loses control of the venture that he or she has been nurturing. In the succession exit, as we've seen, it is very useful for the previous generation to stick around and help the incoming generation manage the venture for awhile. This not only provides the incoming generation with valuable mentoring as it comes up to speed on leading the venture, but it also lets the outgoing generation ease slowly into retirement. Many entrepreneurs who sell their companies are not prepared for the idleness that comes with the sales transaction. Most acquiring ventures are not interested in keeping the founder around, and will simply take over the operation as soon as the sales transaction is closed. For some entrepreneurs, the lack of a place to go, challenges to meet, and people to interact with can be upsetting. In fact, many entrepreneurs quickly leap into a new venture because they simply are not ready to be idle.

Another disadvantage of exit by acquisition is that the parties to the deal must come to agreement on the valuation of the venture. This can be a lengthy and potentially expensive process that may take the entrepreneur's attention away from running the business. The due diligence that needs to be conducted can also be disruptive to an operating venture. Employees who are aware of the potential for the acquisition may be concerned about their jobs and perform less well while the veil of uncertainty exists.

The final disadvantage of the acquisition exit strategy that we'll mention here is the challenge of finding a qualified buyer. This can be a major challenge, depending upon the type of venture. Ventures that are primarily based on local customers and local sales will most likely need to find a local buyer, or someone willing to relocate. If the economy is good there may be many potential buyers around. If the economy is not so good, there may be fewer potential buyers. The entrepreneur who wishes to exit via acquisition does not always have control over exactly when that may occur. It can take years to find a qualified buyer, in which case the value of the business may also change between the initiation of a search for a buyer and finally closing a deal. There are **business brokers** who can help entrepreneurs sell their companies, but they are fee based agents who may also require an up-front retainer. In addition to business brokers, there are a number of websites that entrepreneurs can use to list their business for sale. Some websites that offer such as service include:

- www.bizbuysell.com
- www.businessesforsale.com
- www.businessmart.com
- www.businessbroker.net

In the event that a qualified buyer cannot be found in a reasonable time, the entrepreneur can consider exit via merger. A merger will result in a change of ownership, but does not always include cash as part of the transaction. An entrepreneur who can no longer wait for a buyer can exit the venture by merging with another entity and then holding onto the stock of the merged enterprise. Of course, that is not the only time a merger strategy makes sense. Often two or more companies may be able to pursue more, significant new opportunities under a single venture than they could alone. Let's turn next to the exit via merger strategy.

13.6 EXIT VIA MERGER

Another method for exiting a venture is via a **merger**—a transaction involving two (or more) companies in which only one company survives. Acquisitions are similar to mergers and, sometimes, the two terms are used interchangeably. In reality, they are quite different, with the primary differences between mergers and acquisitions centering on the relative size of the entities involved and on who is in control of the combined entity. Mergers generally occur when the relative size of the ventures involved is equal—or at least they are perceived to bring equal value to the merged entity.

When an entrepreneur decides to merge with another company it's usually the case that the two entities are similar in size and offer similar value to the merged entity. For example, it would be quite unexpected for Google to merge with a five-person technology start-up. More likely, Google, with its tremendous assets and global reach, would be in control and simply acquire the smaller venture.

When two entrepreneurial ventures merge, the question of who will control the merged companies is part of the merger negotiations. An entrepreneur who wants to exit a venture may elect to merge with another company, but it may take some time for the entrepreneur to wriggle free of the new company. Often, when a merger occurs the merged company requires that the top executives from each venture stay with the merged entity to ensure its success. Depending on the size of the new company, the entrepreneur may be able to negotiate a deal whereby he or she earns out of the company over a period of time.

An **earn-out strategy** is used for ventures that have begun to generate consistently strong positive cash flow.[18] The management team initiates a monthly or quarterly buyback of common stock from the owners of one of the merged entities. Typically, an earn-out can be accomplished over an agreed-upon period of time and can provide the entrepreneur seeking to exit with a strong return as company sales expand and costs decline because of increased operating efficiencies. The purchase price can be scaled upward incrementally over time because the entrepreneur seeking to exist provides the luxury of time for the management team to complete the deal.

Why should an entrepreneur merge his or her venture with another firm? There are both defensive and offensive strategies for a merger. Merger motivations range from survival to protection to diversification to growth. When some technical obsolescence (loss of market or raw material) or deterioration of the capital structure has occurred in the entrepreneur's venture, a merger may be the only means for survival and exit. The merger can also protect against market encroachment, product innovation, or an unwarranted takeover. It can also provide a great deal of diversification as well as growth in market, technology, and financial and managerial strength.

A successful merger requires sound planning by the entrepreneur. The merger objectives, particularly those dealing with earnings, must be spelled out with the resulting gains for the owners of both companies delineated. Also, the entrepreneur must carefully evaluate the other company's management to ensure that it would be competent in developing the growth of the combined entity. The value and appropriateness of the existing resources should also be determined. In essence, this involves a careful analysis of both companies to ensure that the weaknesses of one do not compound those of the other. Finally, the technology entrepreneur should work toward establishing a climate of mutual trust to help minimize any possible management threat or turbulence.

13.6.1 Advantages of Exit Via Merger

One major advantage for many entrepreneurs to the exit via merger option is that they can exit their venture more slowly and methodically than exit via acquisition. Usually, in exit via acquisition the entrepreneur is out on the close of the deal. In a merger, the entrepreneur (from either of the merging parties) may elect to stay with the new company after the merger deal. Of course, the entrepreneur may need to take on a lesser role in the merged entity. It is simply untenable for both of the former CEOs to retain that title, for example. The entrepreneur may elect to exit the merged entity over time via what is termed an *earn-out strategy*. That is, the entrepreneur will be retained by the new company through a period of time in which the entrepreneur's stock in the merged entity fully vests. At that point the entrepreneur may leave and start another venture.

Another major advantage to exit via merger is that the entrepreneur may realize greater value in his or her stock holdings in the new company than would have been realized in their former venture. The entrepreneur may have been at the limits of his or her capabilities to lead and manage, and an exit via acquisition would have valued the venture at its current value. In contrast, by accepting ownership of the new, merged company, the entrepreneur may realize significantly greater value due to the greater capabilities of the combined leadership team.

13.6.2 Disadvantages of Exit Via Merger

The primary disadvantage of exit via merger is the need for the merging companies to integrate their businesses into a single business. This can be very difficult for a variety of reasons. For example, if the two companies have vastly different information systems, it can be expensive converting the merged entity to a single system. Another factor that is often cited as an impediment to a successful merger is if the merging businesses have different cultures. A venture's "culture" consists primarily of the unspoken understandings that people share about, for example, how work is done and how customers are handled. If these are not aligned, the merged entity may have a difficult time bringing the two workforces into a cohesive unit.

Another disadvantage of exit via merger is that it can be more time consuming and expensive than other exit options. For example, a merger will require that both of the merging parties undergo due diligence and valuation. The reason this is necessary is that shareholders in the separate entities need to be compensated proportionately based on the combined value of the newly merged entity. In order to determine one's share of the merged entity, it is necessary to have agreement on how much of the merged entity's value was contributed by each of the merging companies. That proportion will then be used to allocate new company shares to the combined ownership pool. This process of valuation, new stock creation and distribution, and other things associated with merging can be expensive and time consuming. Both parties need to be highly motivated to complete the deal.

13.7 EXIT VIA INITIAL PUBLIC OFFERING

Going public via what is called an IPO occurs when the entrepreneur and other owners of the venture offer and sell some part of the company to the public through a registration statement filed with the Securities and Exchange Commission (SEC) pursuant to the Securities Act of 1933. The resulting capital infusion to the company from the increased number of stockholders and outstanding shares of stock provide the company with financial resources and a relatively liquid investment vehicle. Consequently, the company will have greater access to capital markets in the future and a more objective picture of the public's perception of the value of the business. However, given the reporting requirements, the increased number of stockholders, and the costs involved, the technology entrepreneur must carefully evaluate the advantages and disadvantages of going public before initiating the process. Here we'll review some of the major factors and documents associated with IPO transactions. We'll begin by examining the timing of the transaction.

13.7.1 Timing

Is this a good time for the venture to initiate an IPO? This is the critical question that technology entrepreneurs must ask themselves before launching this effort.[19] Some critical questions in making this decision follow.

First, is the company large enough? Although it is not possible to establish rigid minimum-size standards that must be met before a venture can go public, New York investment banking firms prefer at least a 500,000 share offering at a minimum of $10 per share. Assuming that the company is willing to sell shares representing not more than 40% of the total number of shares outstanding after the offering is completed, this means that, in order to support this $5 million offering, the company would have to have a post-offering value of at least $12.5 million. This size offering will only occur with past significant sales and earnings performance or a solid prospect for future growth and earnings.

Second, what is the amount of the company's earnings, and how strong is its financial performance? Not only is this performance the basis of the company valuation, but also it determines if a company can successfully go public and the type of firm willing to underwrite the offering.

Third, are the market conditions favorable for an IPO? Underlying the sales and earnings, as well as the size of the offering, is the prevailing general market condition.[20] Market conditions affect both the initial price that the entrepreneur will receive for the stock and the aftermarket—the price performance of the stock after its initial sale. Some market conditions are more favorable for IPOs than others.

Fourth, how urgently is the money needed? The entrepreneur must carefully appraise both the urgency of the need for new money and the availability of outside capital from other sources. Given that the sale of common stock decreases the ownership position of the technology entrepreneur and other equity owners, the longer the time before going public, given that profits and sales growth occur, the less percentage of equity the technology entrepreneur will have to give up per dollar invested.

Finally, what are the needs and desires of the present owners? Sometimes the present owners lack confidence in the future viability and growth prospects of the business or they have a need for liquidity. Going public is frequently the only method for present stockholders to obtain the cash needed. The Mini-Case below discusses the 2012 Facebook IPO and what it meant to shareholders.

MINI-CASE
Facebook IPO Raises $18B

Oh, to be 28 and a multi-billionaire. That's the outcome of Facebook's IPO for founder and 50% shareholder Mark Zuckerberg. Facebook went public on the NASDAQ exchange on May 18, 2012 amidst much fanfare and hype. The stock was priced at $38/share but opened at over $42/share. More than 82 million shares of the Internet giant were traded in the first 30 seconds after it began trading. On opening day, the stock reached a high of $45/share before settling down and closing just above the IPO price of $38/share. Final trading volume for the day was more than 573 million shares. Zuckerberg's stake in Facebook is estimated to be more than $19 billion. The IPO raised $16 billion for the company to use to grow and hunt for ways to monetize its more than 800 million active users. That is roughly one-eighth of the entire world's population.

Source: Adapted from Koba, M. 2012. "Facebook's IPO: What We Know." CNBC.com, http://www.cnbc.com/id/47043815/Facebook_s_IPO_What_We_Know_Now, accessed on May 20, 2012; Deluca, M. 2012. "Facebook IPO by the Numbers: Zuckerberg's Loot and More." The Daily Beast, http://www.thedailybeast.com/articles/2012/05/18/facebook-ipo-by-the-numbers-zuckerberg-s-loot-more.html, accessed on May 20, 2012.

13.7.2 Selecting an Investment Bank

While a public offering of equity can be through direct sales (e.g., over the Internet) to the public (assuming compliance with all securities laws and regulations), public offerings are normally done through an **underwriter** that acts as an intermediary between the venture issuing the stock and the public. The venture enters into an agreement with the underwriter pursuant to which the underwriter distributes the securities to dealers (brokers) for resale to the public.

Once the entrepreneur has determined that the timing for going public is favorable, he or she must carefully select a managing **underwriter**, an investment bank that will take the lead in forming an **underwriting syndicate**.[21] For example, Goldman Sachs is an investment bank with broad and deep experience as an IPO underwriter. The firm selected to perform as the lead investment bank is of critical importance in establishing the initial price for the stock of the company, in supporting the stock in the aftermarket, and in creating a strong following among security analysts. A syndicate is a group of investment banks and other, usually institutional investors who subscribe to the IPO. That means, they designate in advance how many shares they will purchase at the IPO. This is important because it could be disastrous for a venture to have an IPO where no one purchased the stock.

Selecting the investment banker is a major factor in the success of the public offering. So, the entrepreneur should approach one through a mutual contact. Commercial banks, attorneys specializing in securities work, major accounting firms, providers of the initial financing, or prominent members of the company's board of directors can usually provide the needed suggestions and introductions. As the relationship will be ongoing, not ending with the completion of the offering, the technology entrepreneur should employ several criteria in the selection process, such as reputation, distribution capability, advisory services, experience, and cost.

The success of the offering also depends on the underwriter's distribution capability. An entrepreneur wants the stock of his or her company distributed to as wide and varied a base as possible. As each investment banking firm has a different client base, the entrepreneur should compare client bases of possible managing underwriters. Is the client base made up predominately of institutional investors, individual investors, or balanced between the two? Is the base more internationally or domestically oriented? Are the investors long-term or speculators? What is the geographic distribution—local, regional, or nationwide? A strong managing underwriter and syndicate with a quality client base will help the stock sell well and perform well in the **aftermarket**. The aftermarket is the term used to refer to the performance of a stock in the public markets after the excitement of the IPO has subsided. How a stock performs in the long run is often dependent on the ability of the underwriters to gain interest from *their* investors.[22]

The final factor to be considered in the choice of a managing underwriter is cost. Going public is a very costly proposition and costs *do* vary greatly among underwriters. The average gross spread as a percentage of the offering between underwriters can be as high as 10%. Costs associated with various possible managing underwriters must be carefully weighed against the other four factors. The key is to obtain the best possible underwriter and not try to cut corners given the stakes involved in a successful IPO.

13.8 REGISTRATION STATEMENT AND TIMETABLE

Once the managing underwriter has been selected, a planning meeting should be held of company officials responsible for preparing the **registration statement**, the company's independent accountants and lawyers, and the underwriters and their counsel. At this important meeting, frequently called the "all hands" meeting, a timetable is prepared, indicating dates for each step in the registration process. This timetable establishes the effective date of the registration, which determines the date of the final financial statements to be included. The company's end of the year, when regular audited financial statements are routinely prepared, is taken into account to avoid any possible extra accounting and legal work. The timetable should indicate the individual responsible for preparing the various parts of the registration and offering statement. Problems often arise in an IPO due to the timetable not being carefully developed and agreed to by all parties involved.

After the completion of the preliminary preparation, the first public offering normally requires six to eight weeks to prepare, and file the registration statement with the SEC. Once the registration statement has been filed, the SEC generally takes four to eight weeks to declare the registration effective. Delays frequently occur in this process, such as: (1) during heavy periods of market activity, (2) during peak seasons, such as March, when the SEC is reviewing a large number of proxy statements, (3) when the company's attorney is not familiar with federal or state regulations, (4) when a complete and full disclosure is resisted by the company, or (5) when the managing underwriter is inexperienced.

In reviewing the registration statement, the SEC attempts to ensure that the document makes a full and fair disclosure of the material reported. The SEC has no authority to withhold approval of or require any changes in the terms of an offering that it deems unfair or inequitable as long as all material information concerning the company and the offering are fully disclosed. The National Association of Securities Dealers (NASD) will review each offering, principally to determine the fairness of the underwriting compensation and its compliance with NASD bylaw requirements.

13.8.1 The Prospectus

The prospectus portion of the registration statement is almost always written in a highly stylized narrative form because it is the selling document of the company. Although the exact format is decided by the company, the information must be presented in an organized, logical sequence and in an easy-to-read, understandable manner in order to obtain SEC approval. Some of the most common sections of a prospectus include: the cover page, prospectus summary, the company, risk factors, use of proceeds, dividend policy, capitalization, dilution, selected financial data, the business, management and owners, type of stock, underwriter information, and the actual financial statements.

The cover page includes such information as company name, type and number of shares to be sold, a distribution table, date of prospectus, managing underwriter(s), and syndicate of underwriters involved. There is a preliminary prospectus and then a final prospectus once approved by the SEC.

The preliminary prospectus is used by the underwriters to solicit investor interest in the offering while the registration is pending. The final prospectus contains all of the changes and additions required by the SEC and blue sky examiners and the information concerning the price at which the securities will be sold. The final prospectus must be delivered with or prior to the written confirmation of purchase orders from investors participating in the offering.

13.8.2 The Red Herring

Once the preliminary prospectus is filed, it can be distributed to the underwriting group. This preliminary prospectus is called a **red herring** because a statement printed in red ink appears on the front cover which states that the issuing company is not attempting to sell its shares. The red herring for Mahindra Holidays and Resorts, an India company, is shown in Exhibit 6.3. The registration statements are then reviewed by the SEC to determine if adequate disclosures have been made. Some deficiencies are almost always found and are communicated to the company either by telephone or a *deficiency letter*. This preliminary prospectus contains all the information contained in the final prospectus except that which is not known until shortly before the effective date: offering price, underwriters' commission, and amount of proceeds.

13.8.3 Reporting Requirements

Going public requires a complex set of reporting requirements. The first requirement is the filing of a Form SR sales report, which the company must do within 10 days after the end of the first three-month period following the effective date of the registration. This report includes information on the amount of securities sold and still to be sold, and the proceeds obtained by the company and their use. A final Form SR sales report must be filed within 10 days of the completion or termination of the offering.

The company must file annual reports with the SEC on Form 10-K, quarterly reports on Form 10-Q, and specific transaction reports on Form 8-K. The information in Form 10-K on the business, management, and company assets is similar to that in Form S-1 of the registration statement. Of course, audited financial statements are required.

The quarterly report on Form 10-Q contains primarily the unaudited financial information for the most recently completed fiscal quarter. No 10-Q is required for the fourth fiscal quarter.

A Form 8-K report must be filed within 15 days of such events as the acquisition or disposition of significant assets by the company outside the ordinary course of the business, the resignation or dismissal of the company's independent public accountants, or a change in control of the company.

The company must follow the proxy solicitation requirements regarding holding a meeting or obtain the written consent of security holders. The timing and type of materials involved are detailed in the Securities and Exchange Act of 1933.

These are but a few of the reporting requirements of public companies. All the requirements must be carefully observed, because even inadvertent mistakes can have negative consequences on the company. The reports required must be filed on time.

13.8.4 Advantages of Exit Via IPO

There are three primary advantages of going public: obtaining new equity capital, obtaining value and transferability of the organization's assets, and enhancing the company's ability to obtain future funds. Whether it is first-stage, second-stage, or third-stage financing, a venture is in constant need of capital. The new capital provides the needed working capital, plant and equipment, or inventories and supplies necessary for the venture's growth and survival. Going public is often the best way to obtain this needed capital on the best possible terms.

Going public also provides a mechanism for valuing the company and allowing this value to be easily transferred among parties. Many family-owned or other privately held companies may need to go public so that the value of the company can be disseminated among the second and third generations. Venture capitalists view going public as the most beneficial way to attain the liquidity necessary to exit a company with the best possible return on their earlier-stage funding. Other investors, as well, can more easily liquidate their investment when the company's stock takes on value and transferability. Because of this liquidity, the value of a publicly traded security sometimes is higher than shares of one that is not publicly traded. In addition, publicly traded companies often find it easier to acquire other companies by using their securities in the transactions.

The third primary advantage is that publicly traded companies usually find it easier to raise additional capital, particularly debt. Money can be borrowed more easily and on more favorable terms when there is value attached to a company and that value is more easily transferred. Not only debt financing but future equity capital is more easily obtained when a company establishes a track record of increasing stock value.

13.8.5 Disadvantages of Exit Via IPO

Although going public presents significant advantages for a new venture, the numerous disadvantages must be also carefully weighed. Some technology entrepreneurs want to keep their companies private, even in times of a hot stock market. Why do technology entrepreneurs avoid the supposed gold rush of an IPO?

One of the major reasons is the public exposure and potential loss of control that can occur in a publicly traded company. To stay on the cutting edge of technology, companies frequently need to sacrifice short-term profits for long-term innovation. This can require reinvesting in technology, which in itself may not produce any bottom-line results, particularly in the short run.

Some of the most troublesome aspects of being public are the resulting loss of flexibility and increased administrative burdens. The company must make decisions in light of the fiduciary duties owed to the public shareholder, and it is obliged to disclose to the public all material information regarding the company, its operations, and its management. One publicly traded company had to retain a more expensive investment banker than would have been required by a privately held company in order to obtain an "appropriate" fairness opinion in a desired acquisition. The investment banker increased the expenses of the merger by $150,000, in addition to causing a three-month delay in the acquisition proceedings. Management of a publicly traded company also spends a significant amount of additional time addressing queries from shareholders, press, and financial analysts.

If all these disadvantages have not caused the technology entrepreneur to look for alternative financing other than an IPO, the expenses involved may. The major expenses of going public include accounting fees, legal fees, underwriter's fees, registration and blue sky filing fees, and printing costs. The accounting fees involved in going public vary greatly, depending in part on the size of the company, the availability of previously audited financial statements, and the complexity of the company's operations. Generally, the costs of going public are around $300,000 to $600,000, although they can be much greater when significant complexities are involved. Additional reporting, accounting, legal, and printing expenses can run anywhere from $50,000 to $250,000 per year, depending on the company's past practices in the areas of accounting and shareholder communications.

The underwriters' fees include a cash discount (on commission), which usually ranges from 7% to 10 % of the public offering price of the new issue.

In some IPOs, the underwriters can also require some compensation, such as warrants to purchase stock, reimbursement for some expenses—most typically legal fees—and the right of first refusal on any future offerings. The NASD regulates the maximum amount of the underwriter's compensation and reviews the actual amount for fairness before the offering can take place. Similarly, any underwriter's compensation is also reviewed in blue sky filings.

There are also other expenses in the form of SEC, NASD, and state blue sky registration fees. The final major expense—printing costs—typically ranges from $50,000 to $200,000. The registration statement and prospectus discussed later in this chapter account for the largest portion of these expenses. The exact amount of expenses varies depending on the length of the prospectus, the use of color or black and white photographs, the number of proofs and corrections, and the number printed. It is important for the company to use a good printer because accuracy and speed are required in the printing of the prospectus and other offering documents.

Regardless of how much preparation occurs, almost every entrepreneur is unprepared and wants to halt it at some time during the makeover process. Yet for a successful IPO, each entrepreneur must listen to the advice being given to make the recommended changes swiftly.

13.9 CHAPTER SUMMARY

This chapter examined the topic of valuing and exiting your technology venture. We began by reviewing the due diligence process that usually precedes an exit event. Due diligence is the process used by those who will be acquiring or investing in the venture to ensure that it meets their guidelines and standards. Entrepreneurial ventures can prepare for the due diligence process by running the business efficiently and effectively from the beginning. That means establishing management systems that are aligned with industry best practices and abiding by the accounting rules governing the industry in which the venture competes.

Next we discussed the concept of enterprise value and explored some of the ways in which venture valuation is determined. For example, using a recent valuation within the industry as a comparison, one can then get a rough estimate of the value of another venture. In some industries, a rough valuation can be determined based on a key metric. For example, if a social media company with one million registered users was recently acquired for $5 million, then the value of each registered user was $5. Using simple math, one then can calculate the value of another social media company by multiplying the number of registered users by $5.

More sophisticated valuation techniques are based on future cash flow projections of a company to derive a net present value. Various techniques use different discount rates to account for the risks associated with particular industries.

Next we examined the various strategies that technology entrepreneurs can use to exit their ventures, including succession, acquisition, merger, and IPO. Each of these exit strategies represents a liquidity event for the entrepreneur, and each has its advantages and disadvantages. In particular, the strategies of acquisition, merger, and IPO will be preceded by in-depth due diligence which can be costly and time consuming. Succession, on the other hand, is fraught with its own challenges because it involves handing down ownership of a venture to one's heirs. Entrepreneurs should build their companies from the beginning with some idea in mind of how they would like to exit. The manner in which a company is built and positioned can affect the ease and likelihood of fulfilling a particular exit strategy.

KEYTERMS

Due diligence Involves examining a number of key elements of a target venture, including its financial health, the potential for synergy, the market position and future potential of the venture, the research and development history and roadmap for the venture, legal considerations, and plans for managing the acquired entity.

Valuation The term used to refer to the process and calculations used to establish a dollar value for a venture.

Exit strategy The technology entrepreneur's strategic withdrawal from ownership or operation of his or her venture.

Investment horizon The time an investor's money will be tied up in a technology venture until the time of exit.

Liquidity event The exit from a private venture for shareholders that represents their opportunity to covert shares to cash.

Synergy A concept where the phrase, "the whole is greater than the sum of its parts" is a good definition.

Enterprise value The value of the enterprise as a whole.

Market capitalization The value of public companies calculated as a function of the number of outstanding shares times the price per share.

Present value The value in today's dollars of some future cash flow; because money today can earn interest, future cash flows must be discounted by at least as much as that interest amount to convert to its present value.

Discount rate A discount rate where in conducting a present value analysis, a future cash flow must be discounted by the rate of interest that otherwise could be earned by present dollars.

Risk-adjusted discount rate A discount rate used in the discounted cash flow valuation technique that includes both the alternative investment risk and the venture risk.

Acquisition A transaction involving two companies in which one company purchases the other; the company that is purchased no longer exists, but is incorporated into the other company.

Value drivers Strategic advantages that lead to value creation in an industry and that are attractive elements of an acquisition candidate in an industry.

Vertical strategy A marketing strategy that targets clients by industry type.
Direct purchase A purchase of the firm in which the acquiring company often obtains funds from an outside lender or the seller of the company being purchased.
Bootstrap purchase A purchase where the acquiring company can acquire a small amount of a firm, such as 20% to 30%, for cash; the acquiring company then purchases the remainder of the target venture by a long-term note that is paid off over time out of the acquired company's earnings.
Serial entrepreneur A type of entrepreneur that starts multiple ventures over a career.
Business broker Third-party agents who assist entrepreneurs in the exit process.
Merger A transaction involving two or more companies in which the companies join to form a new bigger company.
Earn-out An strategy used for ventures that have begun to generate consistently strong positive cash flow; the management team initiates a monthly or quarterly buyback of common stock from the owners of one of the merged entities.
Initial public offering A process whereby a private company qualifies to sell its shares on a public stock exchange.
Underwriter A company that administers the public issuance and distribution of securities; a company will utilize one when filing an IPO.
Underwriting syndicate A group of investment banks and other, usually institutional investors who subscribe to an IPO.
Aftermarket The term used to refer to the performance of a stock after the excitement of the IPO has subsided; how a stock performs in the long run is often dependent on the ability of the underwriters to gain interest from their investors.
Registration statement Document filed with the SEC before a company can go public; it consists of two parts: the prospectus and the registration statement.
Red Herring The name given to the preliminary prospectus document due to the red ink used in the printing of the front cover.

ADDITIONAL READING

Goedhart, Marc, Wessels, David, 2010. Valuation: Measuring and Managing the Value of Companies, fifth ed. John Wiley & Sons, Hoboken, NJ.

Lyons, Thomas W., 2010. Exit Strategy: Maximizing the Value of Your Business. Sales Gravy Press, Los Angeles, CA.

Blowers, Stephen C., Griffith, Peter H., Milan, Thomas L., 1999. The Ernst & Young Guide to the IPO Value Journey. John Wiley & Sons, Hoboken, NJ.

WEB RESOURCES

http://www.nvca.org/ This is the website for the National Venture Capital Association. It has many resources and helpful guides to conducting rational venture valuations.

http://www.inc.com/tools/due-diligence-checklist.html: This website provides a due diligence checklist for ventures, which helps them prepare for a thorough due diligence prior to an exit event.

http://venturebeat.com/: This is a good site for tracking recent acquisitions, mergers, IPOs, and other news associated with technology ventures.

ENDNOTES

1 Capon, A. 1997. "Exit Strategies for Entrepreneurs," *Global Investor*, 107: 16–19.

2 Quinlan, R. 2008. "Ideal Acquisitions Need Effective Evaluation." *Financial Executive*, 24(5): 17.

3 Watson, D.G. 2007. "Acquisitions: How to Avoid Implementation Pitfalls and Improve the Odds of Success." *Global Business & Organizational Excellence*, 27(1): 7–19.

4 Kvesic, D.Z. 2008. "Product Lifecycle Management: Marketing Strategies for the Pharmaceutical Industry." *Journal of Medical Marketing*, 8(4): 293–301.

5 Wang, C. and F. Xie. 2009. "Corporate Governance Transfer and Synergistic Gains from Mergers and Acquisitions." *Review of Financial Studies*, 22(2): 829–858.

6 Morrison, N.J., G. Kinley, and K.L. Ficery. 2008. "Merger Deal Breakers: When Operational Due Diligence Exposes Risks." *Journal of Business Strategy*, 29(3): 23–28.

7 Hering, T. and M. Olbrich. 2006. "Valuation of Startup Internet Companies". *International Journal of Technology Management*, 33(4): 409–419.

8 Smith, J.K. and R.L. Smith. 2004. Entrepreneurial Finance 2E. (Hoboken, NJ: John S. Wiley and Sons.)

9 Fitts, J.A., and M.G. Rowe. 2012. "Family Business Transition Planning." *Journal of Accountancy*, 212(5): 22.

10 Giamarco, J. 2012. "The Three Levels of Family Business Succession Planning." *Journal of Financial Services Professionals*, 66(2): 59–69.

11 Scharfstein, A.J. 2011. "Making a Graceful Exit." *Family Business*, 22(3): 20–22.

12 Daley, J. 2011. "Family Affair." *Entrepreneur*, 39(12): 97–105.

13 Nolop, B. 2007. "Rules to Acquire By." *Harvard Business Review*, 85(9): 129–139.

14 Andriole, S.J. 2007. "Mining for Digital Gold: Technology Due Diligence for CIOs." *Communications of AIS*, 2007(20): 371–381.

15 Mero, J. 2005. "People are His Bottom Line." *Fortune*, 155(7): 30.

16 Dalziel, M. 2008. "The Seller's Perspective on Acquisition Success: Empirical Evidence from the Communications Equipment Industry." *Journal of Engineering & Technology Management*, 25(3): 168–183.

17 Adams, M. 2004. "Exit Pay-Offs for the Entrepreneur," *Mergers & Acquisitions: The Dealmaker's Journal*, March; 24–28.

18 Vaughan, B. 2008. "Earn Outs Come Back into Fashion." *Buyouts*, 21(17): 24.

19 Yoon-Jun, L. 2008. "Strategy of Startups for IPO Timing Across High Technology Industries." *Applied Economics Letters*, 15(11): 869–877.

20 Beales, R., R. Cox, and L. Silva. 2008. "Master of Market Timing." *The Wall Street Journal, Eastern Edition*, 251(55): C12.

21 Kulkarni, K., and T. Sabarwal. 2007. "To What Extent are Investment Bank Differentiating Factors Relevant for Firms Floating Moderate Sized IPOs?" *Annals of Finance*, 3(3): 297–327.

22 Gleason, K., J. Johnston, and J. Madura. 2008. "What Factors Drive IPO Aftermarket Risk?" *Applied Financial Economics*, 18(13): 1099–1110.

Appendix I: Example of a Generic Confidentiality Agreement

AGREEMENT REGARDING CONFIDENTIAL INFORMATION AND TECHNOLOGY

This is an Agreement between _____ and _____.

The parties each have particular expertise and have developed or otherwise obtained certain confidential information.

The parties wish to enter into a business relationship which is likely to require that certain confidential information of one party be disclosed to the other. However, the parties do not wish to diminish their rights in their confidential information and require assurances that their rights therein will not be adversely affected by the business relationship.

Accordingly, the parties agree as follows:

1. "Technology" as used in this Agreement means processes, machines, manufactures, compositions of matter, improvements, technological developments, methods, techniques, systems, mask works, software, documentation, data and information (irrespective of whether in human or machine-readable form), works of authorship, and products, whether or not patentable, copyrightable, or susceptible to any other form of protection and whether or not reduced to practice.
2. "Providing Party" means, with respect to an item of information or technology, that party which provides, reveals, or discloses that item to the other party.
3. "Recipient Party" means, with respect to an item of information or technology, that party to which that item is provided, revealed, or disclosed.
4. "Confidential Information" of a Providing Party means any and all Technology and/or information which: (1) is created, developed, or otherwise generated by or on behalf of the Providing Party; or (2) is otherwise marked Confidential by the Providing Party AND (3) is provided to the Recipient Party during the term of this Agreement, EXCEPT

such information which (i) at the time of this Agreement is clearly publicly and openly known and in the public domain; (ii) after the date of this Agreement becomes publicly and openly known and in the public domain through no fault of the Recipient Party; (iii) comes into the Recipient Party's possession and lawfully obtained by the Recipient Party from a source other than from the Providing Party, and not subject to any obligation of confidentiality or restrictions on use; or (iv) is approved for release by written authorization of the Providing Party.

5. The Providing Party reveals Confidential Information to the Recipient Party in strict confidence and solely for the purpose of performing in connection with the business relationship with the Providing Party. The Recipient Party shall not use, or induce others to use, such Confidential Information for any other purpose whatsoever, and shall take all reasonable measures to maintain the confidentiality of such Confidential Information. At a minimum, the Recipient Party shall not: (1) directly or indirectly print, copy, or otherwise reproduce, in whole or in part, any such Confidential Information, without prior consent of the Providing Party and (2) disclose or reveal any such Confidential Information to anyone except those of the Recipient Party's employees who have a specific need to know, are directly involved in the business relationship, and have agreed to be bound by the terms of this Agreement.

6. Upon the Providing Party's request, the Recipient Party will deliver over to the Providing Party all of the Providing Party's Confidential Information, as well as all documents, media, items and comprising, or embodying, such Confidential Information as well as any other documents or things belonging to the Providing Party that may be in the Recipient Party's possession. The Recipient Party shall not retain any copies.

7. *This Agreement shall be governed by and construed in accordance with the laws of the State of ___ without reference to principles of conflicts of laws.* It may be amended only in a writing signed by the parties, and there are no other understandings, agreements, or representations, express or implied. If any clause or provision is or becomes illegal, invalid, or unenforceable, the remaining provisions shall be unimpaired, and the illegal, invalid, or unenforceable provision shall be replaced by a provision, which, being legal, valid, and enforceable, comes closest to the intent of the parties underlying the illegal, invalid, or unenforceable provision.

8. This Agreement shall become effective upon execution by the parties and shall remain in effect until terminated. The minimum term of this Agreement shall be _____. Thereafter, either party may terminate this

agreement upon 30 days' written notice. The parties obligations with respect to Confidential Information disclosed hereunder, however, shall survive the termination of the Agreement.

Date: By: Name: Title:	Date: By: Name: Title:

Appendix II: Example Executive Summary

Phoenix, Arizona www.tidalpoint.com
Contact Varun Kapur Varun.kapur@global.t-bird.edu
Stage
  Pre-launch

Industry
  Software Product
  NAICS Code: 511210
  Software Consulting
  NAICS Code: 541512

MISSION STATEMENT

To empower our customers by providing them with better control of their IT systems through efficient, rational, and cost-effective delivery of IT products and services. And do so by becoming our client's most trusted advisor and partner by sharing knowledge and best practices.

PROBLEM BEING SOLVED

Nature of problem: When it comes to replacement of their legacy IT systems, insurance companies struggle with cost and effort overruns, which makes the cost of transformation and cost of ownership significantly higher.

Market segment: There are approximately 300 mid-sized to large insurance companies that need to replace their legacy IT systems. The segment growth rate is estimated to be 7% compound annual growth rate (CAGR), and the total IT budget by year 2015 for commercial off-the-shelf (COTS) products and services is estimated to be US$ 17.5 billion.

Importance: There is an unmet need for insurance software product providers who combine follow-up services required for product integration and implementation into insurance companies' operational landscape. We will offer a suite of services to the companies along with consultants and experts who can enable a smooth transition for insurance companies.

THE SOLUTION

We will provide a customized off-the-shelf product along with the services required to integrate this product into an insurance company's operational landscape. The product will be called Tidal Point Policy Administration system, which will cover all major classes/business lines and provide an end-to-end processing capability from which are distribution, new business, underwriting including rules engines, policy servicing, claims, and reinsurance. The services provided will include data migration, implementation, business analysis, process consulting, and IT strategy consulting.

UNIQUE SALES PROPOSITION

- Provide an integrated suite of services to the clients without having them organize different activities related to legacy system transformation.
- Overall cost and time will be reduced for the clients due to:
  - Reduced time and efforts in issuing multiple RFPs and RFIs
  - Speed to start improving because one vendor will provide all the services
- More effective approach because all the services will be provided by a single vendor, which will give clients better control and ease of managing the transformation
- Lower cost of ownership for the customers

COMPETITION

IBM, CSC, Guidewire, Accenture (Duck Creek), MajescoMastek, Camilion, Exigen, Insurity, AQS.

MARKET

- 400 Property & Casualty (P&C), Specialty and Life Insurance companies in the U.S. market.

MARKET SEGMENT

- Software (NAICS 511210): Insurance COTS product market estimated to be $17.5 billion by year 2015. Insurance software services including consulting market is estimated at $40.9 billion by 2015.
- Software Consulting (NAICS 541512): Insurance software consulting services are estimated to be $41.0 billion by year 2015.
- Approximately 300 insurance companies (direct carriers) need to replace their legacy IT systems for at least one class of business.

MARKETING PLAN

- Advertisements through industry publications such as *Insurance & Technology*
- Participation in industry conferences and sponsorship of these events
- Personal contacts and outside sales

PRICE

- COTS product: $1,250,000
- Business analysis: $650,000
- Implementation: $700,000
- Process consulting: $200,000
- Data migration: $200,000
- IT strategy: $150,000

Financial summary

Year	2014	2015	2016	2017	2018
Revenue	1,950,000	4,641,000	5,949,762	7,603,796	11,953,167
Cost of goods sold	1,072,500	2,552,550	3,272,369	4,182,088	6,574,242
Gross margin	877,500	2,088,450	2,677,393	3,421,708	5,378,925
Operating expenses	779,500	1,738,590	2,050,380	2,392,271	2,923,929
Operating profit	98,000	349,860	627,013	1,029,437	2,454,996

Appendix III: Sample Development Agreement

DEVELOPMENT AGREEMENT

Preamble

This is a development agreement, effective as of 1 June 2014, between **ClientCo LLC** with offices at 111 AirPark Street, Scottsdale AZ, 85254 (the "Company") and DevCo, Inc with offices at 123 Industry Street, Scottsdale Az, 85254 ("Developer").

Recitals

The company is in the business of access control, has developed certain IP in that field and has developed and markets an access control and security system (ACSS) known as the Ovtcharka ACSS.

Developer has special expertise in facial recognition systems, and, in particular, has extensive experience with 3dP facial recognition systems. Developer is generally familiar with the Ovtcharka ACSS.

The company has committed to deliver Ovtcharka ACSS to various third parties, modified to employ 3dP facial recognition systems as a component thereof; failure to deliver such modified systems on or before 31 March 2015 will result in loss of profitable contracts by the company.

The company wishes to engage developer to perform certain tasks and provide certain services in connection with modifying its Ovtcharka ACSS to employ 3dP facial recognition systems. However, in order to do so, it is necessary that developer be made aware of the proprietary materials and information (PMI) belonging to the company, as well as PMI belonging to 3dP held by company under a confidentiality agreement with 3dP. The company does not wish to lose the confidentiality, or diminish its rights in or control over its PMI, or compromise the confidentiality of the 3dP PMI and requires assurances that its rights therein will not be diminished or impaired by virtue of the dealings with developer.

Further, before engaging Developer, the company also requires assurances that the modifications can be completed within a time frame allowing delivery of the modified systems by 31 March 2015.

Developer is willing to provide, and hereby provides, such assurances. Accordingly, in consideration of the mutual promises contained herein, the parties agree:

1. Definitions. As used herein, the term:
    1.1. "Deliverables" means the items identified in Schedule 1 and the attachments thereto.
    1.2. "Intellectual property" means concepts, inventions, data and information (irrespective of whether in human or machine-readable form), works of authorship, and products, in each case whether or not patentable, copyrightable, or susceptible to any other form of protection and whether or not reduced to practice.
    1.3. "Proprietary materials and information" means any and all software, plans, drawings, or models provided to developer in connection with this development agreement, as well as any and all IP and/or information which: (i) is provided to developer by the company, (ii) is created, developed, or otherwise generated by or on behalf of the company, (iii) concerns or relates to any aspect of the company's business or products or those of 3dP, or (iv) is, for any reason, identified by the company as confidential; except such information which developer can show, clearly and convincingly: (a) is at the time of disclosure, publicly and openly known, (b) becomes publicly and openly known through no fault of developer, or (c) is in developer's possession and documented prior to this agreement, lawfully obtained by developer from a source other than from the company, and not subject to any obligation of confidentiality or restrictions on use.
    1.4. "3dp confidentiality agreement" means the confidentiality agreement between the company and 3dP, dated 12 April 2013, a copy of which is attached to this development agreement.
2. Performance
    2.1. Developer shall present the deliverables to the company, in a form reasonably acceptable to the company by the dates set forth with respect to the deliverables in the "performance schedule" attached hereto as Schedule 2. In connection therewith, the parties shall perform such tasks, and provide such services at such times as set forth in Schedule 2. Time is of the essence.
    2.2. Where acceptance of a deliverable (e.g., "accept or reject" deliverable) is called for in Schedule 2, unless otherwise agreed in writing by the parties, the standard for acceptance shall be "reasonable satisfaction," and the party from whom acceptance is required shall be obligated with respect to subsequent tasks/events in Schedule 2 (including payments) only after the deliverable is accepted.

- 2.3. A notice of rejection shall specify, with particularity, deficiencies in the deliverable. If a deliverable is initially rejected by a specified party, the responsible party will, within 10 business days from receipt of the rejection, or such other time period as may be agreed upon by the parties, make the necessary corrections and re-present the deliverable to the specified party. If the deficiencies are not cured in the re-presented task or deliverable, then the party from whom acceptance is required may, solely at its option:
  - 2.3.1. provide a further written notice of rejection, in which case the responsible party shall, within 10 business days from receipt of the rejection, or such other time period as may be agreed upon by the parties, make the necessary corrections and re-present the deliverable;
  - 2.3.2. elect to accept the task or deliverable subject to the deficiencies; or
  - 2.3.3. immediately terminate this agreement without further obligation on its part.

3. Payment
   - 3.1. In consideration of developer's performance hereunder, the company shall make payment to developer in such amounts and at such times as set forth in Schedule 2. Developer shall be responsible for the payment of all taxes on work performed or deliverables provided pursuant to this development agreement (excepting any tax based on the company's net income). Developer shall maintain proper records in respect of its performance hereunder, adequate to support any charges to the company for any work performed under this development agreement, and to make such records available during normal working hours to customer and/or its nominee.

4. Intellectual property
   - 4.1. All works of authorship and IP that developer, alone or jointly with others, creates, conceives or first makes or fixes in tangible media in connection with developer's engagement by the company (hereinafter "resultant IP") shall be the sole and exclusive property of the company. Developer hereby assigns, and shall assign, to the company all right, title, and interest in and to all resultant IP. All works of authorship created in connection with developer's engagement by the company shall be deemed "works for hire" commissioned by the company to the fullest extent permitted by the copyright laws of the United States. Developer shall at all times take all proper and legal actions in support of (including signing further documentation), and shall not take or induce any action or omission inconsistent with or tending to diminish or impair, the rights of the company in the resultant IP.

4.2. The deliverables shall be:
  4.2.1. the original work of developer, and no portion of the deliverables shall originate with other than developer, except as may be identified in Schedule 4.2.1
  4.2.2. specifically developed for the company, and created, conceived or first made or fixed in tangible media, in connection with developer's engagement by the company except as may be identified in Schedule 4.2.2.
4.3. To the extent that any portion of a deliverable is identified in Schedule 4.2.1 or 4.2.2, developer shall obtain for, or grants to, the company a perpetual royalty free license (with right of sublicense) to make, have made, use, sell, offer for sale, and otherwise market and distribute or have distributed, such portion of the deliverable, or such other license as may be agreed to in writing by the parties.

5. Representations and warranties
  5.1. Each party represents and warrants that it has the ability to enter into this development agreement without seeking the approval or consent of any third party.
  5.2. Developer represents and warrants that neither the company's use nor sale of any deliverable provided by developer, or any product embodying the deliverable provided by developer, or any aspect thereof, constitutes an infringement or violation of any patent, copyright, or other proprietary rights of any third party.
  5.3. Developer represents and warrants that: the tasks and services performed pursuant to the engagement shall be, performed in a good and workmanlike manner in accordance with any established professional standards for such tasks and services, and the best practices in developer's industry; the tasks and services and deliverables shall comply with all applicable laws, regulations, codes, and ordinances; and developer is aware of the purpose for which the deliverables are to be used by the company, and that the deliverables are fit for that purpose and merchantable.

6. Indemnity

The parties shall each indemnify and hold the other harmless from and against any and all claims, liabilities, loss, expense (including reasonable attorneys' fees), or damages arising out of any breach of this development agreement, provided that the indemnified party shall, with reasonable promptness, notify the indemnifying party of any such claim, demand, or suit and shall fully cooperate in the defense thereof. The indemnifying party shall have the right to designate counsel to defend against such claims and suits; however, at the indemnified party's option, the indemnified party shall have the right to participate in the defense with its own counsel at its own

expense. In no event shall any such claims or suits affecting the rights of a party be settled without the prior written consent of that party.
7. Confidentiality
   7.1. PMI is revealed to developer in strict confidence, and solely for the purpose of performing under this development agreement. Developer shall not use, or induce others to use, any PMI for any other purpose whatsoever, nor shall it disclose or reveal any PMI to anyone except those of developer's employees directly involved in the business relationship, with a specific need-to-know, and who have first agreed to be bound by the terms of this development agreement. Upon the company's request, developer will deliver over to the company all PMI, as well as all documents, media, and items comprising, embodying, or relating to the PMI as well as any other documents or things belonging to the company that may be in developer's possession. Developer shall not retain any copies.
   7.2. Developer shall comply in all respects with the 3dP confidentiality agreement.
8. TerminationThis development agreement may be terminated by either party in the event of a breach by the other party, immediately upon the end of a 20-day period after written notice of such breach to the breaching party, if such breach is not cured within the 20-day period. Section 7 hereof shall survive the completion and/or termination of this development agreement for any reason.
9. Miscellaneous

This development agreement may be amended only in a writing signed by both parties. The rights of the company and the obligations of developer under this development agreement shall inure to the benefit of the company's nominees, successors, and assigns. The failure of either party demand strict performance of the other party in a given instance or instances shall not be deemed a waiver with respect to other instances. If any clause or provision of this development agreement is, or becomes, illegal, invalid, or unenforceable, such clause or provisions shall be interpreted to reflect the intent of the parties to the greatest extent which is legal, valid, and enforceable, unless such clause or provision cannot be so interpreted, or a court of competent jurisdiction declines to permit such clause or provision to be so interpreted, in which case such clause or provision shall be severed and the remaining provisions of this development agreement shall continue in full force and effect. This development agreement shall be governed by and construed in accordance with the laws of the United States of America, and State of Arizona without reference to the principles of conflict of laws. The federal and state courts within the State of Arizona will have exclusive jurisdiction to adjudicate any dispute arising out of this development agreement, and both parties consent to the personal jurisdiction of those courts.

Appendix III: Sample Development Agreement

DevCo, Inc	ClientCo LLC
Date:	Date:
Signature:	Signature:
Name (typed or printed)	Name (typed or printed)
Title	Title
Address:	Address:
DevCo, Inc	ClientCo LLC,
123 Industry Street, Scottsdale Az, 85254	111 AirPark Street, Scottsdale Az, 85254
	With a Copy to:

Schedule 1 Deliverables

Deliverable Description	Further Description Attached	Delivery Date
Proposed development schedule	X	
Deliverable 1	X	
Deliverable 2	X	

Schedule 2 Performance Schedule

Time Reference	Developer Action	Company Action
Phase I		
Upon execution of the development agreement		Provide initial draw-down retainer of $xx
Within 10 days of execution of the agreement		Provide documentation on existing Ovtcharka ACSS and 3dP system, including detailed specification and source code
Within 10 working days of the receipt of the Ovtcharka and 3dP documentation	Present proposed specification for deliverables and development schedule	
Within five working days of the receipt of the proposed development schedule		Provide a written notice of acceptance or rejection
Phase II		
Upon commencement of phase II		Provide to developer an additional draw-down retainer of $xx
Within __ working days of the commencement of phase II	Present deliverable 1 for acceptance test	
Within __ working days of the presentation of deliverable 1		Perform acceptance tests; written notice of acceptance or rejection
Within __ working days of the acceptance of deliverable 1	Present deliverable 2 for acceptance test	
Within __ working days of the presentation of deliverable 2		Perform acceptance tests; written notice of acceptance or rejection
Within 30 working days of the acceptance of deliverable x		Make payment of $xx

Appendix IV: Example of an Employment Agreement

AGREEMENT REGARDING TECHNOLOGY AND CONFIDENTIAL INFORMATION

I, Ima Employee of 123 Employee Street, Phoenix Arizona 85254, make the following agreement with ClientCo LLC (the "company"), having offices at 111 AirPark Street, Scottsdale AZ, 85254. This agreement is effective as of the following date: <EffectiveDate.

I acknowledge that I am or will be employed by the company in a capacity in which I am expected to generate technology of value to the company, and under conditions in which I shall have access to confidential information belonging to the company.

Accordingly, in consideration of my employment, and other good and valuable consideration, the company and I agree as follows:

1. During my employment, I will assist the company in all possible ways in the discovery, perfection and development of "intellectual property." The term "intellectual property" includes such things as, for example: concepts, developments, improvements, inventions, data and information (irrespective of whether in human or machine-readable form), works of authorship, mask works; trademarks, goodwill, and products, in each case whether or not patentable, copyrightable, or susceptible to any other form of protection and whether or not reduced to practice.
2. I understand and agree that all IP which I, alone or jointly with others:
    (a) create, conceive, fix in a tangible medium, make, or reduce to practice during the period of my employment (including any periods of leaves of absence); or
    (b) first disclose to others, fix in a tangible medium, make, or reduce to practice, within a period of 1 year after the termination of my employment by the company (except such IP which I can prove by clear and convincing evidence were conceived by me after the termination of my employment and not under circumstances contrary to any provision of this agreement);

and which:
- (a) directly or indirectly results from tasks which have been or may be assigned to me by the company; or
- (b) relates to the existing business of the company or any "affiliated company" at the time of conception or reduction to practice of the IP, actual and/or demonstrably anticipated research or development of the company, or to the fields which I have been or may be directed to investigate;

is "relevant IP" owned by the company. I understand that the term "affiliated company" as used in this agreement means any business entity: (i) which is owned in whole or in part by the company; (ii) which is owned by a business entity which is owned in whole or in part by the company; (iii) which owns a controlling interest in the company; or (iv) in which controlling interest is owned by a business entity which in turn owns the company.

3. I will, at all times, promptly and fully disclose to the company all relevant IP.
4. All relevant IP will be the exclusive property of the company or its nominee, whether or not reduced to practice, published, or patented, copyrighted or licensed to others. I hereby assign (and will assign without further consideration, except as may be provided by statute) to the company or its nominee, all rights to all relevant IP (whether or not patentable, copyrightable, or susceptible to any other form of protection) in the United States and all foreign countries. This assignment includes, among other things:
   - (a) The full and exclusive right, title and interest to such relevant IP, in the United States and all other countries;
   - (b) The right of priority and all other rights under any and all international agreements to which the United States of America adheres;
   - (c) The right to file and prosecute applications in any and all countries for patents, copyright registrations, design registrations, mask work protection, and/or other protection; and
   - (d) All applications for patents, copyright registrations, design registrations, mask work protection and/or other protection, and all patents, registrations, and the like that result from such applications.
5. Any such relevant IP which is copyrightable shall be considered a "work made for hire" under the copyright laws of the United States with respect to all of the rights comprised in the copyright, including, but not limited to, any separate contributions to collective works.
6. I understand that any and all IP and Information which:
   - (a) concerns or relates to any aspect of the business of the company or is otherwise owned or used by the company; and
   - (b) derives independent economic value, actual or potential, from not being generally known to, and not being readily ascertainable by

proper means by, other persons who can obtain economic value from its disclosure or use; and

  (c) is the subject of efforts by the company to maintain its secrecy that are reasonable under the circumstances;

is "confidential information" belonging to the company. All IP and information which concerns or relates to any aspect of the business of the company or is otherwise owned or used by the company shall be considered confidential information, except such particular items of the IP and information which I can clearly and convincingly show are:

  (a) openly known by the general public prior to the date of this agreement, or
  (b) subsequent to the date of this agreement, became openly known by the general public through no fault of mine.

I also understand and agree that the fact that a particular item of the IP and information may not be confidential information does not indicate or imply that any other item of the IP and information is not confidential information or that use of such item in combination with other items, or in a particular context is not confidential information.

7. I acknowledge that any unauthorized disclosure or use of confidential information to which I shall have access by virtue of my position in the company, and particularly IP that is confidential information, would cause the company irreparable injury or loss. Accordingly, I shall not, at any time during my employment, or thereafter, use any confidential information except as required in the performance of my duties on behalf of the company or in a manner otherwise expressly authorized by the company. Unless I have prior written authorization from the company, I shall not disclose any confidential information to others. I understand and agree that in view of the nature of confidential information, the duration (for the lifetime of the confidential information), geographical limitation (universal), and scope (total ban on unauthorized use and disclosure) of this provision are in all aspects entirely reasonable.

8. I also understand that the company respects the proprietary rights of others and agree that I will not disclose to the company any confidential IP or information of others if that disclosure would violate any obligation of confidentiality to which I am subject.

9. I shall not at any time during the term of my employment, or thereafter, take or cause any action which would be inconsistent with, or tend to impair or diminish, the rights of the company in relevant IP or in confidential information, and I will always assist the company in every proper and legal way to obtain, maintain and protect its rights in relevant IP, and its rights in confidential information to which I had access during the term of my employment. I will supply evidence and testimony, and sign all papers deemed necessary by the company or its nominee to

obtain, maintain and protect its rights. However, obtaining, maintaining and protecting the company's rights will be entirely at the expense of the company or its nominee.

10. During the term of my employment with the company I shall devote my full-time efforts to the company and shall not take other employment without prior written consent by the company, nor shall I directly or indirectly engage in any endeavor other than for the company in which I may reasonably be expected to call upon, refer to, or use any confidential information. I acknowledge that these restrictions are necessary for the protection of the company.

11. Further, acknowledging that it would be essentially impossible to refrain from unconsciously using confidential information under some circumstances, I hereby agree that *for a period of 1 year after termination of my employment*, for any reason, I shall not:
    (a) directly or indirectly engage in any endeavor, other than on behalf of the company, in which I may reasonably be expected to call upon, refer to, or use any confidential information; or
    (b) take any employment or engagement with any company which directly or indirectly engages in competition with the company in a capacity in which I might call upon, refer to or use any confidential information

I understand and agree that in view of the nature of confidential information, the duration (1 year), and geographical limitation (universal) of, and activities precluded by, this provision are in all aspects entirely reasonable and are necessary for the protection of the company.

12. Upon termination of my employment with the company, for whatever reason I will return to the company all confidential information, and any other documents relating to the business of the company, and all documents, equipment, and supplies owned by the company that may be in my possession. I will return to the company all copies of documents, drawings, software and programs, including all recordings on magnetic, optical or other media, and all listings, and not take or retain any copies thereof.

13. I understand and agree that this is not an agreement of employment; the company is not required to continue to employ the employee for any particular period. My employment can be terminated at will by the company at any time, with or without cause. The provisions of this agreement shall survive any such termination for any reason.

14. This agreement inures to the benefit of, and is enforceable by, the company, its nominees, successors, and assigns, and shall be binding upon me, my heirs, and legal representatives.

15. I represent that I have not entered, and will not enter, into any agreement or obligation which will prevent my full compliance with the terms

of this agreement and that I do not claim any rights to any IP other than that, if any, which is described in detail in an attachment to this agreement.
16. I understand that this agreement may be amended only in a writing signed by the company, and there are no other understandings, agreements, or representations, express or implied.
17. If any clause or provision of this agreement is or becomes invalid, or unenforceable, I agree that the clause or provision should be interpreted to call for the protection of the company's rights to the greatest extent which is legal, valid, and enforceable, and if the clause or provision cannot be so interpreted, or a court of competent jurisdiction declines to permit such clause or provision to be so interpreted, the clause or provision will be severed and the remaining provisions of this agreement shall continue in full force and effect.
18. **This agreement shall be governed by and construed in accordance with the laws of the United States, and the State of Arizona, without resort to principles of conflicts of law.** The federal and state courts within the State of Arizona will have jurisdiction to adjudicate any dispute arising out of this agreement. Both parties expressly consent to: (a) the personal jurisdiction of the federal and state courts within the State of Arizona; and (b) be bound by orders of such courts.

Employee: Ima Employee Signature Date:	ClientCo LLC By: Signature Name: Title: Date:

Index

Note: Page numbers followed by *f* indicate figures, *b* indicate boxes and *t* indicate tables.

A

Accelerator, 6
Acceptance, contract, 257
    revocation of, 258–259, 258*b*
    seeking, 305
Access control security system (ACSS), 347
Accredited investors, 193–194
Acquisition, 321–325
    advantages of, 323–324
    bootstrap purchase, 323
    deal, 323
    direct purchase, 323
    disadvantages of, 324–325
    preparation for, 322*b*
ACSS. *See* Access control security system
Adoption curve, 234, 234*f*
Advertising, 242
    as promotion, 242–243
Advertising-based business model, 52
Aeroglide Corporation, 131
Affordable method, 246
Aftermarket, 330
Aging research, 9
Agreements, contract, 272–280
    assignment, 276
    confidentiality, 274–275
    consulting, 275
    development, 275
        definitions in, 348
        deliverables in, 352
        indemnity in, 350–351
        intellectual property in, 351
        payment in, 349
        performance in, 348
        performance schedule in, 352
        preamble, 347
        recitals in, 352
        representations in, 350
        warranties in, 350
    distribution, 279
    employment, 273, 353–358
    franchise, 278–279
    know-how licenses, 277–278
    license, 276–280
    maintenance, 275
    manufacturing, 275–276
    noncompete, 273–274
    OEM, 279–280
    operating, 272–273
    patent licenses, 277
    purchase, 280
    sample development, 347–352
    support, 275
    technical service, 279
    trademark licenses, 278
    VAR, 279–280
AIA. *See* American Invents Act
Alternative literature, 57
Amazon, 52–53, 75
Amazon Web Services, 43
American Invents Act (AIA), 97
American Red Cross, 26
Amoruso, Sophia, 39
Analytical skills, 294–295
Angel financing, 198–199
Another Broken Egg Cafe, 72
AOL, 85–86
App Store, 19
Apple, 57, 75, 85–86, 220*b*, 235, 237, 243–244, 289, 297
Applicable law, 272
Apps, 10*b*
Appy Food and Drink Co., 238
Arbitrary determination method, 245
Arbitrary trademarks, 117*t*
Arby's, 72
Arizona Technology Investment Forum (ATIF), 198
Arnolite Pallet Corporation, 77, 218
Articles of organization, 146
Aspirin, 118*f*
Assignment, 272
    agreements, 276
ATIF. *See* Arizona Technology Investment Forum
Attitudes, 302–303
Attribute listing, 226
Authorized stock, 151

B

B2B. *See* Business to business
B2C. *See* Business to consumer
B2G. *See* Business to government
Baby Boomers, 7
Baker v. Selden, 109
Baldwin, Ben, 50
Band-Aid, 107
Banks, investment, 329–330
Barchas, Myron, 127
Basche, Todd, 289
Baseline Ventures, 167
Bell, Alexander Graham, 94
Benefit segmentation, 79
Benton, Angela, 49–50
Berkeley Enterprises, 259
Best Buy, 238–239
Big, hairy, audacious goals (BHAGs), 297, 298*f*
Big data, 7
Big-dream approach, 226
Billable hours, 52

Biogenomics, 71*b*
Biosimilars, 71*b*
Birchbox, 48–49, 48*b*
Bitcoin, 8
Blank, S., 23
Blogging, 244
Bohm, Peter, 127
Boilerplates, 271–272
Bonds, 155
Bootstrap financing, 210
Bootstrap purchase, 323
Bosch & Lomb, 236–237
Bound notebook, 92–93
Box.com, 8
BPO. *See* Business process outsourcing
Brain science, 9
Brainstorming, 224
Brainwriting, 225
Branding, 237
  in marketing, 236–237
Breach, 261–265, 262*f*
  material, 261
Brichter, Loren, 10*b*
Bricklin, Dan, 83
Brilliance, showcasing, 305
Brin, Sergey, 304
Brokers
  business, 325
  in business models, 53
  in distribution, 241
Build-Measure-Learn Feedback Loop, 21, 22*f*
Business brokers, 325
Business formation, 156*f*
Business model
  advertising-based, 52
  billable hours, 52
  brokers, 53
  definition, 51
  distributors, 52
  freemium, 52
  in idea generation process, 51–53
  landlord-leaser-licensor, 53
  pay as you go, 52
  repeatable, 26
  resellers, 52
  retailers, 52–53
  scalable, 26
  subscription pricing, 52
  wholesalers, 52–53
Business model canvas, 26–29, 27*f*
Business partners, 34–35

Business plan
  development, 180–181
  elements of, 171–181, 172*f*
    business description, 174
    executive summary, 171–173
    financial plan, 176–177
    financial statements, 177*t*
    industry description, 174
    marketing plan, 175–176, 176*f*
    operational plan, 178
    organizational plan, 177–178
    production plan, 177
    revenue and expense forecasting, 179
    Section 1, 171–173
    Section 2, 174–178
    Section 3, 179–180
    technology plan, 174
  purpose of, 169–171
  software, 170*b*
  update, 180–181
  writing, 168–171
Business process outsourcing (BPO), 44
Business to business (B2B), 43–44, 69
Business to consumer (B2C), 44–46
Business to government (B2G), 46–47
Busque, Leah, 167*b*, 173, 217

C

Call the loan, 204
Cam'ron, 56*b*
Capital, 34. *See also* Venture capital
  management plan, 188–191
Card reader, 231
Causal logic, 4, 4*f*
C-corporation, 142
  characteristics, 144*t*
  distribution of profits and losses, 150
  Shareholder options, 149
  taxation of, 149
Change, 226–227
Channel members, 241*f*
Checklist method, 225
Check-the-box regulations, 146
China, 114*b*
Church & Dwight, 127
CIP. *See* Continuation-in-part
CISG. *See* Contracts for the International Sale of Goods

Claiming priority, 99
Claims, patent, 102
ClearFit, 50
Cloud computing, 8
Coffee Joulies, 49
Cognitive dissonance, 234
Cohen, David, 49
Colgate-Palmolive, 127
Collaborative commerce, 7
Collateral, 203
Colleran, Kevin, 50
Collins, Jim, 297
Color Lines Clothing, 219
Combination locks, 289
Combination use-based registration systems, trademarks, 113
Commercial impracticability test, 263
Common law
  fraud in, 260
  trademark, 113
Common stock, 152
Communication skills, 296
Comparative, 305
Compensatory damages, 263
Competent evidence, 94*b*
Competition, 239–240
Competitive parity, 217–218
  method of budgeting, 246
Complete performance, 261
Concept stage, product planning, 67–68
Conceptual skills, 297–298
Confidentiality agreements, 96, 351
  in contracts, 274–275
  generic, 339–342
  3dp, 348
ConnectU, 251
Consideration, 259
  in contract anatomy, 268–269
Constitution, U.S., 97
Constructive reduction, 90
Consulting agreements, 275
Consumer Electronics Show, 58
Consumers. *See also* Business to consumer
  markets, 75–76
  newness to, 71–72
  in pricing, 240
Consumption chain, 47, 48*f*
Context awareness, 7–8
Contingency plan, 222
Continuation application, 100
Continuation-in-part (CIP), 100

Continuous innovation, 51
Contracts
  agreements, 272–280
    assignment, 276
    confidentiality, 274–275
    consulting, 275
    development, 275, 347, 348, 349, 350, 351, 352
    distribution, 279
    employment, 273, 353–358
    franchise, 278–279
    know-how licenses, 277–278
    license, 276–280
    maintenance, 275
    manufacturing, 275–276
    noncompete, 273–274
    OEM, 279–280
    operating, 272–273
    patent licenses, 277
    purchase, 280
    sample development, 347–352
    support, 275
    technical service, 279
    trademark licenses, 278
    VAR, 279–280
  amended, 262
  anatomy of, 265–272
    consideration in, 268–269
    definitions, 266
    indemnity, 350–351, 270
    intellectual property in, 267–268
    miscellaneous provisions, 271–272
    performance, 266–267
    preamble, 265, 347
    recitals, 265–266
    representations, 269–270
    termination, 271
    terms, 271
    warranties, 269–270
  breach, 261–265
    material, 261
    remedies for, 262f
  conditions, 262
  damages, 263–265
    compensatory, 263
    punitive, 263
    quasi-contract, 265
    reformation, 265
    rescission, 264
    restitution, 264
    specific performance and, 264

enforcement, 262
formation, 254–259
  acceptance, 257, 258–259, 258b, 305
  consideration, 259
  counteroffer in, 255–257
  offer in, 254–255, 258–259, 258b
gambling, 260
law, 252–253
mutual recision, 262
novation, 262
obligations, 31
performance, 261–265
  complete, 261
  in contract anatomy, 266–267
  deliberate, 33
  in development agreements, 348
  excuses for less than complete, 262–263
  inadequate, 261
  schedule, 267, 352
  specific, 264
  substantial, 261
surety, 260
Contracts for the International Sale of Goods (CISG), 253
  fraud and, 260
  UCC and, 254f
Conversational interview, 67–68
Convertibility, 153
Copyright, 88t, 108–112
  Internet and, 110–111
  law, 109b
  notice, 111
  ownership, 111–112
  registration, 111
  software and, 109–110
Corporate entrepreneurship, 222–223
Corporate stocks, 151
  authorized, 151
  issued shares, 151
  unissued shares, 151
Corporate veil, 132
Corporation. See also specific corporations
  C-corporation, 142
    characteristics, 144t
    distribution of profits and losses, 150
    shareholder options, 149
    taxation of, 149

corporate status maintenance, 145–146
as legal entity, 141–146
legal structure, 141–146
  payroll tax, 141
  self-employment tax, 141–142
  structure of, 141–142
limited liability, 130, 146–147
  advantages and disadvantages of, 147t
  characteristics, 146–147
  distribution of profits and losses, 150
  expenses, 148–149
  formalities, 150
  shareholder options, 149
  taxation of, 149
S-corporation, 142
  characteristics, 143t
  distribution of profits and losses, 150
  shareholder options, 149
  taxation of, 149
self-employment tax in, 141–142
small business, 142–143
structure of, 141–142
tax laws, 142–143
  dividends, 142
  double taxation, 142
Costco, 46
Counteroffer, 255–257
Creativity, 224–226
Crowdfunding, 194
Cumulative preferred stock, 153
Customers
  acquiring, 56–59, 321
  buying process, 233f
  development process, 24f
  discovery and validation, 23–26
    channels, 28
    cost structure, 28
    customer relationships, 28
    customer segments, 28
    definition, 25
    key activities, 29
    key partners, 29
    key resources, 29
    revenue streams, 28
    value proposition, 28
  due diligence and, 313
  feedback, 22

Customers *(Continued)*
  in idea generation process, 56–59
  orientation, 218
  validation, 50*f*, 51
Cyber security, 8

D

Damages, 263–265
  compensatory, 263
  punitive, 263
  quasi-contract, 265
  reformation, 265
  rescission, 264
  restitution, 264
  specific performance and, 264
DCAA. *See* Defense Contract Audit Agency
Debt financing, 189, 202–205
  alternatives to, 207–210
  institutional lender requirements, 203
  lender types, 203–205
  loan rates, 203–205
  payment methods, 203–205
Decision-making skills, 295
Defective creation of limited liability entity, 133
Defense Contract Audit Agency (DCAA), 46–47
Defensive, 305
Definitions, contract, 266
Deliberate performance, 33
Deliberate practice, 17–18, 32–35
  business partners in, 34–35
  capital in, 34
  feedback in, 32–33
  fit in, 33
  motivation in, 32
  repetition in, 33
  starting, 34–35
  understandability in, 32
  youth and, 34
Deliberate speed, 55–56
Deliverables, 352
Dell, 42
Demographic segmentation, 78
Denholtz, Jeffrey, 127
Denholtz, Melvin, 127
Department of Commerce, U.S., 84
Derivation proceeding, 90
Descriptive trademarks, 117*t*
Design patent, 88*t*, 98

Development agreements, 275
  definitions in, 348
  deliverables, 352
  indemnity in, 350–351
  intellectual property in, 351
  payment in, 349
  performance in, 348
  performance schedule in, 352
  preamble, 347
  recitals in, 352
  representations in, 350
  warranties in, 350
Development stage, product, 68
Deviation, management by, 170
Digital Millennium Copyright Act (DMCA), 110
Diligence, 90
Direct purchase, 323
Discount rate, 316
  risk-adjusted, 317
Discounted cash flow technique, 316–318
Discovery Channel, 58
Dispute resolution process, 272
Disruptive technologies, 72, 75, 76*f*
  Kickstarter and, 77*b*
Distribution
  agreements, 279
  in marketing, 240–242
    brokers in, 241
    channels, 240–241, 321
    physical, 242
    supply chain management in, 242
  newness to, 72–73
Distributors, 52
Diversification, 72
Dividends
  corporate taxes and, 142
  in preferred stock distributions, 153
DMCA. *See* Digital Millennium Copyright Act
Dolbear, Amos, 94
Dorf, B., 23
Double taxation, 142
Drawbaugh, Daniel, 94
Dropbox, 8
Due diligence, 311–314
  customers and, 313
  finances in, 312
  management and key personnel, 314

  market and, 313
  operations, 314
  product and service line in, 312
  research and development, 313–314
  synergy in, 312–313

E

Early stage, 198
Earn-out strategy, 326
EatFeastly.com, 7
eBay, 167*b*
Economic buyer, 25
Economies of scale, 322
The Edge (website), 58
Edison, Thomas, 75, 94
Effectual logic, 4, 4*f*
Effectuation, 4, 13
Elevator pitch, 207
Eller, Karl, 303
Elmi, Steele, 127
E-medical records, 8
Emerging growth companies, 194
Employee stock options (ESO), 149, 159–160
  restrictive clauses, 160
  tax issues, 157–161
  vesting in, 159–160
Employment agreements, 273, 353–358
Enterprise capital, 169–171
Enterprise value, 315
Entrepreneurial expertise, 4
Entrepreneurial leadership, 290–293
  ethics and, 303–305
  influence in, 292–293
  skills, 293–303
    analytical, 294–295
    attitudes, 302–303
    communication, 296
    conceptual, 297–298
    decision-making, 295
    personality, 301
    resilience, 300–301
    self-awareness, 300–301
    team-building, 299–300
    values, 301–302
Entrepreneurial management in Lean Startups, 23
Entrepreneurial method, 29–32
  contractual obligations, 31
  private property respect and, 31

rebounding in, 30–31
value creation in, 30
Entrepreneurship, 9f
corporate, 222–223
technology
big data, 7
cloud computing, 8
collaborative commerce, 7
context awareness, 7–8
global economy and, 9
opportunities in, 6–11
quantified self, 7
trends in, 6–11
value creation and, 30
Equitrac, 85–86
Equity, 151–155
financing, 189
alternatives to, 207–210
costs of, 196–197
seed stage, 197
sources of, 197–198
in startup ventures, 157–161
ESOs in, 159–160
vesting in, 160
sweat, 158
ESO. *See* Employee stock options
Esprit de corps, 297
Estate debts, 260
Estée Lauder, 244
Ethics, 303–305
Evident Corporation, 127
Exclusive right, 102–103
Execution, 291f
Executive summary, 171–173, 343–346
competition, 344
market, 344
market segment, 345
mission statement in, 343
price, 345
problem solving in, 343–344
solutions in, 344
USP, 344
Exit event, 202
Exit strategies, 149
via acquisition, 321–325
advantages of, 323–324
bootstrap purchase, 323
deal, 323
direct purchase, 323
disadvantages of, 324–325
preparation for, 322b
definition, 310–311
earn-out, 326

via IPO, 328–330
advantages of, 333–334
disadvantages of, 334–335
investment bank selection, 329–330
timing, 328–329
via merger, 325–327
advantages of, 326–327
disadvantages of, 327
registration statement, 331–335
prospectus in, 331–332
red herring, 332
reporting requirements, 332–333
via succession, 318–321
advantages of, 320
disadvantages of, 320–321
timetable, 331–335
Expenses, estimating, 179, 180
Express warranties, 269–270
Expression, 109
Extensive due diligence, 198

F

Fab, 52–53
Facebook, 41, 85–86, 242b, 244
IPO, 329b
Family businesses, 318, 318f
Fanciful trademarks, 117t
Fancy hands, 167
Federal Drug Administration (FDA), 23
FedEx, 41, 298
Feedback, 32–33
Ferriss, Tim, 167, 173
FICA tax, 141
Final product, development of, 70
Finances, due diligence, 312
Financial plan, 176–177
Financial statements, 177t
Financing
angel, 198–199
bootstrap, 210
debt, 189, 202–205
alternatives to, 207–210
institutional lender requirements, 203
lender types, 203–205
loan rates, 203–205
payment methods, 203–205
equity, 189
alternatives to, 207–210
costs of, 196–197

seed stage, 197
sources of, 197–198
regulation of, 192–202
First mover advantage, 220–221
Fit, 33
Fixed costs, 239
Floodgate, 167
Fora TV, 58
Force majeure, 263, 272
Forced relationships, 225
Founders Fund, 167, 173
Founder's stock, 154–155
Franchise agreements, 278–279
Fraud
in CISG, 260
in common law, 260
in inducement, 260
limited liability and, 133
statute of, 260
in UCC, 260
Free association, 225
Freemium, 52
Frends, 75
Fundamental venture types, 41–47, 42f
business to business, 43–44, 69
business to consumer, 44–46
business to government, 46–47
single-product, 46
Fundraising
elevator pitch, 207
PPM, 206
subscription agreement, 206–207
tools and techniques, 205–207

G

Galvin, Robert, 24–25
Gambling contracts, 260
Geek Squad, 238–239
General partnership, 199
advantages and disadvantages of, 138t
expense of, 148–149
as legal entity, 137
Generic trademarks, 117t
GENIAC, 259
Gesellschaft mit beschrankter Haftung (GmbH), 130
Gigwalk, 167
Gilt, 52–53, 244
Global markets, 11–12, 13–15
GmbH. *See* Gesellschaft mit beschrankter Haftung

GMO labeling, 235*b*
Goods, 252–253
Google, 18, 85–86, 107, 300, 322, 325
Gordon method, 225
Gourmet to Go, 79
Government information, 218
Government markets, 75–76
Government regulations, 44
Gray, Elisha, 94
Gretzky, Wayne, 12
Griffith, Scott, 167
Growing beyond startup, 222–227
    change in, 226–227
    creativity in, 224–226
    innovation, 223–224
    ownership and, 224

H
Hangouts, 3
Harvard Connection, 251
Hawking, Stephen, 58
Health informatics, 9
Hewlett-Packard, 289
High fidelity versions, 25
Hike, 3
Hisrich, Robert D., 222–223
Holley, Jim, 50
Horowitz, Andreesen, 56*b*
Humility, 304, 305*f*

I
Idea development process, 68–70, 69*f*
    competition, 68–69
    description in, 68–69
    final product, 70
    launch, 70
    marketing plan, 70
    modify and validate, 69
    need determination, 69
    product planning, 67
Idea generation process, 47–54
    business model testing in, 51–53
    customers in, 56–59
    innovation, 49–51
    nontraditional, 56–59
        alternative literature, 57
        new places and experiences, 58
        team up, 59
        thought leaders, 58–59
    opportunity register, 54–56, 55*t*
    point of pain identification in, 47–49

Ideas, 9, 9*f*
IKEA, 167*b*
Implied warranties, 270
IMVU, 20, 21*b*
    MVP in, 21
Inadequate performance, 261
Inappropriate confidentiality markings, 96*b*
Indemnity, 270
    in development agreements, 350–351
Index Ventures, 39
Inducement, fraud in, 260
Industrial markets, 75–76
Influence, 292–293
Informal agreements, 251*b*
InfoSys, 44
Initial Public Offering (IPO), 192, 311
    exit via, 328–330
        advantages of, 333–334
        disadvantages of, 334–335
        investment bank selection, 329–330
        timing, 328–329
    Facebook, 329*b*
    registration statement, 331–335
        prospectus in, 331–332
        red herring, 332
        reporting requirements, 332–333
    timetable, 331–335
Innovation, 9*f*
    accounting, 23
    continuous, 51
    forms of, 223*f*
    growing beyond startup, 223–224
    in idea generation process, 49–51
    of products, 49–51
    of services, 49–51
Innovators, 9
Institutional lender requirements, 203
Intangible value, 85*f*
Intel, 85–86
Intellectual property (IP), 11, 84, 158, 348, 353
    consideration in, 86
    in contract anatomy, 267–268
    contributions in, 86
    in development agreements, 351
    license agreements in, 268
    protection, 87
        mechanisms, 88*t*

    recognizing, 89–90
    record keeping, 90–95
        guidelines, 92–95
        procedures, 91–92
    technology ventures and, 84–87
InterDigital, 85–86
Interference proceeding, 90
Intermediaries, 240–241
Internal metrics, 22
Internal Revenue Service (IRS), 135
Internal skills, 322
International protection, trademark, 116
Invention, 84
Investment
    banks, 329–330
    horizon, 310–311
    return on, 199
    transactional analysis of, 157*f*
IP. *See* Intellectual property
IPO. *See* Initial Public Offering
IRS. *See* Internal Revenue Service
Issued shares, 151
Iterative way, 29

J
Jackson, David, 49
Jackson, Peter, 215
Jannard, Jim, 215*b*
JOBS. *See* Jumpstart Our Business Startups Act
Jobs, Steve, 75, 243–244, 297
Joint work, 112
Jointly liable, 129–130
Jolie, Angelina, 245
Jumpstart Our Business Startups (JOBS) Act, 194, 195

K
Katrak, Bela, 219
Kearney, Claudine R., 222–223
Kelleher, Herb, 298
Kentucky Fried Chicken, 72
Key personnel, 314
Kickstarter, 76, 195
    disruptive technologies and, 77*b*
Kimberley-Clark, 238
Kleenex, 107, 118*f*
Kleiner Perkins Caufield & Byers, 76
Know-how, 84
    licenses, 277–278
    nonproprietary, 95
Koch Industries, 318

Kodak, 85–86
Kohls, 244
Kyron Global Accelerator, 6

L

Landlord-leaser-licensor business model, 53
Launching, 70
Law
　applicable, 272
　common
　　fraud in, 260
　　trademark, 113
　contract, 252–253
　copyright, 109*b*
　corporation tax, 142–143
　　dividends, 142
　　double taxation, 142
　patent, 91
　trademark, 114*b*
Leadership skills, 293–303
　analytical, 294–295
　attitudes, 302–303
　communication, 296
　conceptual, 297–298
　decision-making, 295
　personality, 301
　resilience, 300–301
　self-awareness, 300–301
　team-building, 299–300
　values, 301–302
Lean manufacturing, 20, 43–44
Lean Startups, 20–23, 21*b*
　entrepreneurial management in, 23
　innovation accounting in, 23
　MVP in, 21
　validated learning in, 23
Legal structure
　choice of, 135–136
　corporation, 141–146
　　payroll tax, 141
　　self-employment tax, 141–142
　　structure of, 141–142
　general partnership, 137, 199
　　advantages and disadvantages of, 138*t*
　　expense of, 148–149
　　as legal entity, 137
　　limited partnership, 139
　sole proprietorship, 135–136
　　advantages and disadvantages of, 136*t*
　　expense of, 148–149
　　as legal entity, 135–136
　　wrong choice of, 127*b*
Lehman, Tom, 56*b*
Liability, 129–134. *See also* Limited liability
　joint, 129–130
　severe, 129–130
　unlimited, 129–130
　vicarious, 130
License agreements, 89
　in contracts, 276–280
　in intellectual property, 268
Limited liability, 129–130
　defective creation of, 133
　extent of, 131–134
　fraud and, 133
　procedural formalities, 133
　statutory basis and, 133
Limited liability corporations (LLCs), 130, 146–147
　advantages and disadvantages of, 147*t*
　characteristics, 146–147
　distribution of profits and losses, 150
　expenses, 148–149
　formalities, 150
　shareholder options, 149
　taxation of, 149
Limited partnership, 139
　advantages and disadvantages of, 140*t*
Limited partnership agreement (LPA), 199–200
Lincoln Motor Company, 222*b*
LinkedIn, 244
Liquidity event, 310–311
Litronic, 256
LivePlan, 170*b*
LLCs. *See* Limited liability corporations
Loans
　calling, 204
　rates, 203–205
　SBA, 205
　terms of, 204
LPA. *See* Limited partnership agreement
Luxottica, 215

M

Mackey, John, 298
Maintenance agreements, 275
Maintenance plans, 41–42
Management
　capital, 188–191
　by deviation, 170
　in due diligence, 314
　key personnel, 314
　in lean startups, 23
　supply chain, 242
Manufacturer agent, 241
Manufacturing agreements, 275–276
Market, 75–79
　adding, 321
　capitalization, 315
　consumer, 75–76
　creation, 31–32
　due diligence and, 313
　entry positioning, 219
　government, 75–76
　industrial, 75–76
　penetration pricing, 240
　penetration strategy, 219–221
　　first mover advantage, 219–221
　segmentation, 77–79
　　benefit, 79
　　demographic, 78
　　by market type, 78*t*
　　psychological, 79
　target, 47
　　positioning, 79
　　selection, 232–233
Marketing
　branding and, 236–237
　distribution in, 240–242
　　brokers in, 241
　　channels, 240–241, 321
　　physical, 242
　　supply chain management in, 242
　mix, 235
　online, 244
　packaging in, 237
　plan, 175–176
　　development of, 70
　　elements of, 176*f*
　pricing in, 239–240
　　channel members and, 241*f*
　　competition in, 239–240
　　consumer in, 240
　　costs in, 239
　　market penetration, 240
　　marketing in, 239–240
　　subscription, 52

Marketing *(Continued)*
  product mix in, 235–239
  promotion in, 242–245
    advertising as, 242–243
    affordable method, 246
    arbitrary determination
      method, 245
    budget, 245–246
    competitive parity method, 246
    mix, 243f
    objective and task
      method, 246
    percent-of-sales method, 246
    personal selling as, 243
    publicity as, 243–244
    sales, 238–239, 242, 243
    social media as, 244
  services in, 238–239
  technology, 234–235
Market-skimming, 240
Marks & Spencer, 238
Marriage, 260
Maskwork, 88t, 112
Material breach, 261
McKelvey, Jim, 231
McNealy, Scott, 8
Medicare, 141
Menlo Park, 18
Mergers, 325–327
  advantages of, 326–327
  disadvantages of, 327
Microsoft, 75, 85–86
Milliken & Company, 318
Minimal viable product (MVP), 6
  in IMVU, 21
Min/max, 206
Mirror image rule, 255
Mittal, Kevin, 3
Modify and validate, 69
Moghadam, Mahbod, 56b
Motivation, 32
Motorola, 24b
Multiples technique, 315–316
Mutual rescission, 262
MVP. *See* Minimal viable product

N
NAB. *See* National Association of
  Broadcasters
NAICS. *See* North American Industry
  Classification System
Nano-technology, 8
Narendra, Divya, 251
Nas, 56b

NASD. *See* National Association of
  Securities Dealers
Nasty Gal, 39
National Association of Broadcasters
  (NAB), 215
National Association of Securities
  Dealers (NASD), 331, 334–335
Natural Products Expo West, 237
NDA. *See* Nondisclosure agreement
Neostem, 70b
Ness Computing, 309
New product classification, 73f
Newness, 70–73
  to consumer, 71–72
  to distribution system, 72–73
  to organization, 72
Next, 57
Nickelodeon, 238
Nike, 7
99dresses.com, 7
Noncompetition agreement, 97, 260
  in contracts, 273–274
Nondisclosure agreement
  (NDA), 274
Nonobviousness, 105–106
Nonparticipating preferred
  stock, 154
Nonproprietary know-how, 95
Nontraditional idea sources, 56–59
  alternative literature, 57
  new places and experiences, 58
  team up, 59
  thought leaders, 58–59
Nortel, 85–86
North American Industry
  Classification System (NAICS),
  68–69, 73, 175
Northrop Corporation, 256
Nosek, Luke, 167
Notice symbols, trademarks, 116
Novation, 262
Novelty, 105–106
Nuance, 85–86

O
Oakley, 215b
Objective and task method, 246
OEM. *See* Original equipment
  manufacturer agreements
Offer, contract formation, 254–255
  revocation of, 258–259, 258b
Office Action, 106
Online marketing, 244
Online Service Providers (OSP), 110

Open source licensing, 87
OpenTable, 309b
Operating agreement, 146
  in contracts, 272–273
Operational plan, 178
Operations, 314
Opportunity assessment plan,
  73–75, 74t
Opportunity register, 54–56, 55t
Organizational plan, 177–178
Original equipment manufacturer
  (OEM) agreements, 279–280
Original Soup Man, 72
Original work of authorship, 108
Originality, 108
OSP. *See* Online Service Providers
Osterwalder, Alexander, 26
O'Tierney, Tristan, 231
Ouya, 76
Overcapitalization, 191
Ownership, 129–134
  copyright, 111–112
  growing beyond startup and, 224
  interest terminology, 151t
  of patents, 107

P
Packaging, 237
Page, Larry, 304
Panavision, 216
Parameter analysis, 226
Participating preferred stock, 154
Patent Cooperation Treaty (PCT), 107
Patents, 85–86, 97–108
  application, 99–100
  claims, 102
  design, 88t, 98
  eligibility, 104–105
  examination process, 106
  exclusive right, 102–103
  first page of, 101f
  granted in U.S., 99f
  international, 107–108
  licenses, 277
  nonobviousness in, 105–106
  novelty in, 105–106
  ownership of, 107
  patentability, 103–106
  pending, 106–107
  petty, 98
  plant, 98–99
  U.S. law, 91
  utility, 88t, 97–98
  written description in, 100–102

Index

Pay as you go, 52
Payment, 349
PayPal, 167, 231
Payroll tax, 141
PB Loco, 72
PCT. *See* Patent Cooperation Treaty
Pebble, 195*b*
Percent-of-sales method, 246
Perfect tender rule, 261
Performance, 261–265
    complete, 261
    in contract anatomy, 266–267
    deliberate, 33
    in development agreements, 348
    excuses for less than complete, 262–263
    inadequate, 261
    schedule, 267
        in development agreements, 352
    specific, 264
    substantial, 261
Peroxydent Group, 127
Perrier, 237
Persevering, 21
Personal guarantee, 203
Personal holding companies, 142
Personal selling, 242
    as promotion, 243
Personal Software, 83
Personality, 301
Petty patent, 98
Phantom income, 149
Pillars, 35
    business model canvas, 26–29
    customer discovery and validation, 23–26
        channels, 28
        cost structure, 28
        customer relationships, 28
        customer segments, 28
        definition, 25
        key activities, 29
        key partners, 29
        key resources, 29
        revenue streams, 28
        value proposition, 28
    entrepreneurial method, 29–32
        contractual obligations, 31
        private property respect and, 31
        rebounding in, 30–31
        value creation in, 30

Lean Startup, 20–23, 21*b*
    entrepreneurial management in, 23
    innovation accounting in, 23
    MVP in, 21
    validated learning in, 23
    value creation, 18–19
Pivoting, 21
Pixar, 57, 297
Plant patent, 98–99
Pliant Technology, 85–86
PMI. *See* Proprietary materials and information
Point of pain, 47–49
Polymer Technology, 236–237
Porras, Jerry, 297
Post-money valuation, 196–197
Postpurchase behavior, 234
PowerPoint, 207
PPM. *See* Private Placement Memorandum
Practical problems, 50
Preamble, contract, 265
    in development agreements, 347
Preferred stock, 152–155
    convertibility of, 153
    cumulative, 153
    distributions, 153
        dividends in, 153
    founder's stock, 154–155
    nonparticipating, 154
    participating, 154
    raising capital, 155–157
    voting rights, 154
Pre-money valuation, 196
Present value, 316
Pricing
    channel members and, 241*f*
    competition in, 239–240
    consumer in, 240
    costs in, 239
    market penetration, 240
    marketing in, 239–240
    subscription, 52
Primary objectives, 12
    global focus, 12
    research and theory, 13
    rules and principles, 12
Prime rate, 203
Printing costs, 335
Prior art, 105
Private placement, 194

Private Placement Memorandum (PPM), 191, 195
    as fundraising tool, 206
Private property, 31
Procter & Gamble, 19
Production plan, 177
Products
    innovation, 49–51
    lifecycle, 313*f*
    lines
        broadening, 321
        in due diligence, 312
    mix, 235–239
    planning and development, 66–68, 67*f*
        concept stage, 67–68
        development stage, 68
        idea stage, 67
        test marketing stage, 68
    service, 233–234
Profit interest, 156
Promotion, 242–245
    advertising as, 242–243
    affordable method, 246
    arbitrary determination method in, 245
    budget, 245–246
    competitive parity method, 246
    mix, 243*f*
    objective and task method, 246
    percent-of-sales method, 246
    personal selling as, 243
    publicity as, 243–244
    sales, 238–239, 242, 243
    social media as, 244
Proprietary materials and information (PMI), 347, 348
Prospectus, 331–332
Provisional application, 99
Psychological segmentation, 79
Publicity, 242
    as promotion, 243–244
Punitive damages, 263
Purchase agreements, 280
Purchasing product service, 233–234
    product adoption curve, 234*f*
Pure registration systems, trademarks, 113–114

Q

Quantified self, 7
Quantive, 85–86
Quasi-contract, 265

R

Radio, 24b
Raising capital, 155–157
Rap Genius, 56b
Real estate, 260
Recitals, contract, 265–266
    in development agreements, 352
Record keeping, 90–95
    guidelines, 92–95
    procedures, 91–92
Red herring, 332
redBus, 65b, 66
Reformation, 265
Regenerative medicine, 71
Registration statement, 331–335
    prospectus in, 331–332
    red herring, 332
    reporting requirements, 332–333
Regulation of financing, 192–202
Repeatable business model, 26
Repetition, 33
Representations, 269–270
Request for proposals (RFP), 46
Rescission, 264
Research and development, 313–314
Resellers, 52
Resilience skills, 300–301
Restitution, 264
Restrictive clauses, 160
Restrictive covenants, 204
Retailers, 52–53
Return on investment, 199
Revenue, estimating, 179
Revocation, 258–259, 258b
RFP. *See* Request for proposals
R&H Safety Sales, 218
Ries, Eric, 20, 22
Right of reversion, 111
RIM, 85–86
Risk-adjusted discount rate, 317
Robotics, 9
Rokeach, Milton, 302, 302t
Rovio, 19
Rudy, Jerome, 127

S

Sales
    estimating, 179–180
    in marketing, 238–239
    promotion, 242, 243
SalesForce.com, 43
Sama, M. Phanindra, 65b
Sample development agreement, 347–352
Samsung, 235
Samuel Adams, 244
Sandia National Labs, 218
Satisficing, 295
SBA. *See* Small Business Administration Loans
SBIR. *See* Small Business Innovation Research
Scalable business model, 26
Schilowitz, Ted, 215
Schwarz, Jesse, 299
Scientific method, 29
S-corporation, 142
    characteristics, 143t
    distribution of profits and losses, 150
    shareholder options, 149
    taxation of, 149
Search, 291f
SEC. *See* Securities and Exchange Commission
Securities, 192
Securities and Exchange Commission (SEC), 192, 328, 335
Security interest, 203
Seed stage, 197
Self-awareness skills, 300–301
Self-employment tax, 141–142
Self-funding, 188
Self-sufficiency, 190
Serial entrepreneurs, 323–324
Services
    in due diligence, 312
    innovation, 49–51
    in marketing, 238–239
    products, 233–234
    technical, agreements, 279
Set asides, 46
Severability, 272
Severely liable, 129–130
Shareholder options
    C-corporation, 149
    LLC, 149
    S-corporation, 149
Shareholders agreement, 150
SIC. *See* Standard Industrial Classification
Silicon Valley, 18
Silicon Valley Bank, 203–204
Single-product venture, 46

Six Sigma, 43–44
Skype, 85–86
Small Business Administration (SBA) Loans, 205, 218
Small business corporation, 142–143
Small Business Innovation Research (SBIR), 191, 207–208
Small Business Technology Transfer (STTR), 208–209
Smith, Fred, 298
SMS messages, 3
Social media, 242
    as promotion, 244
Social security, 141
Socialcast, 171b
Soderbergh, Steven, 216
Softbank, 3
Software, business planning, 170b
Software Arts, Inc., 83
Sole proprietorship
    advantages and disadvantages of, 136t
    expense of, 148–149
    as legal entity, 135–136
South by Southwest, 58
Southwest Airlines, 41, 298
S&P 500, 85f
Spaly, Brian, 50
Specific performance, 264
Spreadsheet programs, 83b
Square Inc., 231
Square Wallet, 231
Squishable, 242b
Staeger Clear Packaging, 238
Stage-gate process, 22, 22f
Standard Industrial Classification (SIC), 68–69, 73, 175
Stanford University eCorner, 58
Startup ventures, 5
    equity distribution in, 157–161
    ESOs in, 159–160
    vesting in, 160
    growing beyond, 222–227
    change in, 226–227
    creativity in, 224–226
    innovation in, 223–224
    ownership and, 224
StartupValley, 169b
Statute of frauds, 260
Statutory bars, 105
Statutory basis, 133

Index

Stem cell therapy, 71
Stock
   authorized, 151
   common, 152
   corporate, 151
      authorized, 151
      issued shares, 151
      unissued shares, 151
   ESO, 149, 159–160
      restrictive clauses, 160
      tax issues, 157–161
      vesting in, 159–160
   founder's, 154–155
   preferred, 152–155
      convertibility of, 153
      cumulative, 153
      distributions, 153
      founder's stock, 154–155
      nonparticipating, 154
      participating, 154
      raising capital, 155–157
      voting rights, 154
   treasury, 151
STTR. *See* Small Business Technology Transfer
Styrofoam, 107
Subconscious, 50
Subscription agreement, 206–207
Subscription pricing, 52
Substantial performance, 261
Succession, 318–321
   advantages of, 320
   disadvantages of, 320–321
Suggestive trademarks, 117*t*
Sun Microsystems, 8
Super Glue, 107
Supply chain management, 242
Support agreements, 275
Surety contracts, 260
Sustainability programs, 44
Sustainable Minds, 44
Swatch, 238
Sweat equity, 158
Synergy, 312–313

T

Target market, 47
   positioning, 79
   selection, 232–233
Taser, 107
TaskRabbit, 167*b*, 173, 217
Tata Group, 44

Taxation
   of C-corporation, 149
   double, 142
   ESOs and, 157–161
   FICA, 141
   laws
      corporation types and, 142–143
      dividends, 142
      double taxation, 142
   of LLC, 149
   payroll, 141
   of S-corporation, 149
   self-employment, 141–142
Team-building skills, 299–300
Technical service agreements, 279
Technology base, 321
Technology entrepreneurship
   big data, 7
   cloud computing, 8
   collaborative commerce, 7
   context awareness, 7–8
   global economy and, 9
   opportunities in, 6–11
   quantified self, 7
   trends in, 6–11
   value creation and, 30
Technology escrow, 92*b*
Technology marketing, 234–235
Technology plan, 174
Technology ventures, 84–87
TED, 58
Temporary organization, 23
Termination of contract, 271
Terms of contracts, 271
Terms of loans, 204
Tesla Motors, 41
Test marketing stage, 68
Test sales, 25
Tetra Pak, 238
Thiel, Peter, 167, 173
Thought leaders, 58–59
3dp confidentiality agreement, 348
Tichy, Noel M., 292–293
Timetable, 331–335
Toyota Production System (TPS), 20
TPS. *See* Toyota Production System
Trade secrets, 88*t*, 95–97
   procedures for, 95–97
Trademarks, 88*t*, 112–119
   arbitrary, 117*t*
   categories of, 117*t*
   choosing, 119

   combination use-based registration systems, 113
   common law, 113
   descriptive, 117*t*
   examiners, 102
   fanciful, 117*t*
   generic, 117*t*
   international protection of, 116
   law in China, 114*b*
   licenses, 278
   notice symbols, 116
   pure registration systems, 113–114
   registering, 114–115, 115*f*
   rights acquisition, 113–114
   strength of, 116–117
   suggestive, 117*t*
Transactional analysis, 87*f*
   of business formation, 156*f*
   of investment, 157*f*
Transparency, 304
Treasury stocks, 151
TrunkClub, 50
Twitter, 41, 244

U

U Mix, 50
UCC. *See* Uniform Commercial Code
Uhrman, Julie, 76
Unconscionable, 265
Understandability, 32
Underwriter, 329–330
Underwriting syndicate, 329–330
Uniform Commercial Code (UCC), 252–253
   CISG and, 254*f*
   fraud under, 260
   perfect tender rule, 261
   role of, 256–257
Unilever, 238
Unique selling propositions, 217
Unissued shares, 151
Units, 206
Unlimited liability, 129–130
USPTO, 119
Usurious agreements, 260
Utility patent, 88*t*, 97–98

V

Validated learning, 23
Valuation, 315–318
   definition, 310
   post-money, 196–197

Valuation *(Continued)*
  pre-money, 196
  techniques, 315
    discounted cash flow, 316–318
    multiples, 315–316
Value chain analysis, 221–222
Value creation, 18–19, 40
  definition of, 18–19
  in entrepreneurial method, 30
  technology entrepreneurs and, 30
Value drivers, 321
Value propositions, 19
Value-added reseller (VAR) agreements, 279–280
Values, 301–302
VAR. *See* Value-added reseller agreements
Variable costs, 239
Venture capital (VC), 199–201
  firm structure, 200f
  returns, 201f
  specifics sought by, 201–202
Venture Partner, 50
Vertical strategy, 321
Vesting
  requirements, 155
  in startup ventures, 160
Viacom v. YouTube, 110b
Vicariously liable, 130
Video games, 45f
VisiCalc, 83b
Viva, 238
Voting rights, 154

W
Waiver, 272
Walkie-Talkies, 24b
Wal-Mart, 46, 49
Wal-Mart Labs, 46, 49
Warranties
  in contract anatomy, 269–270
  express, 269–270
  implied, 270
Western Union, 94
WhatsApp, 3
Whole Foods, 235b, 237
Wholesalers, 52–53
Winklevoss, Cameron, 251
Winklevoss, Tyler, 251
Wordlock, 289
Work for hire, 111
Written description, 100–102

X
Xerox, 107

Y
Youth, 34
YouTube, 18, 244

Z
Zechory, Ilan, 56b
Zirtual, 167
Zuckerberg, Mark, 251, 329

CPSIA information can be obtained
at www.ICGtesting.com
Printed in the USA
FFOW02n1432290216
21956FF